Cardiovascular Considerations in Hematopoietic Stem Cell Transplantation

Azin Alizadehasl · Ardeshir Ghavamzadeh ·
Amir Hossein Emami · Ghasem Janbabaei ·
Davood Khoda-Amorzideh

Editors

Cardiovascular Considerations in Hematopoietic Stem Cell Transplantation

Springer

Editors
Azin Alizadehasl
Cardio-Oncology Research Center, Rajaie
Cardiovascular Medical and Research
Center
Iran University of Medical Sciences
Tehran, Iran

Amir Hossein Emami
Hematology-Oncology Department, Cancer
Institute, Imam Khomeini Hospital, School
of medicine
Tehran University of Medical Sciences
Tehran, Iran

Davood Khoda-Amorzideh
Cardio-Oncology Research Center, Rajaie
Cardiovascular Medical and Research
Center
Iran University of Medical Sciences
Tehran, Iran

Ardeshir Ghavamzadeh
Cancer and Cell Therapy Research Center
Tehran University of Medical Sciences
Tehran, Iran

Ghasem Janbabaei
Hematologic Malignancies Research Center
Tehran University of Medical Sciences
Tehran, Iran

ISBN 978-3-031-53661-8   ISBN 978-3-031-53659-5   (eBook)
https://doi.org/10.1007/978-3-031-53659-5

This Springer imprint is published by the registered company Springer Nature Switzerland AG
The registered company address is: Gewerbestrasse 11, 6330 Cham, Switzerland

Paper in this product is recyclable.

Contents

HSCT at a Glance

Ardeshir Ghavamzadeh and Maryam Barkhordar

Abstract Hematopoietic stem cell transplantation (HSCT) is the treatment of choice for various pediatric and adult malignant and non-malignant diseases. Many advancements have been achieved to ensure that HSCTs are available and secure for every individual who has a clinical reason for performing it since its discovery as a curative option for either hereditary or inherited hematological illnesses at the beginning of the 1960s. Professor Ghavamzadeh and colleagues performed the first HSCT in Iran on March 13, 1991, at the Bone Marrow Transplant Department of Shariati Hospital in Tehran. Autologous and allogeneic stem cell transplants were conducted at Shariati Hospital's HORCSCT, using bone marrow, peripheral blood, or umbilical cord blood stem cells as the graft sources. In the present chapter, we'll briefly consider the hematopoietic stem cell transplantation process and then look into how transplants have been conducted in Iran over time.

Keywords Hematopoietic stem cell transplantation · HORCSCT · Professor Ghavamzadeh

Abbreviations

HSCT	Hematopoietic stem cell transplantation
MAC	Myeloablative conditioning
RIC	Reduced-intensity conditioning
TBI	Total body irradiation
NRM	Non-relapse mortality
HLA	Human leukocyte antigen

M. Barkhordar
Hematology Oncology and Stem Cell Transplantation Research Center, Tehran University of Medical Sciences, Tehran, Iran

A. Ghavamzadeh (✉)
Cancer and Cell Therapy Research Center, Tehran University of Medical Sciences, Tehran, Iran
e-mail: ghavamza@gmail.com

  1
A. Alizadehasl et al. (eds.), *Cardiovascular Considerations in Hematopoietic Stem Cell Transplantation*, https://doi.org/10.1007/978-3-031-53659-5_1

NMA   Nonmyeloablative conditioning
PBSC   Peripheral blood stem cells
BMSC   Bone marrow Stem Cell
HORCSCT   Hematology, Oncology, and Stem Cell Transplantation Research
             Center
APCC   Asian Pacific Cancer Center
EBMT   European Blood and Marrow Transplantation
ASH   American Society of Hematology
ISH   International Society of Hematology
ESMO   European School of Medical Oncology
ASCO   American Society of Clinical Oncology

1   An Overview of Hematopoietic Stem Cell Transplantation

Hematopoietic stem cell transplantation (HSCT) is the preferred therapeutic option in various pediatric and adult malignant and non-malignant diseases. It originally served as salvage therapy after intensive chemotherapy or radiotherapy and for the treatment of serious hematopoietic impairments; however, stem cell therapy has also developed as an immunotherapy for many immune-related disorders. Thus far, the only established therapeutic application of stem cells has been HSCT. Many advancements have been achieved to ensure that HSCTs are available and secure for every individual who has a clinical reason for performing it since its discovery as a curative option for either hereditary or inherited hematological illnesses at the beginning of the 1960s (Barriga et al. 2012).

Restoration of hematopoiesis in individuals with impaired bone marrow or an immune system could be achieved through HSCT. Autologous transplants use cells from the patient themselves, whereas allogeneic transplants use stem cells from a related or unrelated donor. Peripheral blood, bone marrow, and umbilical cord blood are all available sources of stem cells. HSCT has been progressively employed for treating a wide range of cancers and other disorders during the last five decades.

Autologous and allogeneic HSCT are similar in that patients take "conditioning" therapies followed by intravenous CD34+ stem cells. In autologous HSCT, stem cells are harvested primarily to address the myelosuppression conditioning regimen which is the main method to destroy the primary disease. Autologous HSCT could be used as first-line consolidation for multiple myeloma after the first remission; however, it is typically conducted after the second or subsequent remission in patients with malignant lymphoma. Autologous HSCT may be utilized to treat autoimmune disorders and recurrent germ-cell cancers (Parrondo et al. 2020; Zahid et al. 2017; Mousavi et al. 2018; Swart et al. 2017).

Allogeneic HSCT has also been used to treat various non-cancer diseases such as hemoglobinopathies (thalassemia, sickle cell disease), congenital or acquired bone marrow failure syndromes, primary immunodeficiencies, and inherited metabolic disorders. Repopulating the blood cells and immunological systems or passage of enzymes or other key cellular elements are two ways in which HSCTC might treat the primary problem in each condition (Ghavamzadeh et al. 2018, 2019).

2 Conditioning Regimens

There are three types of transplant preparation protocols that are employed in particular circumstances (Gagelmann and Kroger 2021; Scott et al. 2017):

- Myeloablative conditioning (MAC): The most intensive conditioning regimens utilized in hematopoietic stem cell transplantation are called MAC. In order to completely ablate the bone marrow, they comprise substantial doses of alkylating compounds, either with or without whole-body irradiation (TBI). The most frequently used alkylators are fludarabine, cyclophosphamide, and busulfan. Moreover, melphalan, etoposide, cytarabine, and thiotepa are also mostly used. The goal of MAC is to eliminate the recipient's hematopoietic system and make space for the engraft of the donor's stem cells. A substantial risk of toxicity, including organ damage, graft-versus-host disease (GVHD), and severe infections, has been associated with MAC regimens.
- Reduced-intensity conditioning (RIC): RIC policies tend to be milder compared to MAC protocols. They involve a reduction in TBI or myeloablative chemotherapy doses. RIC aims to reduce the toxicity associated with MAC while still allowing donor stem cell engraftment. A higher chance of recurrence could offset the less toxic nature of RIC regimens. RIC regimens are usually given to elderly or medically compromised patients who are unable to sustain MAC regimens.
- Nonmyeloablative conditioning (NMAC): In hematopoietic stem cell transplantation, NMAC regimens are the least intensive conditioning regimens used. They do not require stem cell support and enable the patient's bone marrow to recover. NMAC aims to allow donor stem cells to engraft while preserving the hematopoietic system of the recipient. NMAC regimens are associated with a minimal risk of toxicity, but they may also increase the risk of relapse. NMAC regimens are commonly used in patients with nonmalignant diseases or in elderly or medically unstable patients who cannot tolerate MAC or RIC regimens.

3 Donor Types and Human Leukocyte Antigen (HLA) Matching in Allogeneic HSCT

High-resolution typing at (HLA-A, -B, -C, HLA-DRB1, and HLA-DQB) is used for matching of HLA. The ideal level of HLA matching for siblings is a 10 of 10. Due to numerous unmeasurable factors not accounted for with HLA typing, related donors get priority versus unrelated ones. Matched unrelated donor transplants have increased in popularity because of the declining size of families and the development of donor database registries (Barker et al. 2019; Rappazzo et al. 2021).

Unfortunately, not all patients can find a compatible donor in the ever-expanding donor registries. Disparities in access to allogeneic transplants exist as minority groups are not adequately represented in donor registries. Mismatched related and unrelated donors may be used when matched donors are not readily accessible. Donors with a 6/8 or 7/8 match have traditionally been used. There is a higher chance of graft rejection and GVHD with these transplants (Rappazzo et al. 2021).

Another mismatch, "haploidentical" transplants, reflects a half-match and often comes from a relative, such as a kid or parent. Global acceptance of haplo-HSCT is growing because of advances in GVHD preventive regimens such as post-transplantation cyclophosphamide. These mismatched transplants may be reality-saving for patients without a matched sibling or unrelated donor (Barkhordar et al. 2021, 2022a, b).

4 Stem Cell Collection

The three stem cell sources were bone marrow, mobilized peripheral blood stem cells (PBSC), and umbilical cord blood. Typically, bone marrow transplants were obtained by a difficult procedure of collecting marrow from the donor under general anesthesia. However, as procedures for stem cell mobilization advance, PBSC is growing more common.

Mobilization usually requires high-dose filgrastim (or occasionally plerixafor) and five consecutive days of apheresis until a good harvest is obtained. Compared to stem cells derived from bone marrow, PBSCs implant more rapidly and fail less frequently; however, GVHD is more likely to occur. PB collection does not need general anesthesia or invasive removal from the donor; however, over a day is often required.

Umbilical cord blood is the third source of stem cells stored early following delivery. Cord blood databases are separate from registries of bone marrow donors (Pagliuca et al. 2017). Regarding matching, CBT's benefits include quick access and relative flexibility. Umbilical blood cells are suitable for usage if they are compatible with around six alleles (two sets of HLA-A, -B, and -DRB1). Disadvantages include a lack of supply, a higher price, a delay in engraftment, and a rise in infections, all of which contribute to greater transplant-related mortality than other donor sources.

5 History of Stem Cell Transplantation in Iran

Known as the "Father of Bone Marrow Transplantation" in Iran, Professor Ardeshir Ghavamzadeh was born in Tehran in 1940. In 1971, he graduated as a doctor from the Medical Faculty of the University of Vienna. He then attended the University of Basel in Switzerland for a fellowship after finishing his residency in internal medicine there in 1991. When he went back to Iran in 1975, he set up the city's first hematology ward.

Professor Ghavamzadeh and colleagues performed the first HSCT in Iran on March 13, 1991, at the Bone Marrow Transplant Department of Shariati Hospital in Tehran. In 1998, Professor Ghavamzadeh established the Hematology-Oncology-Stem Cell Transplantation Research Center (HORCSCT) of Tehran University in Shariati Hospital (Ghavamzadeh et al. 2008, 2013), where Stem Cell Transplantation cured numerous malignant and non-malignant hematologic and non-hematologic disorders.

The HORCSCT collaborated with several of the world's leading scientific and research organizations, including the European Blood and Marrow Transplantation (EBMT) and the International Blood and Marrow Transplantation Registry (IBMTR), to collect information on HSCT recipients. In addition, the HORCSCT joined the EBMT and Asian Pacific Cancer Center and collaborated with the European School of Medical Oncology, the American Society of Clinical Oncology, the American Society of Hematology, and the International Society of Hematology (Ghavamzadeh et al. 2008).

The evolution and establishment of the Hematology-Oncology-Stem Cell Transplantation Research Center of Tehran University, one of the most important transplant institutions in the Middle East (Lida et al. 2019), followed the sequence of events:

- Opening of the Hematology Ward in 1975.
- Establishment of the Hematology and Oncology Department of Shariati Hospital in 1978.
- First HSCT Using Bone marrow Stem Cell (BMSC) 1991.
- First HSCT Using Peripheral Blood Stem Cell (PBSC) 1996.
- Establishment of Hematology, Oncology and Stem Cell Transplant Research Center (HORCSCT) in 1998.
- Establishment of the Iranian Cord Blood Bank in1999.
- Establishment of Data Management Group in 1999.
- First unrelated donor transplant in 2002.
- First haploidentical donor transplant in 2002.
- Establishment of the Iranian Stem Cell Donor Registry in 2009.
- Establishment of the Institute for Regenerative Medicine in 2015.
- Establishment of Oncology Research Center in 2017.
- Establishment of Cell Therapy and HSCT Research Center in 2017.
- Establishment of the Research Institute for Oncology, Hematology, and Cell Therapy, Tehran University of Medical Sciences in 2018.

- Receiving many national and international awards (17th Congress of the Asia Pacific 2012; 10th Asia Oceania Congress of Nuclear Medicine and Biology 2012; 16th Congress of the Asia Pacific Blood and Marrow Transplantation 2011; Shariati Hospital on Teacher's Appreciation 2011; Iranian Unrelated Bone Marrow Donor Registry 2010; ASBMT Young Investigator Award 2010).

Autologous and allogeneic stem cell transplants were conducted at Shariati Hospital's HORCSCT, using bone marrow, peripheral blood, or umbilical cord blood stem cells as the graft sources. Moreover, all kinds of allogeneic HSCT concerning the donor type and match statues have been performed at HORCSCT, including HSCT from relatives (siblings or other relatives) or un-relatives' donors, with full-matched, one locus mismatched or mismatched (Haploidentical) HLA typing. The general results of autologous and allogeneic HSCT recipients (considering the primary diseases, donor types, match status, and other underlying conditions) have been relatively similar or comparable to the main transplant centers in other parts of the world (Barkhordar et al. 2021, 2022a, b).

The first bone marrow transplant in Iran was conducted more than 30 years ago, and since then, more than 20 other bone marrow transplant institutes have been established within our country. The Iranian stem cell donor registry bank was established in 2009 in Shariati Hospital. It is the only stem cell registry bank in the Middle East connected to the international EMDIS system, where all donors have an international GRID code (19-digit code).

References

ASBMT Young Investigator Award, ASBMT Tandem meetings, Feb. 24–28, 2010, Orlando, Florida

Awarded Eleventh Avicenna Festival for developing innovative procedures in the delivery of clinical services "Establish the Iranian Unrelated Bone Marrow Donor Registry". Tehran University of Medical Sciences, 2010

Barker JN, Boughan K, Dahi PB, Devlin SM, Maloy MA, Naputo K, Mazis CM, Davis E, Nhaissi M, Wells D, Cooper C, Ponce DM, Kernan N, Scaradavou A, Giralt SA, Papadopoulos EB, Politikos I. Racial disparities in access to HLA-matched unrelated donor transplants: a prospective 1312-patient analysis. Blood Adv. 2019;3(7):939–44. https://doi.org/10.1182/bloodadvances.201802 8662.

Barkhordar M, Kasaeian A, Tavakoli S, Vaezi M, Foumani HK, Bahri T, Babakhani D, Mirzakhani L, Mousavi A, Mousavi SA, Ghavamzadeh A. Selection of suitable alternative donor in the absence of matched sibling donor: a retrospective single-center study to compare between haploidentical, 10/10 and 9/10 unrelated donor transplantation. Int J Hematol-Oncol Stem Cell Res. 2021;1;51–60. Ijhoscr.tums.as.ir

Barkhordar M, Kasaeian A, Janbabai G, Kamranzadeh Fumani H, Tavakoli S, Rashidi AA, Mousavi SA, Ghavamzadeh A, Vaezi M. Modified combination of anti-thymocyte globulin (ATG) and post-transplant cyclophosphamide (PTCy) as compared with standard ATG protocol in haploidentical peripheral blood stem cell transplantation for acute leukemia. Front Immunol. 2022a;13: 921293. https://doi.org/10.3389/fimmu.2022.921293.

Barkhordar M, Kasaeian A, Janbabai G, Mousavi SA, Fumani HK, Tavakoli S, Bahri T, Ghavamzadeh A, Vaezi M. Outcomes of haploidentical peripheral stem cell transplantation with combination of post-transplant cyclophosphamide (PTCy) and anti-thymocyte globulin (ATG)

compared to unrelated donor transplantation in acute myeloid leukemia: a retrospective 10-year experience. Leuk Res. 2022b;120: 106918. https://doi.org/10.1016/j.leukres.2022.106918.

Barriga F, Ramírez P, Wietstruck A, Rojas N. Hematopoietic stem cell transplantation: clinical use and perspectives. Biol Res. 2012;45(3):307–16. https://doi.org/10.4067/S0716-976020120003 00012.

Gagelmann N, Kröger N. Dose intensity for conditioning in allogeneic hematopoietic cell transplantation: can we recommend "when and for whom" in 2021? Haematologica. 2021;106(7):1794–804. https://doi.org/10.3324/haematol.2020.268839.

Ghavamzadeh A, Alimogaddam K, Jahani M, Mousavi S, Iravani M, Bahar B, et al. HSCTin Iran: 1991 to 2008. Hematol Oncol Stem Cell Ther. 2008;1:231–8.

Ghavamzadeh A, Jalili M, Rostami S, Yaghmaie M, Aliabadi LS, Mousavi SA, Vaezi M, Fumani HK, Jahani M, Alimoghaddam K. Corrigendum to: comparison of induction therapy in non-high risk acute promyelocytic leukemia with arsenic trioxide or in combination with ATRA leukemia research 66 (2018) 85–88. Leuk Res. 2018;67:116. https://doi.org/10.1016/j.leukres.2018.03.003.

Ghavamzadeh A, Kasaeian A, Rostami T, Kiumarsi A. Comparable outcomes of allogeneic peripheral blood versus bone marrow hematopoietic stem cell transplantation in major thalassemia: a multivariate long-term cohort analysis. Biol Blood Marrow Transp: J Am Soc Blood Marrow Transp. 2019;25(2):307–12. https://doi.org/10.1016/j.bbmt.2018.09.026.

Ghavamzadeh A, Alimoghaddam K, Ghaffari F, Derakhshandeh R, Jalali A, Jahani M. Twenty Years of Experience on Stem Cell Transplantation in Iran. Iran Red Crescent Med. 2013;15(2)

Iida M, Kodera Y, Dodds A, Ho AYL, Nivison-Smith I, Akter MR, Wu T, Lie AKW, Ghavamzadeh A, et al. Registry Committee of the Asia-Pacific Blood and Marrow Transplantation Group (APBMT). Advances in hematopoietic stem cell transplant Asia-Pacific region: The second report from APBMT 2005–2015. Bone Marrow Transplant. 2019;54(12):1973–1986. https://doi.org/10.1038/s41409-019-0554-9

Mousavi A, Abedinzadeh N, Taj L, Kasaeian A, Alimoghaddam K, Vaezi M, Zokaasadi M, Kamranzadeh H, Ardeshir G. Successful outcome of autologous stem cell transplantation for relapsed or refractory germ cell tumors. Int J Hematol-Oncol Stem Cell Res. 2018;12:191–5.

Outstanding poster award for "treatment of inborn errors of metabolism with allogeneic hematopoietic stem cell transplantation" presented at The 16th Congress of the Asia Pacific Blood and Marrow Transplantation October 30–2 November 2011,Sydney, Australia

Pagliuca S, Peffault de Latour R, Volt F, Locatelli F, Zecca M, Dalle JH, Comoli P, Vettenranta K, Diaz MA, Reuven O, Bertrand Y, Diaz de Heredia C, Nagler A, Ghavamzadeh A, Ruggeri A, et al. Long-term outcomes of cord blood transplantation from an HLA-identical sibling for patients with bone marrow failure syndromes: a report from eurocord, cord blood committee and severe aplastic anemia working party of the European society for blood and marrow transplantation. Biol Blood Marrow Transp: J Am Soc Blood Marrow Transp. 2017;23(11):1939–48. https://doi.org/10.1016/j.bbmt.2017.08.004.

Parrondo RD, Ailawadhi S, Sher T, et al. Autologous stem-cell transplantation for multiple myeloma in the era of novel therapies. JCO Oncol Pract. 2020;16:56–66.

Rappazzo KC, Zahurak M, Bettinotti M, Ali SA, Ambinder AJ, Bolaños-Meade J, Borrello I, Dezern AE, Gladstone D, Gocke C, Fuchs E, Huff CA, Imus PH, Jain T, Luznik L, Rahmat L, Swinnen LJ, Wagner-Johnston N, Jones RJ, Ambinder RF. Nonmyeloablative, HLA-mismatched unrelated peripheral blood transplantation with high-dose post-transplantation cyclophosphamide. Transp Cell Therapy. 2021;27(11):909.e1-909.e6. https://doi.org/10.1016/j.jtct.2021.08.013.

Scott BL, Pasquini MC, Logan BR, et al. Myeloablative versus reduced-intensity hematopoietic cell transplantation for acute myeloid leukemia and myelodysplastic syndromes. J Clin Oncol. 2017;35:1154–61.

Selected as an "exemplary professor" of Shariati Hospital on Teacher's Appreciation Day, 2011

Swart JF, Delemarre EM, van Wijk F, et al. Haematopoietic stem cell transplantation for autoimmune diseases. Nat Rev Rheumatol. 2017;13:244–56.

Third prize for best oral presentation at 17th Congress of the Asia Pacific. Blood and Marrow Transplantation October 26–28 2012, Hyderabad, India

Young Investigator Award for Best Scientific Presentation in Clinical Nuclear Medicine for "Autologous HSCTin patients with high risk neuroblastoma treated with/without 131I-MIBG" presented at 10th Asia Oceania Congress of Nuclear Medicine and Biology (AOFNMB).16–20 May 2012, Tehran, Iran

Zahid U, Akbar F, Amaraneni A, et al. A review of autologous stem cell transplantation in lymphoma. Curr Hematol Malig Rep. 2017;12:217–26.

Burden of Cardiovascular Disease in HSCT

Amir Hossein Emami, Azin Alizadehasl, Feridoun Noohi Bezanjani, and Hanieh Hajiali

Abstract Cancers inflict the largest burden in the world. Burden of Cancer incidence and mortality is also growing. Hematopoietic stem cell transplantation (HSCT) is a curative therapeutic method for malignant hematologic disorders and is associated with early and late treatment related mortality. Cardiac toxicity and cardiovascular consequences are not common in post HSCT patients compared to other HSCT related complications. However, these complications play an important role in survivors' outcome. It is necessary to identify and decrease these life-threatening complications to improve patients' outcome and decline related cost burden.

Keywords Cancer · Global cancer burden · Hematopoietic stem cell transplantation · Cardiovascular disease · Cardio-oncology

Abbreviations

CAD	Coronary artery disease
CHF	Congestive heart failure
CNS	Central nervous system
CVRF	Cardiovascular risk factor
CVD	Cardiovascular disease
DALYs	Disability adjusted life years
GBD	Global burden of disease
GVHD	Graft versus host disease
HM	Hematological malignancy
HSCT	Hematopoietic stem cell transplantation

A. Hossein Emami
Hematology-Oncology Department, Cancer Institute, Imam Khomeini Hospital, School of medicine, Tehran University of Medical Sciences, Tehran, Iran

A. Alizadehasl (✉) · F. N. Bezanjani · H. Hajiali
Cardio-Oncology Research Center, Rajaie Cardiovascular Medical and Research Center, Iran University of Medical Sciences, Tehran, Iran
e-mail: alizadeasl@gmail.com

A. Alizadehasl et al. (eds.), *Cardiovascular Considerations in Hematopoietic Stem Cell Transplantation*, https://doi.org/10.1007/978-3-031-53659-5_2

PE   Pericardial effusion
SVTA Supraventricular tachyarrhythmia
TBI   Total body irradiation
WHO World health organization

1 Introduction

Cancers inflict the largest burden in the world based on world health organization (WHO) reports (Mattiuzzi and Lippi 2019). It is a main cause of mortality and morbidity. Burden of Cancer incidence and mortality is growing through the world, parallel to aging and growth of population (Sung et al. 2021). Cancer is increasing as a second leading cause of death (Mattiuzzi and Lippi 2019; Sung et al. 2021). On the other hand, there is a reduction in mortality rates of coronary heart disease and stroke in recent years (Sung et al. 2021). WHO reported epidemiologic trend of more common causes of death from 2016 to 2060. Cancer is expected to become the first cause of death worldwide in 2060 with a 2.08 fold increase compared to 1.76 fold increase for ischemic heart disease (Mattiuzzi and Lippi 2019). Risk of developing malignancies is about 20.2% for lifetime (22.4% in men, 18.2% in women). Cancer associated burden is non significantly higher in men than women (9.6% versus 8.6%) (Mattiuzzi and Lippi 2019). About 19.3 million new cancer patients and 10 million cancer deaths were reported through the world, in 2020. There is estimated 50% of all patients and 58.3% of cancer deaths occurred in Asia in 2020 where about 59.5% of world population is resided (Sung et al. 2021). It is also estimated 1,918,030 new cancer cases and 609,360 cancer deaths including 350 deaths per day for lung cancer in the United State in 2022 (Siegel et al. 2022). Global cancer burden is anticipated to be 28.4 million patients in 2040, about 47% increase compared to 2020 (Sung et al. 2021). Five-year cancer survival in low and middle income countries is much lower than high income countries due to variations in prevention, screening, diagnostic and treatment methods. Despite the net benefits of cancer treatment are small in low income countries because of poor outcome condition of malignancies, investment on patients' survival improvement has long term benefits and substantial returns for countries with all income groups. Global benefit worth is estimated about 3 trillion dollars in 2020–2030 if patients are comprehensively diagnosed and achieved appropriate treatment (Ward et al. 2021).

Hematological malignancies (HMs) are main contributors of cancer global burden. They are classified to four common subtypes: Hodgkin lymphoma (HL), non-Hodgkin lymphoma (NHL), multiple myeloma (MM) and leukemias. Hodgkin lymphoma, non-Hodgkin lymphoma and multiple myeloma are reported for 1.4, 7 and 2.3 million disability adjusted life years (DALYs) in 2017, respectively (Keykhaei et al. 2021). It is estimated about 1.3 million new HM patients with 700,000 deaths related to it through the world, in 2020. Global burden of disease (GBD) study showed increase in incidence of non Hodgkin lymphoma and leukemias about 45% and 26%

from 2006 to 2016, respectively. Incidence of hematological malignancies often rises with aging (Chen et al. 2022). Old aged patients with hematological malignancies have significant mortality and morbidity. Half of HM patients and about 70% of cancer deaths occur in those 65 years and older (Krok-Schoen et al. 2018). Advances in therapeutic strategies have resulted in improvement of patients' outcome. These cancer survivors are at risk of late complications such as therapeutic methods' side effects (Chen et al. 2022).

Hematopoietic stem cell transplantation (HSCT) is a curative therapeutic method for both malignant hematologic disorders or non malignant diseases including bone marrow failure syndromes, immune deficiencies, aplastic or sickle cell anemia, and some metabolic disorders. It may be the only treatment for patients with advanced hematologic malignancies when other therapeutic strategies such as chemo-radiotherapy have been failed (Zhao et al. 2022). HSCT has continuously increased worldwide with more than 7% per year and average number of 90,000/ year. In 2019, statistical review showed one and half million cases received HSCT through the world. The use of HCT technology has been expanded in last 30 years across Europe. There were 48,512 HSCTs including 41% allogenic and 59% autologous which increased 2.2% compared to 2018 in Europe (Zhao et al. 2022; Passweg et al. 2021). Alongside the risk of relapse, HSCT is also associated with early and late treatment related mortality. Infections, toxicity after allogenic HSCT and graft versus host disease(GVHD) are essential causes of death in these patients (Styczyński et al. 2020). Survival rate in HSCT patients is reported 24–76% (Alizadehsal et al. 2023). Improvement in patient's survival has occurred in recent years due to advances in transplantation protocols and post transplantation supportive care. It is reported post HSCT patients who have survived more than 5 years with no relapse have a high probability of surviving for 15 years more. However, they have 4 to ninefold risk of mortality compared to general population within the next 30 years post HSCT (Zhao et al. 2022). HSCT survivors can experience severe acute or chronic complications such as relapsing, graft failure, infections, CNS disorders, hormonal disturbances and cardiovascular toxicity that affect patients' mortality and morbidity (Zhao et al. 2022). In a study, incidence of late complications was estimated 93.2% in post HSCT patients during a follow up range of 2–21 years. In addition, severe or life threatening complications are reported in 18% of HSCT survivors (Zhao et al. 2022).

2   Cardiovascular Disease in Post HSCT Patients

First HSCT induced cardiotoxicity was reported in 1976 in a case study including 29 patients with bone marrow transplantation. Two patients died due to congestive heart failure and autopsy showed extensive myocardial necrosis, fibrin deposition and extravasation of red blood cells (RBCs) (Zhao et al. 2022). Cardiac toxicity and cardiovascular consequences are not common in post HSCT patients compared to other HSCT related complications (for both allogenic and autologous transplantation) (Zhao et al. 2022; Tichelli et al. 2008). Most studies have evaluated clinical cardiac

Table 1 Prevalence of CVD and other complications related to CVD in a meta-analysis in 2022

CVD and complications related to it in post HSCT patients	Prevalence rate (%)
CVD	16.84
Pericardial effusion	19.72
Arrhythmia	3.91
CHF	3.66
Stroke	0.22
CAD	1.36
Death	1.53

CAD coronary artery disease. *CHF*, congestive heart failure. *CVD*, cardiovascular disease (Alizadehsal et al. 2023)

complications. Subclinical involvement appears earlier in post transplantation period and can occur more frequent (Tichelli et al. 2008). Alizadehasl et al. reported 16.84% prevalence of cardiovascular disease in these patients as a systematic review and meta analysis in 2022. This prevalence was higher in patients who received allogenic transplantation than autologous HSCT (Table 1) (Alizadehsal et al. 2023).

Acute and late cardiac complications are important limitations for patients undergoing HSCT and are associated with high mortality rate, morbidity and decline in quality of life in long term HSCT survivors (Zhao et al. 2022). Long term post allogenic HSCT survivors have tenfold risk for mortality compared to matched general population (Tichelli et al. 2008). Cardiac toxicity contributes to 2.4% and 3% of late death causes in post autologuos and allogenic HSCT, respectively (Tichelli et al. 2008). Risk of late deaths related to cardiac complications is fourfold after autologous HSCT in females and 2.3 fold after allogenic HSCT in both males and females, compared to general population (Tichelli et al. 2008). Relative mortality reduces with time from HSCT, although remain high even at 15 years after transplantation in post allogenic HSCT patients (Tichelli et al. 2008; Auberle et al. 2023). There is a high burden of chronic health in HSCT survivors. However, mortality related to cardiovascular disease (stroke, heart failure, myocardial infarction) in these patients can underestimate true burden of CVD after HSCT (Armenian et al. 2017).

Cardiac complications are related to multi factors such as previous cardiotoxic chemotherapy and radiation, HSCT conditioning and patients' cardiovascular risk factors (hypertension, diabetes, dyslipidemia) and it can be expected to increase of incidence with longer follow up after HSCT (Tichelli et al. 2008; Armenian et al. 2017).

2.1 Cardiovascular Risk Factors in HSCT Survivors

HSCT survivors have increased risk for developing cardiovascular risk factors because of pre transplantation and conditioning related therapeutic exposure, GVHD and its management. Armenian and colleague reported 10 year incidence of hypertension, diabetes, dyslipidemia and multi risk factors ($\geq$ 2) about 37.7%, 18.1%, 46.7% and 31.4% in HSCT survivors, respectively. Allogenic HSCT survivors had higher risk for cardiovascular risk factors (CVRFs) than autologous HSCT recipients. These patients who developed high grade graft versus host disease (GVHD) had additional risk for CVRFs with incidence of 25% for diabetes, 50% for hypertension and dyslipidemia within 10 years after HSCT which may be due to using of immunosuppressants for prevention/treatment of acute GVHD. Some studies have shown abdominal radiation may lead to insulin resistance or metabolic syndromes due to radiation induced pancreatic and/or liver injury. However, certain mechanisms by which total body irradiation (TBI) increases risk of diabetes or dyslipidemia are not clear (Armenian et al. 2012).

2.2 Cardiovascular Disease

Cardiovascular disease is associated with premature death and it is also irreversible and debilitating (Armenian et al. 2012). Both young and old patients who receive HSCT, have increased and significant risk for development of cardiovascular disease in first decade after transplantation accompanied with worsening cardiovascular risk factors at more time passes after HSCT (Armenian et al. 2017). The incidence of arterial events such as coronary artery disease (CAD) or stroke in allogenic HSCT recipients is 10% at 15 years that exceeds 20% at 20 years. The highest risk is reported in patients with multiple CVRFs with cumulative incidence of 10–12% at 10 years or in pre HSCT exposure to cardiotoxic therapeutic methods with increased risk of 15% at 10 years (Armenian and Chow 2014). The median age at first presentation of cardiovascular disease in allogenic HSCT survivors is as low as 53 years (range of 35–66 years),earlier than would be expected for general population or autologous HSCT survivors (Armenian et al. 2017; Armenian and Chow 2014). Cardiovascular related mortality in HSCT recipients is estimated 2 to fourfold higher than general population (Armenian et al. 2017).

2.3 Heart Failure

Congestive heart failure (CHF) can occur during immediate or late post HSCT period. Risk factors for early hear failure after transplantation consist of pre HSCT left ventricle ejection fraction (LVEF),conditioning regimen with high dose cyclophosphamide and TBI with morbidity rate of 5–43%. Mortality related to early CHF also ranges from 1 to 9% (Armenian and Chow 2014). Late occurring CHF is primarily associated with pre HSCT exposure to anthracyclines. It often presents 1 year after HSCT and the risk is especially high for autologous transplantation recipients with incidence rate of 5% at 5 years which increases to 10% at 15 years post HSCT.

Congestive heart failure related to HSCT is a poor outcome condition with less than 50% surviving five years after diagnosis. In addition, CVRFs can affect CHF risk. A study showed presence of hypertension or diabetes in HSCT survivors with exposure to high dose anthracyclines led to 35 and 27 fold increased risk for CHF, respectively (Armenian and Chow 2014).

2.4 Arrhythmia

Previous studies showed incidence rate of arrhythmia in patients receive HSCT is 9–27%. Tonorezos and colleague revealed arrhythmia after HSCT is associated with significant mortality and morbidity in a single center study in 2015. They reported post HSCT patients with new arrhythmia had a eightfold risk for intensive care unit (ICU) admission, ninefold risk for in hospital death and a threefold risk for death within 1 year after transplantation (Tonorezos et al. 2015). Turk and colleague evaluated incidence of arrhythmia in HSCT patients in United State between 2008 and 2012. 11% of post HSCT patients developed arrhythmia during their hospital admission that was related to higher mortality rate, increased length of hospital stay and admission cost. Incidence of arrhythmia was also higher in patients with prior cardiovascular disease particularly congestive heart failure (Turk et al. 2019). Post HSCT patients are at moderate risk of developing supraventricular tachyarrhythmia (SVTA). Post transplant SVTA is relatively uncommon. Although, it is clinically an important complication and associated with higher mortality rate. Some studies showed patients with non Hodgkin lymphoma have increased risk for SVTA especially older cases and those with prior cardiac disease (Hidalgo et al. 2004).

2.5 Pericardial Disease

Incidence of Post transplantation pericardial effusion is reported about 0.2–16.9% (Versluys et al. 2018). Significant Post HSCT pericardial effusion has low incidence and is rare but potentially life threatening complication (Liu et al. 2015). These patients have poor clinical outcome (Kubo et al. 2021). Known mechanisms of early pericardial effusion (PE) within first 100 days after transplantation consist of iron overload due to frequent blood infusion, pretransplant conditioning including radiation or high dose cyclophosphamide. Chronic GVHD and age at transplantation play a main role in late onset massive PE (Liu et al. 2015).

3 Conclusion and Future Horizon

The world's population and life expectancy have been increasing. The incidence of most cancer is associated with aging. Novel adequate diagnostic and therapeutic methods of malignancies result in the improvement of cancer patients' outcome (Alizadehsal and Fini 2023). HSCT is commonly used as cellular immunotherapy over 60 years ago and provide cure of thousands of patients with incurable disease (Balassa et al. 2019). Improvement in HSCT strategies has contributed to increase in survival about 10% per decade (Armenian and Chow 2014). Cancer and related therapeutic method including HSCT have known effects on patients' mortality and morbidity. They are associated with difference complications such as cardiac involvement, despite the improvement of patients' survival and life expectancy. These complications play an important role on survivors' outcome, too. It is necessary to identify and decrease these life threatening complications to improve patients' outcome and decline related cost burden.

References

Alizadehasl A, Fini H. Thrombotic events in cancer patients.2023. https://doi.org/10.5772/intech open.109619

Alizadehasl A, Ghadimi N, Hosseinifard H, Roudini K, Emami AH, Ghavamzadeh A, et al. Cardiovascular diseases in patients after hematopoietic stem cell transplantation: systematic review and meta-analysis. Current Res Translational Med. 2023;71(1):103363. https://doi.org/10.1016/j.ret ram.2022.103363

Armenian SH, Sun C-L, Vase T, Ness KK, Blum E, Francisco L, et al. Cardiovascular risk factors in hematopoietic cell transplantation survivors: role in development of subsequent cardiovascular disease. Blood. 2012;120(23):4505–12. https://doi.org/10.1182/blood-2012-06-437178.

Armenian SH, Chemaitilly W, Chen M, Chow EJ, Duncan CN, Jones LW, et al. National institutes of health hematopoietic cell transplantation late effects initiative: the cardiovascular disease and associated risk factors working group report. Biol Blood Marrow Transp. 2017;23(2):201–10. https://doi.org/10.1016/j.bbmt.2016.08.019.

Armenian SH, Chow EJ. Cardiovascular disease in survivors of hematopoietic cell transplantation. Cancer. 2014 Feb 15;120(4):469–79. https://doi.org/10.1002/cncr.28444. Epub 2013 Oct 25

Auberle C, Lenihan D, Gao F, et al. Late cardiac events after allogeneic stem cell transplant: incidence, risk factors, and impact on overall survival. Cardio-Oncol. 2023;9(1). https://doi.org/10.1186/s40959-022-00150-1

Balassa K, Danby, R, Rocha, V. Haematopoietic stem cell transplants: principles and indications. British J Hosp Med.2019;80(1):33–9. https://doi.org/10.12968/hmed.2019.80.1.33.

Chen L, Zheng Y, Yu K, et al. Changing causes of death in persons with haematological cancers 1975–2016. Leukemia. 2022;36:1850–60. https://doi.org/10.1038/s41375-022-01596-z.

Hidalgo J, Krone R, Rich M, et al. Supraventricular tachyarrhythmias after hematopoietic stem cell transplantation: incidence, risk factors and outcomes. Bone Marrow Transp. 2004;34:615–9. https://doi.org/10.1038/sj.bmt.1704623.

Keykhaei M, Masinaei M, Mohammadi E, et al. A global, regional, and national survey on burden and Quality of Care Index (QCI) of hematologic malignancies; global burden of disease systematic analysis 1990–2017. Exp Hematol Oncol. 2021;10:11. https://doi.org/10.1186/s40164-021-001 98-2.

Krok-Schoen L, Fisher JL, Stephens JA, Mims A, Ayyappan S, et al. Incidence and survival of hematological cancers among adults ages ≥ 75 years. Cancer Med. 2018;7(7):3425–33. https://doi.org/10.1002/cam4.1461

Kubo H, Imataki O, Fukumoto T, Oku M, Ishida T, et al. Risk factors for and the prognostic impact of pericardial effusion after allogeneic hematopoietic stem cell transplantation. Transp Cell Therapy.2021;27(11):949.e1–8

Liu Y-C, Chien S-H, Fan N-W, Hu M-H, Gau J-P, et al. Risk factors for pericardial effusion in adult patients receiving allogeneic haematopoietic stem cell transplantation. Br J Haematol. 2015;169:737–45. https://doi.org/10.1111/bjh.13357.

Mattiuzzi C, Lippi G. Current Cancer Epidemiology. J Epidemiol Glob Health. 2019;9(4):217–22. https://doi.org/10.2991/jegh.k.191008.001.

Passweg JR, Baldomero H, Chabannon C, et al. Hematopoietic cell transplantation and cellular therapy survey of the EBMT: monitoring of activities and trends over 30 years. Bone Marrow Transp. 2021;56:1651–64. https://doi.org/10.1038/s41409-021-01227-8.

Siegel RL, Miller KD, Fuchs HE, Jemal A. Cancer statistics, 2022. CA Cancer J Clin. 2022;72(1):7–33. https://doi.org/10.3322/caac.21708.

Styczyński J, Tridello G, Koster L, Iacobelli S, van Biezen A, et al. Infectious Diseases Working Party EBMT. Death after hematopoietic stem cell transplantation: changes over calendar year time, infections and associated factors. Bone Marrow Transplant. 2020;55(1):126–136. https://doi.org/10.1038/s41409-019-0624-z. Epub 2019 Aug 27

Sung H, Ferlay J, Siegel RL, Laversanne M, Soerjomataram I, Jemal Α, Bray F. Global cancer statistics 2020: GLOBOCAN estimates of incidence and mortality worldwide for 36 cancers in 185 countries. CA Cancer J Clin. 2021;71(3):209–49. https://doi.org/10.3322/caac.21660.

Tichelli A, Rovó A, Gratwohl A. Late pulmonary, cardiovascular, and renal complications after hematopoietic stem cell transplantation and recommended screening practices. Hematology Am Soc Hematol Educ Program. 2008;2008(1):125–33. https://doi.org/10.1182/asheducation-2008.1.125.

Tichelli A, Bhatia S, Socié G. Cardiac and cardiovascular consequences after haematopoietic stem cell transplantation. Br J Haematol. 2008;142(1):11–26. https://doi.org/10.1111/j.1365-2141.2008.07165.x.

Tonorezos ES, Stillwell EE, Calloway JJ, Glew T, Wessler JD, et al. Arrhythmias in the setting of hematopoietic cell transplants. Bone Marrow Transp. 2015;50(9):1212–6. https://doi.org/10.1038/bmt.2015.127. Epub 2015 Jun 1.

Turk A, Liu J, Jamshed U, et al. Arrhythmia in bone marrow transplant: why we should worry. J Am Coll Cardiol. 2019;73(9_Supplement_1):500. https://doi.org/10.1016/S0735-1097(19)311 08-8

Versluys AB, Grotenhuis HB, Boelens MJJ, et al. Predictors and outcome of pericardial effusion after hematopoietic stem cell transplantation in children. Pediatr Cardiol. 2018;39:236–44. https://doi.org/10.1007/s00246-017-1747-x.

Ward ZJ, Scott AM, Hricak H, Atun R. Global costs, health benefits, and economic benefits of scaling up treatment and imaging modalities for survival of 11 cancers: a simulation-based analysis. Lancet Oncol. 2021;22(3):341–50. https://doi.org/10.1016/S1470-2045(20)30750-6.

Zhao Y, He R, Oerther S, Zhou W, Vosough M, Hassan M. Cardiovascular complications in hematopoietic stem cell transplanted patients. J Pers Med. 2022;12(11):1797. https://doi.org/10.3390/jpm12111797.

Risk Factors and Mechanisms of Cardiotoxicity in HSCT

Ghasem Janbabai, Mohammad Vaezi, Davood Khoda-Amorzideh, and Maryam Mohseni Salehi

Abstract Cardiotoxicity in individuals undergoing HSCT is influenced by a variety of factors. To begin, it can be said that any previous cardiotoxic therapies, such as chest radiation and anthracyclines, raise the risk of cardiotoxicity in the future. Furthermore, the development of comorbid disorders such as dyslipidemia (DLP), diabetes mellitus (DM), or hypertension (HTN) as a result of HSCT or GVHD is accompanied by an increased risk of cardiovascular disease in the long run (Tuzovic et al. in Curr Oncol Rep 21(3):28, 2019). In addition, the effects of dimethyl sulfoxide, which is employed to preserve stem cells, are likely to have a role in cardiac events (Cox et al. JunCell Tissue Bank 13:203–215, 2012). Other HSCT-related comorbidities, such as sepsis (Ghafoor in Critical care of the pediatric immuno-compromised hematology/oncology patient. Springer, Cham, pp. 211–235, 2009), thrombotic microangiopathy (Pfeiffer et al. in Bone Marrow Transp 52:630–633, 2017; Dandoy et al. in Biol Blood Marrow Transplant 19:1546–1556, 2013; Jodele et al. in Transfus Apher Sci 54:181–190, 2016), and hepatic veno-occlusive disease (Ovchinsky et al. in Biol Blood Marrow Transp 24:207–218, 2016) (Table 1) might also cause cardiac problems.

Keywords Hematopoietic stem cell transplantation · HSCT · Cardiotoxicity · GvHD · Anthracyclines · Chest radiation · Clonal hematopoiesis of indeterminate potential

G. Janbabai
Hematologic Malignancies Research Center, Tehran University of Medical Sciences, Tehran, Iran

M. Vaezi
Hematology-Oncology and Stem Cell Transplantation Research Center, Tehran University of Medical Sciences, Tehran, Iran

D. Khoda-Amorzideh (✉) · M. M. Salehi
Cardio-Oncology Research Center, Rajaie Cardiovascular Medical and Research Center, Iran University of Medical Sciences, Tehran, Iran
e-mail: dkh1359@gmail.com

Abbreviations

ACS	Acute coronary syndrome
AML	Acute myeloid leukemia
ASCVD	Atherosclerotic cardiovascular disease
BMI	Body mass index
CAD	Coronary artery disease
CHF	Congestive heart failure
CHIP	Clonal Hematopoiesis of Indeterminate Potential
CKD	Chronic kidney disease
CVD	Cardiovascular diseases
DLP	Dyslipidemia
DM	Diabetes mellitus
GvHD	Graft versus host disease
HDL	High-density lipoprotein
HF	Heart failure
HSCT	Hematopoietic stem cell transplantation
HTN	Hypertension
MDS	Myelodysplastic syndrome
MI	Myocardial infarction
SLE	Systemic lupus erythematosus
TBI	Total body radiation

1 History of Cardiotoxic Therapies

See Table 1.

Table 1 Cardiotoxicity of Most important transplant complication

Transplant complication	Cardiovascullar effect
Graft versus host disease (GVHD)	Myocardial injury/infarction Pericardial effusion Bradycardia Dysrhythmia
Thrombotic microangiopathy	Systemic hypertension Pulmonary hypertension Pericardial effusion
Hepatic veno-occlusive disease	Fluid overload leading to decompensated heart failure
Sepsis	Myocardial depression Tachyarrhythmia Hypotension

Many therapies which are used prior to HSCT targeting the underlying primary hematologic malignancy, such as chest radiation and anthracyclines, as well as therapies used during the transplant, such as the combination of total body radiation (TBI) and a multi-drug regimen (including melphalan (Olivieri et al. 1998), carmustine (Kanj et al. 1991), cytarabine (Vaickus and Letendre 1984), and cyclophosphamide (Goldberg et al. 1986)) and even post-HSCT maintenance therapies can cause cardiotoxicity in both short- and long-term (Table 2). Chemotherapeutic drugs, for example, high-dose cyclophosphamide (28%) and anthracyclines (3–26%), administered as part of preparative regimens or even primary therapy have been also linked to heart failure (HF) (Yeh and Bickford 2009; Gottdiener et al. 1981; Tichelli et al. 2007; Armenian et al. 2011a).

Monoclonal antibodies and other targeted medicines, which are routinely used as 'maintenance therapy' both before and after HSCT, have lately been associated with a number of distinct cardiac complications. Hypertension (HTN) (both systemic and pulmonary), vascular disease, arrhythmia, cardiometabolic consequences, myocardial ischemia, and QT prolongation are among these complications (Tuzovic et al. 2019; Chang et al. 2017a, b; Moslehi 2016).

1.1 Anthracyclines

Anthracyclines were used in the induction and/or consolidation phases of treatment for the vast majority of individuals with acute leukemia who were transplant candidates. Patients receiving anthracyclines before HSCT have a considerably higher risk of cardiovascular mortality (Armenian et al. 2012). The cumulative anthracycline dosage is also linked to post-transplant cardiac problems (Sakata-Yanagimoto et al. 2004; Peres et al. 2010).

Receiving an anthracycline dosage of more than 250 mg/m^2 was associated with late-onset congestive heart failure (CHF) in a nested case–control study of HSCT survivors, who were treated at City of Hope (Armenian et al. 2008). In acute myeloid leukemia (AML), CPX-351, which is a liposomal form of cytarabine and daunorubicin, has demonstrated a substantial overall survival advantage when compared to normal cytarabine and daunorubicin (7 + 3) treatment (Lancet et al. 2018), enabling a hitherto poor-risk subset of AML patients to undergo HSCT. The forthcoming cardiac adverse effects of the anthracycline liposomal formulation in CPX-351 with/without HSCT have yet to be determined.

1.2 Chest Radiation

Cardiovascular mortality and morbidity are well-known side effects of chest radiation (Heidenreich 2009). On the basis of coronary CT angiography, a study on Hodgkin's lymphoma survivors with previous radiation treatment (48% of whom got 30 Gy

Table 2 Cardiac complication of common drugs used Before HSCT

Drug class	Agent	Mechanisms of cardiotoxicity	Cardiovascular toxicities
Alkylating agents	Cyclophosphamide Ifosfamide Melphalan Busulfan Carmustine Thiotepa	Endothelial capillary Damage promote cell death by induction of DNA instability leading to breakage	Cardiomyopathy Hemorrhagic myocarditis/pericarditis Pericardial effusion/ tamponade Arrhythmias (atrial fibrillation, supraventricular tachycardia) Chest pain Edema
Alkylating agents (platinums)	Carboplatin Cisplatin Oxaliplatin	Inhibits DNA synthesis	Vasculotoxicity Endothelial dysfunction Arterial vasospasm Hypertension Ischemic heart disease Stroke Edema Thromboembolic events Endocardial fibrosis/ pulmonary fibrosis Arrhythmias Pericardial effusion Raynaud's phenomenon Bradycardia, LBBB, Heart failure
Anthracyclines	Doxorubicin Daunorubicin Epirubicin Idarubicin Valrubicin Mitoxantrone	Increased toxic oxygen-free Radical Increased oxidative stress	Cardiomyopathy Atrial arrhythmias Ventricular arrhythmia Coronary artery disease Nonspecific ST-T changes on ECG Pericarditis, myocarditis
Antimetabolites	Fludarabine Cytarabine	Interfere with DNA and RNA synthesis to prevent cell proliferation	Pericarditis Pericardial Effusion/tamponade (cytarabine) Chest pain Hypotension (fludarabine)
Topoisomerase inhibitors	Etoposide	Coronary artery Vasospasm Induction of an Immune response Direct injury to the myocardium	Hypotension Ischemic heart disease Vasospastic angina

(continued)

Table 2 (continued)

Drug class	Agent	Mechanisms of cardiotoxicity	Cardiovascular toxicities
Monoclonal antibodies	Rituximab Bevacizumab	Unknown anaphylaxis and hypersensitivity reaction (Rituximab) Unknown possible exacerbate of preexisting coronary and peripheral vessels owing to antibody of PlGF Infusion hypersensitivity Reaction (Bevacizumab)	Hypertension Anginal pain Arrhythmia (ventricular fibrillation) Peripheral edema Thromboembolic events Decreased left ventricular ejection fraction/ Heart failure Ischemic heart disease
Total body irradiation		RT causes tissue injury primarily through the generation of oxidative stress; inflammation is seen acutely and fibrosis over time	Cardiomyopathy Heart failure Coronary artery disease Valvular heart disease Pericardial disease Autonomic dysfunction

or more) found that coronary artery disease (CAD) was evident in 12 of 31 (39%) individuals (Mulrooney et al. 2014). Chest radiation before transplant is also found to be associated with a 9.5-fold rise in CAD in HSCT survivors (Armenian et al. 2010).

1.3   Cyclophosphamide

Cyclophosphamide, which is an alkylating drug, is often employed to ablate the bone marrow in preparative regimens of HSCT. Cardiotoxicity is a well-known side effect of cyclophosphamide, particularly at larger dosages than 100 mg per kg. Exudative pericarditis, severe myocarditis, malignant arrhythmia, myocardial depression, and CHF are all among these possible side effects (Shanholtz 2001).

2   Graft Versus Host Disease (GvHD)

GvHD is a disease in which immune cells of the donor's bone marrow assault the recipient's tissues because they are recognized as "non-self". GvHD is a frequent consequence of allogeneic HSCT, which involves the implant of bone marrow stem cells from a non-identical donor, while autologous HSCT does not cause it. The increased risk of cardiotoxicity found in patients following allogeneic HSCT compared to autologous HSCT might be due to the development of GvHD (Armenian 2014). Chemotherapy and radiation may also exacerbate this process by causing underlying tissue damage and lead to significant organ damage (Rackley et al. 2005).

There are various reasons explaining why GvHD may increase the risk of cardiovascular events in patients. For starters, having a history of grade II-IV acute GvHD has been linked to the development of several diseases like DM, HTN, and HF, which are well-known cardiovascular risk factors (Armenian et al. 2012). After allogeneic transplantation, acute GVHD is also associated with hypertriglyceridemia and hypercholesterolemia (Rackley et al. 2005). Interestingly, these comorbidities are most likely caused by the adverse effects of calcineurin inhibitors and steroids that are widely used to treat GvHD (Armenian et al. 2012). Ibrutinib, a newly-approved drug for chronic GvHD, has also been shown to be related to atrial fibrillation (AF) (Wang et al. 2013). Ruxolitinib has been introduced to the GVHD therapies as an off-label treatment of steroid-refractory GvHD, despite a documented high incidence of hyperlipidemia (Mesa et al. 2015) (Table 3).

Finally, GvHD may cause inflammation and injury in the vascular endothelium, which can lead to accelerated atherosclerosis and other arterial incidents (Armenian et al. 2012; Tichelli 2008). GvHD is also related to thrombosis and may raise the risk

Table 3 Cardiac complication of common drugs used after HSCT

Drug class	Agent	Cardiovascular toxicities
Anti-thymocyte globulin		Chest pain Peripheral edema Hypertension Hypotension Bradycardia, tachycardia Myocarditis
Calcineuric inhibitors	Tacrolimus Cyclosporine	Hypertrophic cardiomyopathy, Hypertension (tacrolimus) Hypertension, dyslipidemia (cyclosporine)
Corticosteroids	Dexamethasone Methylprednisolone	Hyperglycemia Dyslipidemia Bradycardia Cardiac arrhythmia Hypertension
mTor inhibitor	Sirolimus	Peripheral edema Hypertension Chest pain Hyperglycemia hyperlipidemia
Btk inhibitor	Ibrutinib	Hypertension Arrhythmia [atrial fibrillation, atrial flutter, ventricular arrythmia] Peripheral edema Heart failure Cerebrovascular accident
Tyrosin kinase inhibitor	Ruxolitinib	Hypertension Hyperlipidemia

of microvascular and epicardial diseases, such as myocardial infarction/injury due to its inflammatory environment (Akahori et al. 2003). If the serosal tissue is involved, GvHD might manifest with pericardial effusion or cardiac tamponade, comparable to autoimmune diseases like systemic lupus erythematosus (SLE) (Ferreira et al. 2014).

3 Cardiovascular Comorbidities and Risk Factors

HSCT patients generally have an increased cardiovascular risk because of the common risk factors of cancer and cardiovascular diseases (CVD). Despite the fact that advanced age is a well-known independent CVD risk factor (Yeh and Bickford 2009), advances in HSCT procedures have enabled HSCT to be offered to the elderly. To that aim, patients aged 60 and more accounted for 55% of autologous HSCT recipients and 39% of allogeneic HSCT recipients in 2018 (D'Souza 2020). Because of chemo-radiation history, many elderly HSCT candidates have underlying cardiac dysfunction and/or CVD.

TBI, usually employed in HSCT preparative regimen, has been linked to a higher risk of DM and dyslipidemia (DLP) (Armenian et al. 2012). Furthermore, the 10-year cumulative incidences of hypertriglyceridemia, low high-density lipoprotein (HDL) cholesterol, elevated blood glucose, obesity, and elevated blood pressure were 65.0%, 52.0%, 33.1%, 18.6%, and 8.8%, respectively, in an analysis of 123 HSCT survivors, of whom 95.9% received allogeneic HSCT and 100% were pre-treated with TBI (Friedman et al. 2017).

In previous studies, allogeneic HSCT has been associated with a higher prevalence of cardiovascular risk factors such as smoking, DLP, and HTN (Armenian et al. 2012, 2008; Shanholtz 2001; Khayata et al. 2018; Cardinale et al. 2006; Chow et al. 2011). DM and HTN are accompanied by an increased risk of CHF in autologous HSCT recipients (Armenian et al. 2011b). Having two or more comorbidities (such as chronic lung disease, chronic kidney disease (CKD), DM, and HTN) were linked to the incidence of late CHF in both allogeneic and autologous HSCT in a study by Armenian et al. (Armenian et al. 2008). DLP was found to be prevalent in 36 and 28% of 1196 adult patients undergoing HSCT before autologous and allogenic procedures, respectively, which increased to 62 and 74% within three months (Premstaller et al. 2018) after transplant in a 25-year study on 1196 HSCT survivors.

Cardiac arrhythmias (Peres et al. 2010; Tonorezos et al. 2015; Cardinale et al. 2004), sub-clinical rise of cardiac biomarkers (Cardinale et al. 2000; Snowden et al. 2000; Leger et al. 2018), ischemia episodes (Felker et al. 2000; Chung et al. 2008), and HF are the most common cardiovascular consequences. In comparison to autologous HSCT, allogeneic HSCT survivors had a greater prevalence of cardiovascular risk factors such as HTN, DM, and DLP (Armenian et al. 2012).

The presence of more than one comorbidity is associated with an increased risk of cardiac problems during and after HSCT. A high global cardiovascular risk score, according to Tichelli et al. (2008), is defined as the presence of at least 50% of the following disorders: DLP, arterial HTN, elevated BMI, DM, smoking, or low physical activity. They discovered that having a high global cardiovascular risk score was related to a 9.81 relative risk of having an arterial vascular incident after allogeneic HSCT (Tichelli et al. 2008).

4 Clonal Hematopoiesis of Indeterminate Potential (CHIP)

Clonal Hematopoiesis of Indeterminate Potential (CHIP) is an emerging cardio-vascular risk factor in the HSCT recipients (Gibson and Steensma 2018; Gibson et al. 2017). Somatic mutations may occur in bone marrow cells, just like every other body tissue. The risk of mutations grows with time and, as a result, individual's age. The most typically impacted genes are DNA methyltransferase 3A (DNMT3A) and Tet Methylcytosine Dioxygenase 2 (TET2), which undergo mutations causing them to lose activity, which is consistent with their tumor suppressive identity (Jaiswal et al. 2017). Patients with hematological malignancies, such as AML, have been shown to have mutations in the DNMT3A gene. TET2 mutations are also seen in AML, myelodysplastic syndrome (MDS), and various myeloproliferative disorders (Cimmino et al. 2017). In bone marrow-derived macrophages, deficiency of TET2 promotes the expression of pro- IL-1β through epigenetic processes and processing of pro- IL-1β by regulating the expression and activity of NOD-like" receptor protein3 (NLRP3) inflammasome (Fuster et al. 2017). These cells also insult the blood vessels and cause atherosclerotic plaques to have a much more severe inflammatory milieu. This approach might explain population-based studies that found an eightfold greater risk of myocardial infarction (MI) in individuals who carry TET2 mutations compared to those who do not (Jaiswal et al. 2017). In patients with past acute coronary syndrome (ACS) (i.e. MI) (Ridker et al. 2017a, b), targeted therapy with monoclonal antibodies against IL-1β has been demonstrated to reduce the recurrence of acute ischemia episodes.

Another essential consideration is that different elements, particularly the autonomic nervous system, have an impact on bone marrow. In proliferative disorders like MDS and AML, activation of β2-adrenergic receptors present in the bone marrow promotes progression of the disease at the expense of reduction of β3- adrenergic receptor activity in the regeneration niche (Hanoun et al. 2015). Cardiometabolic risk factors, like DLP, also activate the bone marrow regeneration niche, which is more active in patients with Atherosclerotic Cardiovascular Disease (ASCVD) compared to healthy controls (Seijkens et al. 2014; Dutta et al. 2012, 2015; Heidt et al. 2014; Hoogeveen et al. 2018; Valk et al. 2017). When the bone marrow β3-adrenergic receptors are stimulated, as in the case of MI (Dutta et al. 2012, 2015; Heidt et al. 2014; Hoogeveen et al. 2018; Valk et al. 2017), hematopoietic and myeloid progenitor cells, which seed in the spleen, are released, and feed extramedullary hematopoiesis, and finally reinforce the inflammatory cell population, such as macrophages and monocytes, in atherosclerotic plaques (Dutta et al. 2015; Heidt et al. 2014; Hoogeveen et al. 2018; Valk et al. 2017). This discovery somehow explains the surprising clinical finding of an increased risk of MI recurrency after an MI. It is yet unknown if this axis is changed in individuals with hematological malignancies and to what degree it is altered.

Stress, somatic mutations, and clonal hematopoiesis are hypothesized to raise the cardiovascular risk by raising systemic inflammation (Gibson and Steensma 2018). The frequency of CHIP before HSCT rises with age in adult autologous HSCT recipients and is linked to a higher risk of cardiovascular mortality (Gibson et al. 2017). In the case of allogeneic or pediatric HSCT, however, equivalent outcomes have not been seen (Frick et al. 2019; Collord et al. 2018).

5 Functional Disability

HSCT recipients must remain in the hospital for an average of four weeks or longer. Patients are often severely ill, acquire diminished exercise ability, and lose muscle mass over this lengthy stay (Scott et al. 2016). Although there is a lack of evidence on the effects of longer hospitalization days on cardiovascular complications in HSCT patients, studies on nononcologic patients may be used to estimate this impact. According to one research, 3-weeks of bed rest may cause cardiovascular and pulmonary fitness to deteriorate at the same rate as 40-year of aging (Saltin et al. 1968; McGavock et al. 2009). Given the accompanying increased cardiovascular risks, the effect of extended bed rest on cardiovascular and pulmonary fitness in HSCT recipients is anticipated to be greater than in healthy controls.

References

Akahori M, Nakamae H, Hino M, Yamane T, Hayashi T, Ohta K, et al. Electrocardiogram is very useful for predicting acute heart failure following myeloablative chemotherapy with hematopoietic stem cell transplantation rescue. Bone Marrow Transpl. 2003;31:585–90.

Armenian SHCE. Cardiovascular disease in survivors of hematopoietic cell transplantation. Cancer. 2014;120(4):469–79.

Armenian SH, Sun CL, Francisco L, Steinberger J, Kurian S, Wong FL, et al. Late congestive heart failure after hematopoietic cell transplantation. J Clin Oncol. 2008;26(34):5537–43.

Armenian SH, Sun CL, Mills G, Teh JB, Francisco L, Durand JB, et al. Predictors of late cardiovascular complications in survivors of hematopoietic cell transplantation. Biol BloodMarrow Transpl. 2010;16(8):1138–44.

Armenian SH, Sun CL, Kawashima T, et al. Long-term health-related outcomes in survivors of childhood cancer treated with HSCT versus conventional therapy: a report from the Bone Marrow Transplant Survivor Study (BMTSS) and Childhood Cancer Survivor Study (CCSS). Blood. 2011a;118:1413–20.

Armenian SH, Sun CL, Shannon T, Mills G, Francisco L, Venkataraman K, et al. Incidence and predictors of congestive heart failure after autologous hematopoietic cell transplantation. Blood. 2011b;118(23):6023–9.

Armenian SH, Sun CL, Vase T, Ness KK, Blum E, Francisco L, et al. Cardiovascular risk factors in hematopoietic cell transplantation survivors: role in development of subsequent cardiovascular disease. Blood. 2012;120(23):4505–12.

Cardinale D, Sandri MT, Martinoni A, Tricca A, Civelli M, Lamantia G, et al. Left ventricular dysfunction predicted by early troponin I release after high-dose chemotherapy. J Am Coll Cardiol. 2000;36:517–22.

Cardinale D, Sandri MT, Colombo A, Colombo N, Boeri M, Lamantia G, et al. Prognostic value of troponin I in cardiac risk stratification of cancer patients undergoing high-dose chemotherapy. Circulation. 2004;109:2749–54.

Cardinale D, Colombo A, Sandri MT, Lamantia G, Colombo N, Civelli M, et al. Prevention of high-dose chemotherapy-induced cardiotoxicity in high-risk patients by angiotensin-converting enzyme inhibition. Circulation. 2006;114:2474–81.

Chang HM, Okwuosa TM, Scarabelli T, Moudgil R, Yeh ETH. Cardiovascular complications of cancer therapy: best practices in diagnosis, prevention, and management: part 2. J Am Coll Cardiol. 2017a;70(20):2552–65.

Chang HM, Moudgil R, Scarabelli T, Okwuosa TM, Yeh ETH. Cardiovascular complications of cancer therapy: best practices in diagnosis, prevention, and management: part 1. J Am Coll Cardiol. 2017b;70(20):2536–51.

Chow EJ, Mueller BA, Baker KS, Cushing-Haugen KL, Flowers ME, Martin PJ, et al. Cardiovascular hospitalizations and mortality among recipients of hematopoietic stem cell transplantation. Ann Intern Med. 2011;155:21–32.

Chung T, Lim WC, Sy R, Cunningham I, Trotman J, Kritharides L. Subacute cardiac toxicity following autologous haematopoietic stem cell transplantation in patients with normal cardiac function. Heart (Br Card Soc). 2008;94:911–8.

Cimmino L, et al. Restoration of TET2 function blocks aberrant self-renewal and leukemia progression. Cell. 2017;170:1079-1095.e20.

Collord G, Park N, Podesta M, Dagnino M, Cilloni D, Jones D, et al. Clonal haematopoiesis is not prevalent in survivors of childhood cancer. Br J Haematol. 2018;181(4):537–9.

Cox MA, Kastrup J, Hrubisko M. Historical perspectives and the future of adverse reactions associated with hemopoietic stem cells cryopreserved with dimethyl sulfoxide. Cell Tissue Bank. 2012;13(2):203–15.

D'Souza AF. Current use of and trends in hematopoietic cell transplantation in the United States. Biol Blood Marrow Transplant. 2020;26:e177–82.

Dandoy CE, Hirsch R, Chima R, Davies SM, Jodele S. Pulmonary hypertension after hematopoietic stem cell transplantation. Biol Blood Marrow Transp. 2013;19(11):1546–56.

Dutta P, et al. Myocardial infarction accelerates atherosclerosis. Nature. 2012;487:325–9.

Dutta P, et al. Myocardial infarction activates CCR2+ hematopoietic stem and progenitor cells. Cell Stem Cell. 2015;16:477–87.

Felker GM, Thompson RE, Hare JM, Hruban RH, Clemetson DE, Howard DL, et al. Underlying causes and long-term survival in patients with initially unexplained cardiomyopathy. N Engl J Med. 2000;342:1077–84.

Ferreira DC, de Oliveira JS, Parísio KRF. Pericardial effusion and cardiac tamponade: clinical manifestation of chronic graft versus host disease after allogeneic hematopoietic stem cell transplantation. Rev Bras Hematol Hemoter. 2014;36(2):159–61.

Frick M, Chan W, Arends CM, Hablesreiter R, Halik A, Heuser M, et al. Role of donor clonal hematopoiesis in allogeneic hematopoietic stem-cell transplantation. J Clin Oncol. 2019;37(5):375–85.

Friedman DN, Hilden P, Moskowitz CS, et al. Cardiovascular risk factors in survivors of childhood hematopoietic cell transplantation treated with total body irradiation: a longitudinal analysis. Biol Blood Marrow Transp. 2017;23:475–82.

Fuster JJ, et al. Clonal hematopoiesis associated with TET2 deficiency accelerates atherosclerosis development in mice. Science. 2017;355:842–7.

Ghafoor S, James M, Goldberg J, McArthur JA. Cardiac dysfunction in hematology oncology and hematopoietic cell transplant patients. In: Critical care of the pediatric immunocompromised hematology/oncology patient. Cham: Springer; 2019. pp. 211–235

Gibson CJ, Steensma DP. New insights from studies of clonal hematopoiesis. Clin Cancer Res. 2018;24(19):4633–42.

Gibson CJ, Lindsley RC, Tchekmedyian V, Mar BG, Shi J, Jaiswal S, et al. Clonal hematopoiesis associated with adverse outcomes after autologous stem-cell transplantation for lymphoma. J Clin Oncol. 2017;35(14):1598–605.

Goldberg MA, Antin JH, Guinan EC, Rappeport JM. Cyclophosphamide cardiotoxicity: an analysis of dosing as a risk factor. Blood. 1986;68(5):1114–8.

Gottdiener JS, Appelbaum FR, Ferrans VJ Deisseroth A, Ziegler J. Cardiotoxicity associated with high-dose cyclophosphamide therapy. Arch Intern Med. 1981;141:758–63

Hanoun M, Maryanovich M, Arnal-Estape A, Frenette PS. Neural regulation of hematopoiesis, inflammation, and cancer. Neuron. 2015;86:360–73.

Heidenreich PAKJ. Radiation induced heart disease: systemic disorders in heart disease. Heart. 2009;95:252–8.

Heidt T, et al. Chronic variable stress activates hematopoietic stem cells. Nat Med. 2014;2014(20):754–8.

Hoogeveen RM, et al. Monocyte and haematopoietic progenitor reprogramming as common mechanis underlying chronic inflammatory and cardiovascular diseases. Eur Heart J. 2018;39:3521–7.

Jaiswal S, et al. Clonal hematopoiesis and risk of atherosclerotic cardiovascular disease. N Engl J Med. 2017;377:111–21.

Jodele S, Dandoy CE, Myers KC, El-Bietar J, Nelson A, Wallace G, et al. New approaches in the diagnosis, pathophysiology, and treatment of pediatric hematopoietic stem cell transplantation-associated thrombotic microangiopathy. Transfus Apher Sci. 2016;54(2):181–90.

Kanj SS, Sharara AI, Shpall EJ, Jones RB, Peters WP. Myocardial ischemia associated with high-dose carmustine infusion. Cancer. 1991;68(9):1910–2.

Khayata M, Al-Kindi S, Njoroge L, De Lima MJG, Oliveira GH. Preexisting cardiovascular disease in patients undergoing hematopoietic stem cell transplantation. J Clin Oncol. 2018;36: e19501.

Lancet JE, Uy GL, Cortes JE, Newell LF, Lin TL, Ritchie EK, et al. CPX-351 (cytarabine and daunorubicin) liposome for injection versus conventional cytarabine plus daunorubicin in older patients with newly diagnosed secondary acute myeloid leukemia. J Clin Oncol. 2018;36(26):2684–92.

Leger KJ, Baker KS, Cushing-Haugen KL, Flowers MED, Leisenring WM, Martin PJ, et al. Lifestyle factors and subsequent ischemic heart disease risk after hematopoietic cell transplantation. Cancer. 2018;124:1507–15.

McGavock JM, Hastings JL, Snell PG, et al. A forty-year follow-up of the Dallas bed rest and training study: the effect of age on the cardiovascular response to exercise in men. J Gerontol A Biol Sci Med Sci. 2009;64:293–9.

Mesa RA, Verstovsek S, Gupta V, Mascarenhas JO, Atallah E, Burn T, et al. Effects of ruxolitinib treatment on metabolic and nutritional parameters in patients with myelofibrosis from COMFORT-I. Clin Lymphoma Myeloma Leuk. 2015;15(4):214–21.

Moslehi JJ. Cardiovascular toxic effects of targeted cancer therapies. N Engl J Med. 2016;375(15):1457–67.

Mulrooney DA, Nunnery SE, Armstrong GT, Ness KK, SrivastavaD, Donovan FD, et al. Coronary artery disease detected by coronary computed tomography angiography in adult survivors of childhood Hodgkin lymphoma. Cancer. 2014;120(22):3536–44

Olivieri A, Corvatta L, Montanari M, Brunori M, Offidan M, Ferretti GF, et al. Paroxysmal atrial fibrillation after high-dose melphalan in five patients autotransplanted with blood progenitor cells. Bone Marrow Transp. 1998;21(10):1049–53.

Ovchinsky N, Frazier W, Auletta JJ, Dvorak CC, Ardura M, Song E, et al. Consensus report by the pediatric acute lung injury and sepsis investigators and pediatric blood and marrow transplantation consortium joint working committees on supportive care guidelines for management of veno-occlusive disease in children and adolescents, part 3: focus on cardiorespiratory dysfunction, infections, liver dysfunction, and delirium. Biol Blood Marrow Transp. 2016;24(2):207–18.

Peres E, Levine JE, Khaled YA, Ibrahim RB, Braun TM, Krijanovski OI, et al. Cardiac complications in patients undergoing a reduced-intensity conditioning hematopoietic stem cell transplantation. Bone Marrow Transp. 2010;45(1):149–52.

Pfeiffer TM, Rotz SJ, Ryan TD, Hirsch R, Taylor M, Chima R, et al. Pericardial effusion requiring surgical intervention after stem cell transplantation: a case series. Bone Marrow Transp. 2017;52(4):630–3.

Premstaller M, Perren M, Koçack K, Arranto C, Favre G, Lohri A, et al. Dyslipidemia and lipid-lowering treatment in a hematopoietic stem cell transplant cohort: 25 years of follow-up data. J Clin Lipido. 2018;12:464–63.

Rackley C, Schultz KR, Goldman FD, Chan KW, Serrano A, Hulse JE, et al. Cardiac manifestations of graft-versus-host disease. Biol Blood Marrow Transp. 2005;11:773–80.

Ridker PM, et al. Antiinflammatory therapy with canakinumab for atherosclerotic disease. N Engl J Med. 2017a;377:1119–31.

Ridker PM, et al. Effect of interleukin-1β inhibition with canakinumab on incident lung cancer in patients with atherosclerosis: exploratory results from a randomised, double- blind, placebo-controlled trial. Lancet. 2017b;390:1833–42.

Shanholtz C. Acute life-threatening toxicity of cancer treatment. Crit Care Clin. 2001;17(3):483–502

Sakata-Yanagimoto M, Kanda Y, Nakagawa M, Asano-Mori Y, Kandabashi K, Izutsu K, et al. Predictors for severe cardiac complications after hematopoietic stem cell transplantation. Bone Marrow Transp. 2004;33(10):1043–7.

Saltin B, Blomqvist G, Mitchell JH, Johnson RL Jr., Wildenthal K, Chapman CB. Response to exercise after bed rest and after training. Circulation. 1968;38:VII1–78.

Scott JM, Armenian S, Giralt S, Moslehi J, Wang T, Jones LW. Cardiovascular disease following hematopoietic stem cell transplantation: pathogenesis, detection, and the cardioprotective role of aerobic training. Crit Rev Oncol Hematol. 2016;98:222–34.

Seijkens T, et al. Hypercholesterolemia- induced priming of hematopoietic stem and progenitor cells aggravates atherosclerosis. FASEB J. 2014;28:2202–13.

Snowden JA, Hill GR, Hunt P, Carnoutsos S, Spearing RL, Espiner E, et al. Assessment of cardiotoxicity during haemopoietic stem cell transplantation with plasma brain natriuretic peptide. Bone Marrow Transp. 2000;26:309–13.

Tichelli AGA. Vascular endothelium as "novel" target of graft versus host disease. Best Pr Res Clin Haematol. 2008;21(2):139–48.

Tichelli A, Bucher C, Rovo A, et al. Premature cardiovascular disease after allogeneic hematopoietic stem-cell transplantation. Blood. 2007;110:3463–71.

Tichelli A, Passweg J, Wójcik D, Rovó A, Harousseau JL, Masszi T, et al. Late cardiovascular events after allogeneic hematopoietic stem cell transplantation: a retrospective multicenter study of the late effects working party of the European group for blood and marrow transplantation. Haematologica. 2008;93(8):1203–10.

Tonorezos ES, Stillwell EE, Calloway JJ, Glew T, Wessler JD, Rebolledo BJ, et al. Arrhythmias in the setting of hematopoietic cell transplants. Bone Marrow Transpl. 2015;50:1212–6.

Tuzovic M, Mead M, Young PA, Schiller G, Yang EH. Cardiac complications in the adult bone marrow transplant patient. Curr Oncol Rep. 2019;21(3):28

Vaickus L, Letendre L. Pericarditis induced by highdose cytarabine therapy. Arch Intern Med. 1984;144(9):1868–9.

van der Valk FM, et al. Increased haematopoietic activity in patients with atherosclerosis. Eur Heart J. 2017;38:425–32.

Wang ML, Rule S, Martin P, Goy A, Auer R, Kahl BS, et al. Targeting BTK with ibrutinib in relapsed or refractory mantlecell lymphoma. N Engl J Med. 2013;369(6):507–16.

Yeh ET, Bickford CL. Cardiovascular complications of cancer therapy: incidence, pathogenesis, diagnosis, and management. J Am Coll Cardiol. 2009;53:2231–47.

Cardiotoxicity of Commonly Used Drugs in HSCT

Bita Shahrami and Mohammad Vaezi

Abstract Many drugs used in hematopoietic stem cell transplantation (HSCT) settings have the potential to cause adverse cardiovascular effects. Drug-induced cardiotoxicity is a serious adverse effect with a high rate of morbidity and mortality in patients undergoing autologous or allogeneic HSCT. The mechanism of drug-induced cardiotoxicity can involve direct or indirect damage to cardiac cells. Furthermore, these toxicities may manifest either early or late. There are several risk factors that can increase an individual's susceptibility to cardiotoxicity. Identifying prevention and management strategies for cardiac complications induced by drugs is crucial.

Keywords Adverse drug reaction (ADR) · Cardiac complications · Bone marrow transplantation (BMT) · Hematopoietic stem cell transplantation (HSCT) · Pharmacotherapy

Abbreviations

ADR	Adverse drug reaction
BMT	Bone marrow transplant
Cap	Capsule
CYP450	Cytochrome P450
DR	Delayed-release
ECG	Electrocardiogram

B. Shahrami
Department of Clinical Pharmacy, School of Pharmacy, Tehran University of Medical Sciences, Tehran, Iran

B. Shahrami · M. Vaezi (✉)
Hematology-Oncology and Stem Cell Transplantation Research Center, Tehran University of Medical Sciences, Tehran, Iran
e-mail: m-vaezi@sina.tums.ac.ir

A. Alizadehasl et al. (eds.), *Cardiovascular Considerations in Hematopoietic Stem Cell Transplantation*, https://doi.org/10.1007/978-3-031-53659-5_4

EF	Ejection fraction
ER	Extended-release
GVHD	Graft-versus-host disease
HSCT	Hematopoietic stem cell transplant
IM	Intramuscular
Inj	Injection
IV	Intravenous
Susp	Suspension
Tab	Tablet
TDM	Therapeutic drug monitoring
TdP	Torsade points

1 Introduction

Patients undergoing hematopoietic stem cell transplantation (HSCT) receive complex multiple-drug regimens associated with significant toxicities. These drug toxicities can affect various organ systems, impacting patients' quality of life, duration of hospitalization, and transplant outcomes (Negrin 2020). The cardiovascular system may be adversely affected by several drugs, potentially leading to serious and fatal complications such as heart failure, hypertension, tamponade, and arrhythmia. Drug-induced cardiotoxicity is a serious adverse effect associated with a high rate of morbidity and mortality. Many drugs used in HSCT settings, including before, during, and transplant phases, can have varying degrees of cardiotoxicity. The incidence of drug-induced cardiotoxicity in HSCT patients has been estimated to vary, ranging from 0.9% to 8.9% in different studies (Tuzovic et al. 2019). Chemotherapy agents used as conditioning regimens, immunosuppressive agents for graft-versus-host disease (GVHD) prophylaxis, antimicrobial agents for infection prevention and treatment, as well as supportive care drugs, all have the potential to cause cardiotoxicity. Cardiotoxicity induced by these drugs can develop early or late after HSCT (Raj et al. 2009).

In this chapter, we will review the most commonly used drugs in HSCT that can lead to cardiotoxicity. This topic will also cover the mechanisms, onset, risk factors, and prevention and management strategies for cardiotoxicity associated with each individual drug.

Table 1 summarizes the key characteristics of drug-induced cardiotoxicity in HSCT, while Fig. 1 provides a visual overview of this topic.

Table 1 Characteristics of drugs-induced cardiotoxicity in HSCT

Drug name	Drug class	Indication in HSCT	Cardiotoxicity					
			Very common (> 10%)	Common 1–10%	Uncommon (< 1%)	Rare (< 0.1 %)	Very rare (< 0.01 %)	Not defined
Cy	Alkylating agent	Conditioning	–	–	**Heart failure** **Myocarditis** Tachycardia ECG changes	**Ventricular arrhythmia** Supraventricular arrhythmia Chest pain	Hypotension **Pericarditis**	**Pericardial effusion** **QT prolongation** Heart block **Hemopericardium**
Bu	Alkylating agent	Conditioning	Edema Hypertension Tachycardia Thrombosis Chest pain Vasodilation	**Cardiac Tamponade**	–	–	–	Arrhythmia Cardiomegaly Heart block Hypotension Pericardial effusion Ventricular premature contractions
Mel	Alkylating agent	Conditioning	Peripheral edema **Supraventricular tachycardia** **Atrial fibrilation**	–	–	Ventricular arrhythmia Heart failure Cardiomyopathy	–	Vasculitis
Flu	Antimetabolite	Conditioning	–	Arrhythmia Heart failure Myocardial infarction Cerebrovascular accident	–	**Acute congestive heart failure** Bradycardia	–	–
BCNU	Alkylating agent	Conditioning	–	Chest pain	–	–	**Myocardial ischemia**	Tachycardia Occlusive arterial disease

(continued)

Table 1 (continued)

Drug name	Drug class	Indication in HSCT	Cardiotoxicity					
			Very common (> 10%)	Common 1–10%	Uncommon (< 1%)	Rare (< 0.1 %)	Very rare (< 0.01 %)	Not defined
VP16	Topoisomerase II inhibitor	Conditioning	–	**Hypotension**	–	Myocardial infarction	–	–
ARA-C	Antimetabolite	Conditioning	–	–	–	Cardiomyopathy	–	–
TT	Alkylating agent	Conditioning	–	–	–	**Cardiomyopathy**	–	–
ATG	Immune globulin	GVHD prophylaxis	Hypertension Hypotension Tachycardia Bradycardia Edema	**Cardiopulmonary events** Chest pain	–	–	–	Myocarditis
MTX	Antifolate	GVHD prophylaxis	–	–	–	–	Supraventricular tachycardia Ventricular arrhythmias Myocardial infarction Cardiomyopathy	Chest pain Thrombosis
CsA	CNI	GVHD prophylaxis	**Hypertension Dyslipidemia**	Chest pain Edema	Acute myocardial infarction	–	–	–
Tac	CNI	GVHD prophylaxis	**Hypertension Dyslipidemia Arrhythmia QT prolongation**	Thrombosis Syncope	–	**Myocardial hypertrophy**	–	–

(continued)

Table 1 (continued)

Drug name	Drug class	Indication in HSCT	Cardiotoxicity					
			Very common (> 10%)	Common 1–10%	Uncommon (< 1%)	Rare (< 0.1 %)	Very rare (< 0.01 %)	Not defined
MP	Glucocorticoid	GVHD treatment	–	–	–	–	–	**Fluid retention Hypertension Dyslipidemia Arrhythmias Thromboembolic event Premature atherosclerotic disease**
Vorico	Azole antifungal	Fungal infection	Hypertension	**QT prolongation Torsade de pointes** Myocardial infarction Arrhythmia Cardiomegaly Chest pain Heart failure Orthostatic hypotension	–	–	–	–
Itra	Azole antifungal	Fungal infection	–	Chest pain Peripheral edema Hypertension	–	–	–	**Heart failure** Hypotension
Fluco	Azole antifungal	Fungal infection	–	**Dysrhythmia QT prolongation Torsade de pointes**	–	**Kousin syndrome**	–	–

(continued)

Table 1 (continued)

Drug name	Drug class	Indication in HSCT	Cardiotoxicity					
			Very common (> 10%)	Common 1–10%	Uncommon (< 1%)	Rare (< 0.1 %)	Very rare (< 0.01 %)	Not defined
Posa	Azole antifungal	Fungal infection	Hypertension Hypotension Edema	**QT prolongation Torsade de pointes**	–	–	–	–
Cipro Levo	Fluoroquinolone	Bacterial infection	–	–	**QT prolongation Torsade de pointes Aortic aneurysm Aortic dissection** Myocardial infarction Vasculitis Syncope Bradycardia Tachycardia Hypotension Hypertension	–	–	–

(continued)

Table 1 (continued)

Drug name	Drug class	Indication in HSCT	Cardiotoxicity					
			Very common (> 10%)	Common 1–10%	Uncommon (< 1%)	Rare (< 0.1 %)	Very rare (< 0.01 %)	Not defined
Ondan	5–HT3 antagonist	Nausea/vomiting	–	**QT prolongation Torsade de pointes**	Hypotension	–	–	**Arrhythmia Bradycardia** Tachycardia Angina Peripheral vascular disease
Olanz	Antipsychotic	Nausea/vomiting	**Orthostatic hypotension**	**QT prolongation Torsade de pointes** Chest pain Hypertension Tachycardia Edema	Cerebrovascular accident Syncope			Arrhythmia

The significant adverse drug reaction in the HSCT setting is shown in bolded

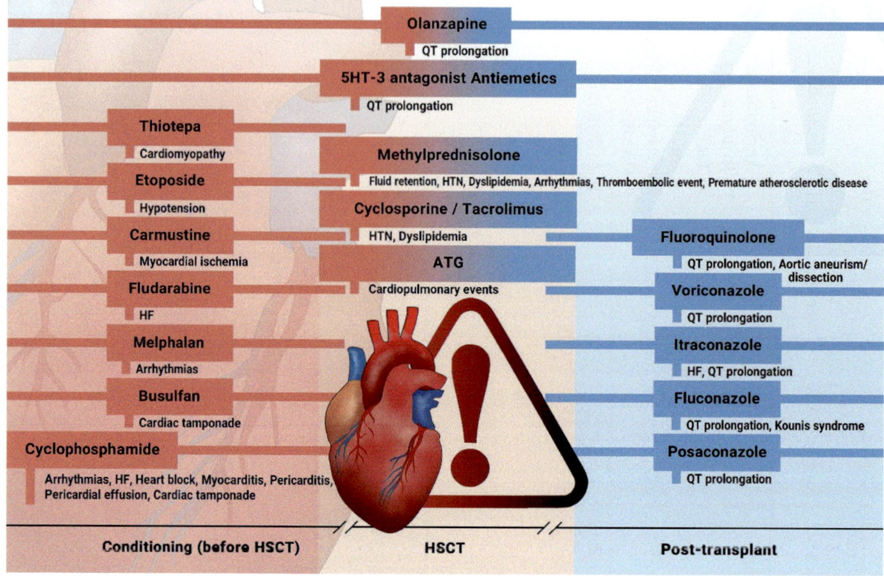

Fig 1 Overview of drug-induced cardiotoxicity in HSCT

Cyclophosphamide (Cy)

Trade name	Endoxan®, Procytox®, Alkyroxan®
Pharmacologic category	Alkylating agent
Dosage forms	Solution, IV: 500 mg/2.5 mL, 1 g/5 mL, 2 g/10 mL Solution reconstituted, IV: 200, 500 mg, 1, 2 g Cap, oral: 25, 50 mg Tab, oral: 25, 50 mg
Route of administration	IV infusion Oral

Cyclophosphamide (Cy) is a nitrogen mustard alkylating agent used as part of the conditioning regimen in autologous and allogeneic HSCTs. Post-transplantation cyclophosphamide has also emerged as a prophylaxis against acute GVHD. This drug is not cell-cycle phase-specific and prevents cell division by cross-linking DNA strands and decreasing DNA synthesis. Additionally, cyclophosphamide possesses potent immunosuppressive and immunomodulatory activities (Ogino and Tadi 2020).

The primary organ affected by cyclophosphamide following HSCT is the heart. Cardiac complications induced by this drug are severe and can limit the use of high-dose therapy. Notable cyclophosphamide-related cardiotoxicities include arrhythmias (supraventricular and ventricular arrhythmia and QT prolongation), heart failure, heart block, hypotension, hemopericardium (secondary to diffuse hemorrhagic myocardial necrosis), myocarditis, exudative pericarditis, and pericardial effusion (Shanholtz 2001; Morandi et al. 2001; Yeh et al. 2021).

Mechanism: The mechanism of cyclophosphamide-induced cardiotoxicity is not clearly established. Cyclophosphamide metabolites, specifically acrolein and/or phosphamide mustard, are believed to cause oxidative stress to the myocardium and direct endothelial injury. It leads to the extravasation of plasma proteinaceous fluid, erythrocytes, and toxic metabolites that cause direct damage to the myocardium and capillary blood vessels. This has been supported by reports of fibrin-fibrin or fibrin-platelet microthrombosis in the capillary and fibrin strands in the interstitium and cardiac myocytes (Dhesi et al. 2013; Appelbaum et al. 1976).

Onset: The onset of cardiotoxicity induced by cyclophosphamide is usually early and within the first 48 h, but may occur up to 10 days after drug initiation.

Risk factors: Higher doses of cyclophosphamide (> 100 mg/kg), older age (> 55 years), preexisting cardiovascular disease, diabetes, prior radiation therapy to the mediastinum or chest wall, prior or concurrent cardiotoxic medications (such as anthracyclines), and a heavy pretreatment are the known risk factors for this adverse effect (Nieto et al. 2000; Bearman et al. 1990). Furthermore, patients with lymphoma are at higher risk for the development of cyclophosphamide cardiotoxicity in comparison to breast cancer patients (Brockstein et al. 2000). Drug interactions that influence the hepatic metabolism of cyclophosphamide or its metabolites may also enhance the cardiotoxic effects of cyclophosphamide. For example, potent inhibition of cytochrome P450 2B6 (CYP2B6) by some azole antifungals decreases cyclophosphamide metabolism and increases conditioning-related toxicities (Marr et al. 2004).

Diagnosis: Cyclophosphamide-induced cardiotoxicity is diagnosed with evidence of clinical and radiological findings. ***Physical exam***: An examination of cardiac percussion, palpation, auscultation, signs of volume overload, and measurement of blood pressure is essential for the detection of cardiac dysfunction and has a prognostic value. ***Echocardiography***: Echocardiography is the most commonly used, non-invasive method for diagnosing cardiotoxicity induced by cyclophosphamide. The earliest change found in the echocardiogram is diastolic dysfunction (Kamezaki et al. 2005). The presence of a reduced ejection fraction (EF) can be an indicator of severe left ventricular dysfunction. In the case of cyclophosphamide-induced hemorrhagic myocarditis, forms of hypertrophic and hypertensive cardiomyopathy may be seen (Birchall et al. 2000). ***ECG***: A 12-lead ECG may be useful for early prediction of acute heart failure. QT prolongation can be the earliest sign of cyclophosphamide cardiotoxicity associated with acute heart failure (Nakamae et al. 2000, 2004). ***Cardiac MRI***: Early left ventricular remodeling and infarction with hemorrhagic transformation can be diagnosed by cardiac MRI (Friedrich et al. 2009). ***Cardiac markers***: Cardiac markers including B-type natriuretic peptide (BNP) and highly sensitive plasma cardiac troponin I or T may be useful in predicting acute heart failure and myocardial damage, respectively (Morandi et al. 2001; Kuittinen et al. 2005).

Prevention: Considering the potential risk factors when planning cyclophosphamide therapy is the most important strategy for preventing fetal cardiotoxicity. Prescribing an alternative treatment is recommended if a patient has significant risk factors or preexisting cardiovascular disease.

Management: Early detection of cyclophosphamide-induced cardiotoxicity along with a timely intervention is crucial to prevent further complications. Cyclophosphamide therapy should be stopped if there are any clinical or laboratory signs of cardiotoxicity. Mild to moderate cardiac toxicities are reversible and expected to be resolved within a few days to weeks following discontinuation of the drug. The management of heart failure and arrhythmia associated with cyclophosphamide is no different from a standard approach. Symptoms may be managed with diuretics, angiotensin-converting enzyme inhibitors (ACEIs) or angiotensin-II receptor blockers (ARBs), beta-blockers, and inotrope agents. These patients may require aggressive monitoring and hemodynamic support (Dhesi et al. 2013). Although supportive care including antioxidants (e.g., ascorbic acid), theophylline, and extracorporeal membrane oxygenation (ECMO) has been proposed, the evidence for these therapies is limited (Kamezaki et al. 2005; Lee et al. 1996).

Busulfan (Bu)

Trade name	Busulfex®, Myleran®
Pharmacologic category	Alkylating agent
Dosage forms	Solution, IV: 6 mg/mL Tab, oral: 2 mg
Route of administration	IV infusion Oral

Busulfan (Bu) is an alkylating agent used with cyclophosphamide as part of the HSCT conditioning regimen. This drug interferes with DNA replication and RNA transcription and also exhibits little immunosuppressive activity (Patel and Busulfan 2021).

Cardiovascular adverse effects of busulfan include edema, tachycardia, hypertension, thrombosis, chest pain, and vasodilation. Furthermore, arrhythmia, cardiomegaly, heart block, pericardial effusion, and ventricular premature contractions may occur with busulfan with a very lower incidence. Cardiac tamponade is a rare but fetal busulfan cardiotoxicity that has been reported with high-dose oral regimens in pediatric patients with thalassemia undergoing HSCT (Angelucci et al. 1992).

Mechanism: The precise mechanism underlying the cardiovascular effects of busulfan has not been established. Pericardial membranes have been suggested as the primary factor in the pathogenesis of cardiac tamponade (Angelucci et al. 1992).

Additionally, long-term busulfan treatment has been associated with endocardial fibrosis of the left ventricle (Weinberger et al. 1975).

Onset: Cardiac tamponade presents with a sudden onset of circulatory shock and cardiac arrest.

Risk factors: Children with thalassemia appear to have a heightened susceptibility to busulfan-induced cardiac tamponade. Moreover, oral regimens of busulfan carry a greater risk of this adverse effect compared to intravenous (IV) administration (Angelucci et al. 1992).

Diagnosis: Patients receiving busulfan should be closely monitored for signs and symptoms of cardiac tamponade. Most patients experience abdominal pain and vomiting as precursors to tamponade, and the diagnosis is confirmed through ultrasonography.

Management: Prompt intervention is required if tamponade is suspected. Immediate fluid removal and emergency pericardiocentesis are the only effective treatments for saving patients.

Melphalan (Mel)

Trade name	Alkeran®
Pharmacologic category	Alkylating agent
Dosage forms	Solution reconstituted, IV: 50 mg Tab, oral: 2 mg
Route of administration	IV infusion Oral

Melphalan (Mel) is a nitrogen mustard alkylating agent used in the conditioning regimen of HSCT for patients with multiple myeloma. It cross-links strands of DNA and inhibits DNA and RNA synthesis by forming carbonium ions. Melphalan is effective against both resting and rapidly diving tumor cells (Jones 2002).

Among HSCT patients, high-dose melphalan is notably associated with arrhythmias, including supraventricular tachycardia (SVT) and atrial fibrillation (AF) (Feliz et al. 2011). Furthermore, high-dose melphalan has been linked to the development of ventricular arrhythmias, heart failure, and cardiomyopathy (Yanamandra et al. 2016; Bleeker et al. 2011; Ritchie et al. 2001; Sampat et al. 2020).

Mechanism: The mechanism of melphalan-induced cardiotoxicity remains relatively unclear. One study reported an increase in the generation of reactive oxygen species (ROS) in response to melphalan treatment (Liu et al. 2020). Additionally, it has been previously identified that direct or indirect damage to cardiac myocytes, possibly acting on an atrial substrate, can lead to arrhythmias following melphalan administration (Nowis et al. 2010).

Onset: The onset of melphalan-induced arrhythmia has been reported to be rapid.

Risk factors: Well-known risk factors for melphalan-induced SVT and AF include older age (> 60 years), renal impairment, the presence of amyloidosis in multiple myeloma, a larger left atrium size, and a history of previous cardiac comorbidities (Feliz et al. 2011; Mileshkin et al. 2005). Additionally, concurrent administration of bortezomib at higher concentrations has been found to increase the cardiotoxicity of melphalan (Nowis et al. 2010).

Prevention: Considering the risk factors when planning high-dose melphalan treatment is an important strategy for preventing cardiotoxic effects.

Management: The standard management of melphalan cardiotoxicity has not been firmly established. Arrhythmia management may necessitate the use of rate- and rhythm-control medications. Amifostine has demonstrated effectiveness in countering the cardiac toxicity associated with higher doses of melphalan (Phillips et al. 2004). Additionally, there is a proposal to use N-Acetyl-L-cysteine (NAC) supplementation to reduce ROS levels and protect cardiomyocytes from cell death, as indicated in an in vitro study (Liu et al. 2020).

Fludarabine (Flu)

Trade name	Fludara®
Pharmacologic category	Antimetabolite
Dosage forms	Solution, IV: 25, 50 mg/2 mL Solution reconstituted, IV: 50 mg Tab, oral: 10 mg
Route of administration	IV infusion Oral

Fludarabine (Flu) is a purine analog antimetabolite that inhibits DNA synthesis, DNA primase, and DNA ligase 1. This drug is the most commonly used agent in conditioning regimens for HSCT.

The cardiovascular adverse reactions listed in the drug package insert include arrhythmia, heart failure, myocardial infarction, and cerebrovascular accidents. In rare instances, the combination of fludarabine with melphalan following HSCT conditioning regimens has been associated with acute congestive heart failure (Ritchie et al. 2001; Newbery et al. 2019).

Mechanism: It remains unclear how the combination of melphalan and fludarabine affects the heart. The indirect targeting of cardiomyocytes and endothelial cells may compromise the protective effects of these chemotherapy agents (Newbery et al. 2019).

Onset: Acute heart failure has been reported during the conditioning with melphalan and fludarabine in documented cases.

Risk factors: As of now, no known risk factors have been identified for the development of fludarabine cardiotoxicity.

Prevention: Predicting the pre-treatment risk for cardiotoxicity is essential.

Management: The treatment of fludarabine-induced heart failure is not well defined, and it may be advisable to apply standard heart failure guidelines.

Carmustin (BCNU)

Trade name	BiCNU®, Gliadel®
Pharmacologic category	Alkylating agent
Dosage forms	Solution reconstituted, IV: 50, 100, 300 mg
Route of administration	IV infusion

Carmustine (BCNU), as a nitrosourea alkylating agent, is widely used as part of BEAM or CBV conditioning regimens for autologous HSCT in patients with lymphoma. This drug alkylates and crosslinks DNA and RNA and may also inhibit enzyme processes through the carbamylation of protein amino acids.

While the cardiac effects of carmustine are extremely rare, there have been reported cases of acute cardiotoxicity associated with high-dose carmustine. Myocardial ischemia (acute angina pectoris) with reversible ECG changes during carmustine infusion has been observed in patients receiving high-dose carmustine, followed by autologous HSCT (Kanj et al. 1991). Additionally, the drug package insert mentions the development of chest pain, flushing, and hypotension with the rapid infusion of carmustine.

Mechanism: The mechanism of carmustine-induced myocardial ischemia remains unknown. Possible mechanisms include increased myocardial oxygen demand, coronary spasm, and alterations in coronary blood flow and distribution—similar to 5-fluorouracil cardiotoxicity (Kanj et al. 1991).

Onset: Chest pain, flushing, and hypotension can occur during carmustine administration, especially with rapid infusion. In some case reports, signs and symptoms of cardiac ischemia have manifested a few minutes after BCNU infusion.

Risk factors: While rapid infusion of carmustine can be associated with flushing, chest pain, and hypotension, the risk factors for carmustine-induced myocardial ischemia are currently unknown.

Prevention: To minimize cardiovascular adverse effects, high-dose carmustine should be infused over at least 2 h. It is also important to monitor blood pressure and vital signs during the infusion.

Management: Supportive care, including intravenous fluid, vasopressors/inotropes, nitrates, and morphine sulfate, has proven effective in managing patients who experience hypotension and myocardial ischemia.

Etoposide (VP16)

Trade name	Toposar®, VePesid®
Pharmacologic category	Topoisomerase II inhibitor
Dosage forms	Solution, IV: 100 mg/5 mL, 500 mg/25 mL, 1 g/50 mL Cap, oral: 50 mg
Route of administration	IV infusion Oral

Etoposide (VP16) is a podophyllotoxin derivative that inhibits DNA topoisomerase II, causing DNA strand breaks. This drug is widely used as part of BEAM or CBV conditioning regimens for autologous HSCT in patients with lymphoma.

The most common cardiac side effect of etoposide is hypotension, typically observed with rapid IV infusion (Yeh and Bickford 2009). The infusion-related hypotension is temporary and has not been associated with cardiac toxicity or ECG changes. While there are a few case reports of etoposide-associated myocardial infarction available, none of them were reported in HSCT settings (Gill et al. 2017; Airey et al. 1995).

Mechanism: Etoposide-induced hypotension results from rapid IV infusion and typically responds promptly to the cessation of the infusion.

Onset: Transient hypotension may occur during the rapid administration of etoposide. However, no delayed hypotension has been reported with this drug.

Risk factors: Etoposide-induced hypotension is infusion-related and is more likely to occur with rapid administration. Geriatric patients may be more susceptible to experiencing etoposide-induced hypotension (McEvoy et al. 2010).

Prevention: To reduce the risk of hypotension, high-dose etoposide should be administered slowly, at a rate of at least 100 mg/m^2/h.

Management: If hypotension occurs during etoposide administration, the infusion should be interrupted. IV hydration or other supportive therapy may be necessary. When reinitiating the infusion, it is recommended to do so at a slower rate.

Cytarabine (ARA-C)

Trade name	Cytosar®
Pharmacologic category	Antimetabolite (Pyrimidine analogue)
Dosage forms	Solution inj: 20, 100 mg/mL
Route of administration	IV infusion SubQ

Cytarabine (ARA-C) is an analog of pyrimidine and is incorporated into DNA. It inhibits DNA polymerase, resulting in decreased DNA synthesis and repair, ultimately leading to cell death during the S phase. Cytarabine is commonly used in the BEAM conditioning regimen for autologous HSCT in patients with lymphoma.

There have been reported cases of cardiomyopathy occurring after high-dose therapy for acute myeloid leukemia (AML) (Yeh and Bickford 2009).

Thiotepa (TT)

Trade name	Tepadina®
Pharmacologic category	Alkylating agent
Dosage forms	Solution reconstituted, IV: 15, 100 mg
Route of administration	IV infusion

Thiotepa (TT) is an alkylating agent commonly used as part of the conditioning regimen for HSCT in patients with solid tumors or hematologic malignancies. This drug produces cross-linking of DNA strands, resulting in disruptions to DNA, RNA, and protein synthesis.

While cardiotoxicity is a known concern with other alkylating agents like cyclophosphamide, there have been no reports of cardiotoxicity associated with the use of thiotepa. However, there have been reported cases of cardiomyopathy linked to high-dose thiotepa (Alidina et al. 1999).

Mechanism: Preclinical models suggest that thiotepa-induced cardiotoxicity may be attributed to damage to cardiac myofibrils and cardiac neural ganglia, similar to the damage induced by doxorubicin (Pannuti and Barboni 1972).

Onset: Thiotepa-induced cardiomyopathy typically presents with an acute onset, often occurring during the immediate post-HSCT period (Alidina et al. 1999).
Risk factors: Proposed risk factors for thiotepa-induced cardiomyopathy include pre-existing cardiac conditions with reduced EF, a history of a second transplant, and female sex (particularly in females undergoing allogeneic HSCT).

Prevention: It is recommended to exercise caution when using high doses of thiotepa in patients with a history of cardiac disease. In some cases, it may be necessary to consider alternative preparative regimens, including the use of less intensive, non-myeloablative regimens.

Management: The management of thiotepa-induced cardiomyopathy does not differ significantly from the standard approach to cardiomyopathy.

Anti-thymocyte Globulin (ATG)

Trade name	Atgam®
Pharmacologic category	Immune globulin
Dosage forms	Solution reconstituted, IV: 25 mg

(continued)

(continued)

Trade name	Atgam®
Route of administration	IV infusion

Anti-thymocyte globulin (ATG) is a polyclonal purified antibody derived from horses or rabbits that targets human lymphocytes, leading to the depletion of T-lymphocytes and subsequent immunosuppression. This drug is used to prevent both acute and chronic GVHD following allogeneic HSCT.

Cardiovascular reactions associated with ATG include myocarditis, cardiac irregularity, chest pain, edema, hypertension, hypotension, tachycardia, and bradycardia. These adverse cardiovascular effects are relatively rare, occurring in less than 5% of cases (Godown et al. 2011; Elazhary and Alawyat 2022).

Mechanism: Cytokine release and activation of monocytes and lymphocytes may contribute to cytokine release syndrome (CRS), which, in turn, can lead to cardiopulmonary events following the administration of ATG.

Onset: Adverse cardiovascular reactions, especially bradycardia, have been reported during or immediately after the infusion of ATG.

Risk factors: Rapid infusion rates of ATG have been associated with CRS. Furthermore, patients with underlying bradycardia may be more susceptible to experiencing severe bradycardia due to ATG (Godown et al. 2011).

Prevention: Currently, there is no standardized preventive strategy. However, ATG should be administered through a high-flow vein over a period of at least 6 h to minimize the risk.

Management: In most reported cases, bradycardia has improved after the completion of ATG treatment. For cases of bradycardia during ATG administration, the use of isoproterenol and glycopyrrolate may be effective (Godown et al. 2011).

Methotrexate (MTX)

Trade name	Trexall®
Pharmacologic category	Antimetabolite (Antifolate)
Dosage forms	Solution, IV: 50 mg/2 mL, 250 mg/10 mL, 1 g/40 mL Solution reconstituted, IV: 1 g Tab, oral: 5, 7.5, 10, 15 mg
Route of administration	IV Oral

Methotrexate (MTX) is an antifolate agent that inhibits dihydrofolate reductase, disrupting DNA synthesis. Short-course methotrexate is the standard prophylaxis against GVHD in allogenic HSCT (Ram and Storb 2013).

Methotrexate-induced cardiotoxicity is extremely rare due to the drug's favorable effects on cardiopulmonary endothelium. However, there have been reported cases of methotrexate-induced cardiac conduction disturbances, including SVT, ventricular arrhythmias, myocardial infarction, and cardiomyopathy (Shah et al. 2022; Bidaki et al. 2017). It's worth noting that none of these reports were associated with HSCT.

Mechanism: Methotrexate can promote cell death, possibly affecting cardiac myocytes.

Risk factors: Most reported cases of methotrexate-associated cardiotoxicity have resulted from methotrexate poisoning or overdose. Some individuals with preexisting morbidities like heart failure have also experienced such toxicity.

Prevention: The most crucial preventive strategy appears to be therapeutic drug monitoring (TDM) of methotrexate to prevent drug overdose.

Management: In cases of methotrexate toxicity, the initiation of leucovorin (LCV) rescue should commence as soon as possible. In some instances, cardiac function has improved after initiating acute cardiomyopathy management, which may include veno-arterial extracorporeal membrane oxygenation (VA-ECMO) (Shah et al. 2022).

Cyclosporine A (CsA)

Trade name	Sandimmune®, Neoral®
Pharmacologic category	Calcinurin inhibitor
Dosage forms	Solution, IV: 50 mg/mL Cap, oral: 25, 50, 100 mg Solution, oral: 100 mg/mL
Route of administration	IV infusion Oral

Cyclosporine A (CsA) is a calcineurin inhibitor (CNI) that inhibits the production and release of interleukin II and several other cytokines in T lymphocytes. This immunosuppressive drug is widely used for the prophylaxis of acute GVHD in allogenic HSCT (Choi and Reddy 2014).

The most common cardiovascular side effect with CNIs is systemic hypertension, which can occur in 13–53% of patients. Cyclosporine, in particular, has a higher incidence of this side effect compared to tacrolimus. Other cardiovascular adverse effects of CNIs include dyslipidemia, characterized by increased low-density lipoprotein (LDL) cholesterol, decreased high-density lipoprotein (HDL) cholesterol, and elevated triglyceride (TG) levels (Kockx and Kritharides 2012). These cardiovascular complications can have significant implications for short-term and long-term transplant outcomes, increasing morbidity and mortality (Miller 2002).

Mechanism: ***Hypertension***: The precise mechanism underlying CNI-induced hypertension remains unknown and appears to be independent of nephrotoxicity. Several factors have been suggested to contribute to the development of hypertension, including increased renal tubular salt reabsorption, activation of neurohormonal vasoconstriction, alterations in vascular reactivity, and the involvement of the sympathetic nervous system. Vasoconstriction is triggered by elevated levels of vasoconstrictor mediators such as sympathetic tone, the renin-angiotensin system, and endothelin-1, coupled with reduced levels of vasodilator mediators like prostaglandins and nitric oxide. Furthermore, cyclosporine activates renal Na–K-Cl cotransporters through both local and systemic mechanisms (Taler et al. 1999). ***Dyslipidemia***: Cyclosporine-induced dyslipidemia is thought to result from various mechanisms. This drug binds to the LDL receptor, leading to increased LDL cholesterol levels. Additionally, it reduces post-heparin lipolytic activity while increasing hepatic lipase activity, ultimately impairing the clearance of VLDL and LDL cholesterol (Kockx and Kritharides 2012).

Onset: ***Hypertension***: CNI-induced hypertension typically develops within the first few weeks of therapy, although the onset can vary. ***Dyslipidemia***: Dyslipidemia tends to occur after approximately three months of treatment with CNIs.

Risk factors: ***Hypertension***: Known risk factors for the development of cyclosporine-induced hypertension include higher doses, longer duration of therapy, preexisting hypertension, an elevated body mass index (BMI), male gender, black ethnicity, concurrent glucocorticoid use, and a history of recurrent disease (Taler et al. 1999). ***Dyslipidemia***: long-term treatment with cyclosporine has been associated with the development of dyslipidemia.

Prevention: ***Hypertension***: To prevent cyclosporine-induced hypertension, it is essential to monitor blood pressure regularly following any changes in the dosage of cyclosporine. Careful monitoring of cyclosporine plasma concentrations is crucial to guide dosing and ensure that therapeutic levels are maintained to maximize effectiveness while minimizing side effects. Blood concentrations should be checked within 2–3 days of starting cyclosporine and after any dosage adjustments. The target cyclosporine level may vary based on factors such as the time of HSCT and the presence of GVHD (Miller 2002). Additionally, because cyclosporine is extensively metabolized by CYP3A4, it can interact with other drugs, potentially leading to adverse reactions like hypertension. Healthcare providers should consider these drug-drug interactions to prevent toxicity. Non-pharmacologic strategies, including weight management, regular exercise, smoking cessation, and control of dyslipidemia and diabetes if present, may also help mitigate the risk of severe hypertension. ***Dyslipidemia***: For the prevention of cyclosporine-induced dyslipidemia, it is recommended to monitor the lipid profile both at baseline and three months after initiating cyclosporine therapy.

Management: ***Hypertension***: In patients with no history of hypertension who develop sustained high blood pressure during cyclosporine therapy, it is advisable to reduce the cyclosporine dosage by 25–50%, or to the lowest effective dose. Typically, blood pressure elevation responds well to dose reduction. However, if adjusting the immunosuppressive dose is not feasible, antihypertensive medications may be necessary. Calcium channel blockers (CCBs) are the preferred choice for antihypertensive treatment during CNI therapy. Dihydropyridine CCBs, such as amlodipine, are usually the first-line treatment. It's essential to note that non-dihydropyridine

CCBs, like diltiazem, can impact cyclosporine metabolism, potentially increasing its serum concentration. Therefore, if diltiazem is initiated, an immediate reduction in the cyclosporine dose and close monitoring of blood levels are warranted. If blood pressure remains uncontrolled with CCBs, other antihypertensive drugs can be added. ***Dyslipidemia***: In patients who develop hyperlipidemia as a result of cyclosporine therapy, lipid-lowering treatment may be necessary to achieve target lipid levels. HMG-CoA reductase inhibitors (statins) are the preferred choice for lipid-lowering therapy. However, it's crucial to consider their potential for serious drug interactions in HSCT patients. Cyclosporine weakly inhibits enzymes such as CYP3A4, CYP2C9, OATPs, and BCRP, which can lead to increased serum concentrations of all statins and an elevated risk of rhabdomyolysis (Ballantyne et al. 2003). To manage cyclosporine-induced dyslipidemia, it's recommended to use the lowest effective statin dose (i.e., atorvastatin 10 mg or an equivalent dose). Additionally, close clinical monitoring and measurement of creatine kinase (CK) levels to detect muscle injury are advised when co-administering cyclosporine and statins. In cases where patients cannot respond to or tolerate statin therapy, conversion from cyclosporine to tacrolimus may be considered.

Tacrolimus (Tac)

Trade name	Prograf®
Pharmacologic category	Calcinurin inhibitor
Dosage forms	Solution, IV: 5 mg/mL Cap, oral: 0.5, 1, 5 mg Tab ER, oral: 0.75, 1, 4 mg
Route of administration	IV infusion Oral

Tacrolimus (Tac or FK506) is a CNI that inhibits T lymphocyte activation by binding to an intracellular protein, FKBP-12. This drug serves as an alternative to cyclosporine for the prophylaxis of GVHD following allogeneic HSCT.

Tacrolimus use carries a risk of hypertension and dyslipidemia, although these risks are generally lower when compared to cyclosporine. Importantly, myocardial hypertrophy is a rare but potentially life-threatening cardiotoxicity associated with tacrolimus. Other cardiovascular adverse effects may include abnormal T waves on ECG, QT interval prolongation, and the potential for torsades de pointes (TdP) arrhythmias. Additional cardiovascular concerns with tacrolimus use encompass heart failure, arrhythmias, thrombosis, and syncope.

Mechanism: ***Hypertension***: The mechanism underlying tacrolimus-induced hypertension is similar to that of cyclosporine (refer to the "Cyclosporine" section). ***Dyslipidemia***: The mechanism of tacrolimus-induced dyslipidemia is similar to that of cyclosporine (refer to the "Cyclosporine" section). ***Myocardial hypertrophy***: The precise mechanism behind tacrolimus-induced myocardial hypertrophy remains unclear. This drug has the potential to increase the thickness of the left ventricular wall, leading to heart failure associated with hypertrophic cardiomyopathy (Atkison et al. 1995). *QT prolongation*: QT prolongation induced by tacrolimus is thought to result from the inhibition of delayed rectifier K+ channels, inward rectifier K+ channels, and L-type Ca2+ currents (Hodak et al. 1998).

Onset: *Hypertension*: The onset of tacrolimus-induced hypertension is similar to cyclosporine (refer to the "Cyclosporine" section). *Dyslipidemia*: The onset of tacrolimus-induced dyslipidemia is similar to cyclosporine (refer to the "Cyclosporine" section). *Myocardial hypertrophy*: Cardiomyopathy associated with tacrolimus treatment typically emerges several months after initiating therapy (Hernández Silva et al. 2022).

Risk factors: *Hypertension*: The risk factors associated with tacrolimus-induced hypertension are similar to those associated with cyclosporine (refer to the "Cyclosporine" section). *Dyslipidemia*: The risk factors associated with tacrolimus-induced dyslipidemia are similar to those associated with cyclosporine (refer to the "Cyclosporine" section). *Myocardial hypertrophy*: Tacrolimus-induced cardiomyopathy is primarily reported in pediatric transplant recipients. Furthermore, maintaining blood levels below 15 ng/mL is crucial to mitigate the risk of this side effect (Nakata et al. 2000).

Prevention: *Hypertension*: The prevention strategies for tacrolimus-induced hypertension are similar to those associated with cyclosporine (refer to the "Cyclosporine" section). *Dyslipidemia*: The prevention strategies for tacrolimus-induced dyslipidemia is similar to those associated with cyclosporine (refer to the "Cyclosporine" section). *Myocardial hypertrophy*: TDM is essential to ensure the effectiveness of tacrolimus and prevent toxicity. *QT prolongation*: Tacrolimus should be avoided in patients with congenital long QT syndrome due to its potential to prolong the QT interval. Before initiating tacrolimus therapy and during its course, it is important to correct any electrolyte abnormalities, such as hypokalemia, hypomagnesemia, and hypocalcemia.

Management: *Hypertension*: The management of tacrolimus-induced hypertension is similar to cyclosporine (refer to the "Cyclosporine" section). *Dyslipidemia*: The mechanism of tacrolimus-induced dyslipidemia is similar to cyclosporine (refer to the "Cyclosporine" section). *Myocardial hypertrophy*: If myocardial hypertrophy is diagnosed, a dose reduction or discontinuation of tacrolimus is necessary. In such cases, cyclosporine or an mTOR inhibitor can be considered alternatives in these patients. *QT prolongation*: For patients experiencing palpitations, lightheadedness, or dizziness during tacrolimus therapy, an ECG should be conducted. If the QTc interval exceeds 450 ms in males or 460 ms in females, tacrolimus and any concomitant medications known to prolong the QT interval should be temporarily withheld. Additionally, it is essential to correct any electrolyte abnormalities, such as hypocalcemia, hypokalemia, or hypomagnesemia (Khatib et al. 2021).

Methylprednisolone Succinate (MP)

Trade name	Solu-Medrol®
Pharmacologic category	Corticosteroid
Dosage forms	Solution reconstituted, IV: 125, 500, 1, 2 g
Route of administration	IV/IV infusion

Methylprednisolone (MP), a systemic corticosteroid, is the preferred treatment for acute GVHD of grade II or higher in patients undergoing allogeneic HSCT. Similar to other corticosteroids, methylprednisolone works by reducing inflammation through the suppression of polymorphonuclear leukocyte migration and the reversal of increased capillary permeability.

Corticosteroids can induce various adverse cardiovascular effects, including fluid retention, hypertension, dyslipidemia, arrhythmias, thromboembolic events, and the development of premature atherosclerotic diseases, such as myocardial infarction, stroke, heart failure, and increased all-cause mortality. Notably, serious cardiovascular toxicities, including sudden death, have been reported following pulse infusions of corticosteroids (Saag and Furst 2012).

Mechanism: The principal cardiovascular side effects of glucocorticoids stem from their impact on various physiological factors, including plasma volume, fluid and electrolyte retention, epinephrine synthesis, and angiotensin levels. Steroids increase glomerular filtration, stimulate the synthesis of angiotensinogen and atrial natriuretic peptide, and elevate peripheral vascular resistance, resulting in systemic hypertension. Moreover, the mineralocorticoid-receptor-mediated effects of glucocorticoids can induce sodium retention and expand plasma volume, further contributing to arterial hypertension. Additionally, they modulate vascular tone by reducing the expression of calcium-activated potassium channels. Glucocorticoid treatment elevates the risk of cardiovascular disease through its impact on insulin resistance, blood glucose levels, lipid profiles, blood pressure, and obesity, which can potentiate atherosclerosis and increase the likelihood of thromboembolic complications (Lorraine and John 2013; Ng and Celermajer 2004).

Risk factors: The risk of corticosteroid-induced cardiotoxicity depends on both the dose and duration of treatment.

Prevention: The use of systemic corticosteroid should be approached with caution in patients who have other risk factors for cardiovascular disease, particularly those with a history of heart failure, hypertension, or recent myocardial infarction. To minimize the risk of significant adverse effects, it is advisable to use the lowest possible dose and limit the duration of corticosteroid treatment. In patients undergoing methylprednisolone therapy for GVHD, it is essential to monitor blood pressure, blood glucose levels, and electrolyte levels. After one week of acute GVHD treatment, the dosage of methylprednisolone should be gradually tapered based on the patient's response (Saag and Furst 2012).

Management: Fluid retention and hypertension should be managed as part of the standard approach. Consideration should be given to withholding corticosteroids if patients develop acute heart failure, myocardial infarction, or stroke during the treatment.

Azole Antifungals

Azole antifungal agents are the backbone of the prevention and treatment of systemic fungal infections in patients undergoing HSCT. The most commonly used azoles in the HSCT setting are fluconazole, itraconazole, voriconazole, and posaconazole.

Voriconazole

Trade name	Vfend®
Pharmacologic category	Antifungal agent (azole derivative)
Dosage forms	Solution reconstituted, IV: 200 mg Tab: 50, 200 mg Susp, oral: 40 mg/mL
Route of administration	IV infusion Oral

Azole antifungals, including voriconazole, have been associated with prolonged QT intervals on ECG, which may lead to TdP, polymorphic ventricular arrhythmias, cardiac arrest, and sudden death (Philips et al. 2007). Other cardiovascular adverse reactions listed in the drug package insert include hypertension, myocardial infarction, arrhythmia, cardiomegaly, chest pain, heart failure, and orthostatic hypotension.

Mechanism: Azole antifungals can block both I_{Kr} and rapidly activate delayed rectifier potassium channels, increasing the risk of TdP (Cleary et al. 2013).

Onset: The onset of voriconazole-induced QT prolongation varries and may occur within the first 24 h up to 23 days after initiation.

Risk factors: Several risk factors have been proposed for the development of azole-induced cardiotoxicties including females, age > 65 years, structural heart disease, congenital long QT syndrome, history of myocardial infarction or heart failure with reduced EF, genetic defects of cardiac ion channels, longer baseline QTc interval (e.g., > 450 ms) or lengthening of the QTc by $\geq$ 60 ms, electrolyte disturbances (e.g., hypocalcemia, hypokalemia, hypomagnesemia), bradycardia, hepatic or kidney impairment, diuretic use, sepsis, critical illness, HIV (especially when CD4 count is less than 200 cells/mm^3), drug-induced QTc prolongation/TdP, a history of drug-induced TdP, and concurrent administration of multiple medications that prolong the QT interval or medications with drug interactions that increase serum concentrations of QT-prolonging medications (Ashley and Perfect 2017). Furthermore, drug-drug interactions play a significant role in the risk related to the cardiovascular toxicity of voriconazole, either through an additive pharmacodynamic effect (e.g., domperidone), reducing the clearance of voriconazole (e.g., amiodarone), or by lowering serum electrolytes (e.g., amphotericin B). QT prolongation may occur independently of drug concentrations (Alkan et al. 2004).

Prevention: Electrolyte abnormalities (e.g., hypokalemia, hypomagnesemia, hypocalcemia) should be corrected before initiating and during azole therapy. Dose adjustment based on serum concentration and TDM are helpful in ensuring efficacy and avoiding toxicity (Elewa et al. 2015).

Management: An ECG should be performed for all patients who experience palpitation, lightheadedness, or dizziness during azole therapy. Voriconazole and any concomitant drugs known to prolong the QT interval should be withheld if QTc is > 450 mesc (in males) or > 460 mesc (in females). Additionally, any electrolyte abnormalities (e.g., hypocalcemia, hypokalemia, hypomagnesemia) should be corrected (Khatib et al. 2021).

Itraconazole

Trade name	Itranom®
Pharmacologic category	Antifungal agent (azole derivative)
Dosage forms	Cap, oral: 100 mg Solution, oral: 10 mg/mL
Route of administration	Oral

Itraconazole can cause a triad of hypertension, hypokalemia, and peripheral edema (Sharkey et al. 1991). The FDA warns that this drug can cause or exacerbate heart failure (Ahmad et al. 2001). When itraconazole was administered IV to animals and healthy human volunteers, negative inotropic effects were observed.

Mechanism: The mechanism of itraconazole-induced heart failure is unknown. It is thought that itraconazole has a direct negative inotropic effect on the heart, leading to decreased cardiac contractility. It may also decrease the heart rate, coronary flow, and prolongation of PR and QRS intervals (Qu et al. 2013). In addition, fluid retention and hypertension induced by itraconazole may be related to the inhibition of 11-β-hydroxysteroid dehydrogenase and mineralocorticoid excess (Hoffmann et al. 2018).

Onset: The time between itraconazole initiation and the presentation of cardiotoxicity has been reported to vary widely (range: 0.3–104 weeks, median: 4 weeks) (Teaford et al. 2020).

Risk factors: In post-marketing experience, heart failure was more frequently reported in patients receiving a total daily dose of 400 mg of itraconazole, although there were also cases reported among those receiving lower total daily doses.

Prevention: Itraconazole should be used with caution in patients with heart failure with reduced EF and in patients with risk factors for heart failure, including those with chronic obstructive pulmonary disease (COPD), renal failure, edematous disorders, ischemic or valvular disease. Signs and symptoms of heart failure should be monitored carefully during treatment. Dose adjustment based on trough serum concentration and TDM are recommended to ensure efficacy and avoid toxicity.

Management: Itraconazole should be discontinued if signs or symptoms of heart failure occur or worsen during administration. The reinitiating of treatment should be reassessed, considering the risk–benefit of use.

Fluconazole

Trade name	Diflucan®
Pharmacologic category	Antifungal agent (azole derivative)
Dosage forms	Solution, IV: 200 mg/100 mL, 400 mg/200 mL Tab, oral: 50, 100, 150, 200 mg Susp, oral: 10 mg/mL
Route of administration	IV infusion Oral

Among azole antifungals, fluconazole has the most arrhythmia adverse effects. This drug has been associated with QT prolongation, which may lead to TdP or polymorphic ventricular arrhythmias. Another rare vascular adverse effect of fluconazole is Kounis syndrome, also called hypersensitivity coronary syndrome or allergic angina (Mahal 2016). This syndrome was first described by Kounis as the "concurrence of acute coronary syndromes with conditions associated with mast cell activation, involving interrelated and interacting inflammatory cells such as allergic or anaphylactoid insults" (Kounis 2006).

Mechanism: ***QT prolongation***: Similar to other azole antifungals, fluconazole may block I_{Kr} and rapidly activate delayed rectifier potassium channels, increasing the risk of TdP. ***Kounis syndrome***: Kounis syndrome is thought to be caused by an allergen mediated-IgE and mast cell activation and degranulation, leading to histamine release.

Onset: ***QT prolongation***: The onset of fluconazole-induced QT prolongation is rapid and usually occurs within the first 24 h to a couple of days after initiation. ***Kounis syndrome***: Kounis syndrome may occur immediately after the first dose of fluconazole.

Risk factors: ***QT prolongation***: Risk factors for the development of fluconazole-induced QT prolongation are similar to voriconazole (refer to the "Voriconazole" section). ***Kounis syndrome***: Type I Kounis syndrome may be seen in patients with no obvious cardiac risk factors and non-obstructive coronary anatomy (Mancano 2017).

Prevention: ***QT prolongation***: Electrolyte abnormalities (e.g., hypokalemia, hypomagnesemia, hypocalcemia) should be corrected before initiating and during azole therapy. Considering drug-drug interactions is crucial to prevent further toxicities. ***Kounis syndrome***: Patients should be monitored for signs and symptoms of chest pain and coronary vasospasm.

Management: ***QT prolongation***: Fluconazole and any concomitant drugs known to prolong the QT interval should be withheld if QTc > 450 mesc (in males) or > 460 mesc (in females). Moreover, any electrolyte abnormalities (e.g., hypocalcemia, hypokalemia, hypomagnesemia) should be corrected (Khatib et al. 2021). ***Kounis syndrome***: The occurrence of an allergic reaction accompanied by chest pain and ECG changes following fluconazole therapy should alert the clinician to potential Kounis syndrome. Patients should be promptly managed with antihistamines, corticosteroids, and anti-Ig E monoclonal antibodies. Nitrates and CCBs with the antivasospastic properties have also been proven effective (Mahal 2016).

Posaconazole

Trade name	Noxafil®, Posanol®
Pharmacologic category	Antifungal agent (azole derivative)
Dosage forms	Solution, IV: 300 mg/16.7 mL Tab, DR: 100 mg Susp, oral: 40 mg/mL
Route of administration	IV infusion Oral

Similar to other azole antifungals, posaconazole can cause QT prolongation and TdP, but the incidence of this adverse effect is very low (Mohr et al. 2008). The mechanism, onset, risk factors, prevention, and management of posaconzaole-induced TdP are similar to voriconazole (refer to the "Voriconazole" section).

Fluoroquinolones (FQ)

Ciprofloxacin

Trade name	Cipro®
Pharmacologic category	Antibiotic, fluoroquinolone
Dosage forms	Solution, IV: 200 mg/100 mL, 400 mg/200 mL Tab, oral: 100, 250, 500, 750 mg Tab, ER: 500, 1000 mg Susp, oral: 250 mg/5 mL, 500 mg/5 mL
Route of administration	IV infusion Oral

Levofloxacin

Trade name	Levaquin®, Tavanex®
Pharmacologic category	Antibiotic, fluoroquinolone
Dosage forms	Solution, IV: 500 mg/100 mL, 750 mg/150 mL Tab, oral: 250, 500, 750 mg Solution, oral: 25 mg/mL Susp, oral: 250, 500 mg/5 mL
Route of administration	IV infusion Oral

Fluoroquinolones (FQs) are highly effective antibiotics used for bacterial infections prophylaxis during neutropenia in high risk HSCT patients. While generally safe and well tolerated, rare but severe cardiovascular adverse effects may occur with their use. Notable cardiotoxicities associated with fluoroquinolones include QT prolongation and aortic aneurysm and dissection. The drug package insert also reports myocardial infarction, vasculitis, syncope, bradycardia, tachycardia, hypotension, and hypertension.

Mechanism: ***QT prolongation***: Fluoroquinolones can prolong the QT interval by inhibiting KCNH2 (Potassium Voltage-Gated Channel Subfamily H Member 2), which may lead to TdP (Kang et al. 2001). ***Aortic aneurysm/dissection***: Fluoroquinolones can cause aortic aneurysm and dissection by up-regulation of matrix metalloproteinase (MMP) enzymes known to play a role in aneurysm development and capable of damaging components of the extracellular matrix. These antibiotics also directly affect the viability of chondrocytes and tenocytes responsible for collagen synthesis, leading to the generation of reactive oxygen species, caspase activation, and apoptosis.

Onset: ***QT prolongation***: The onset of QT prolongation varies and is usually seen at supra-therapeutic doses of fluoroquinolones. ***Aortic aneurysm/dissection***: The onset of fluoroquinolone-associated aortic aneurysm and dissection is time-related and delayed, typically occurring approximately 60 days after initiating fluoroquinolone therapy.

Risk factors: ***QT prolongation***: Risk factors for fluoroquinolone-induced QT prolongations are similar to other drugs, including females, individualized over 65 years old, a history of structural heart disease (e.g., myocardial infarction, heart failure with reduced EF), genetic defects of cardiac ion channels, congenital long QT syndrome, a history of drug-induced TdP, baseline QT prolongation (e.g., > 500 ms) or lengthening of the QTc by ≥ 60 ms, electrolyte disturbances (e.g., hypocalcemia, hypokalemia, hypomagnesemia), bradycardia, kidney or hepatic impairment, the use of loop diuretic, sepsis, concurrent administration of multiple medications (≥2) that prolong the QT interval, and drug-drug interactions leading to QT prolongation. Among fluoroquinolones, moxifloxacin is associated with the highest risk of QT prolongation, arrhythmia, and cardiovascular mortality, followed by levofloxacin and then ciprofloxacin (Chou et al. 2015; Gorelik et al. 2019). ***Aortic aneurysm/ dissection***: Risk factors for the development of fluoroquinolone-induced aortic aneurysm and/or dissection include old age, peripheral vascular disease, a history of aneurysms, atherosclerosis, hypertension, genetic conditions predisposing to aortic aneurysm (e.g., Marfan syndrome, Ehlers-Danlos syndrome), and longer courses of therapy (>14 days). Severe emotional or physical stress has been correlated with the onset of pain following this adverse effect (Lee et al. 2015; Singh and Nautiyal 2017).

Prevention: ***QT prolongation***: Fluoroquinolone use should be avoided in patients taking other QT-prolonging drugs, patients with long QT syndromes, and those with other significant risk factors for arrhythmia. Serum potassium and magnesium levels should be monitored and corrected. ***Aortic aneurysm/dissection***: The FDA recommends avoiding fluoroquinolones in patients with any risk factor for aortic aneurysm and dissection, including patients with Marfan syndrome, Ehlers-Danlos syndrome, peripheral atherosclerotic vascular diseases, uncontrolled hypertension, and/or advanced age.

Management: ***QT prolongation***: An ECG should be offered for all patients presenting palpitation, lightheadedness, or dizziness during fluoroquinolone therapy. Fluoroquinolone and any concomitant drugs known to prolong the QT interval should be withheld if QTc > 450 mesc (in males) or > 460 mesc (in females). Additionally, any electrolyte abnormalities (e.g., hypocalcemia, hypokalemia, hypomagnesemia) should be corrected (Khatib et al. 2021). ***Aortic aneurysm/dissection***: Fluoroquinolone should be stopped immediately if a patient reports side effects suggestive of aortic aneurysm or dissection.

5-HT3 Antagonist

Ondansetron

Trade name	Zofran®, Demitron®
Pharmacologic category	5-HT3 antagonist
Dosage forms	Solution, inj: 4 mg/2 mL Tab, oral: 4, 8 mg Solution, oral: 4 mg/5 mL
Route of administration	IV IM Oral

5-hydroxytryptamine 3 (5-HT3) receptor antagonists are the preferred agents for preventing chemotherapy-induced nausea and vomiting in patients undergoing HSCT. While these medications provide powerful symptom relief, they may have the potential to prolong the QT interval and increase the risk of serious arrhythmias such as TdP.

Mechanism: QT prolongation with 5-HT3 antagonists may occur due to blockade of cardiac HERG K+ channels. The suppression of autonomic reflexes may contribute to bradycardia, hypotension, and tachyarrhythmias (Tricco et al. 2013).

Onset: The onset of QT-prolongation induced by 5-HT3 antagonists is rapid and usually occurs 1 to 2 h after administration. However, QT-prolongation has been recorded within 15 min after administration. This adverse effect may persist more than 2 h after the administered dose.

Risk factors: Known risk factors for 5-HT3 antagonists-induced QT-prolongation include higher doses, IV formulation, concomitant medications that prolong the QT interval, females, hypothermia, concomitant volatile anesthetics or cumulative high-dose anthracycline therapy, underlying heart disease, including heart failure or acute coronary syndrome, electrolyte abnormalities (e.g., hypokalemia, hypomagnesemia), and a history of QT prolongation, bradycardia, tachycardia, or cardiac rhythm disorders (especially ventricular arrhythmia) (Li and Ramos 2017). Among 5-HT3 antagonists, ondansetron has the highest risk of QT prolongation, particularly when administrated in IV single doses of more than 16 mg.

Prevention: ECG monitoring is recommended in high-risk patients, including the elderly, those using other medications known to prolong QT interval, patients with heart failure, bradyarrhythmias, and those receiving cumulative high-dose anthracycline therapy. Serum potassium and magnesium levels should also be monitored and corrected. Ondansetron and dolasetron should be avoided in patients with congenital long QT syndrome, hypokalemia, hypomagnesemia, and complete heart block. Single IV doses > 16 mg of ondansetron are no longer recommended due to the potential for QT prolongation.

Management: An ECG should be offered to all patients who present with palpitation, lightheadedness, or dizziness during 5-HT3 antagonists (particularly ondansetron) therapy. Ondansetron and any concomitant drugs known to prolong the QT interval should be withheld if QTc > 450 mesc (in males) or > 460 mesc (in females). Additionally, any electrolyte abnormalities (e.g., hypocalcemia, hypokalemia, hypomagnesemia) should be corrected (Khatib et al. 2021).

Olanzapine

Trade name	Zyprexa®
Pharmacologic category	Antipsychotic
Dosage forms	Solution reconstituted, IM: 10 mg Tab: 2.5, 5, 7.5, 10, 15, 20 mg
Route of administration	IM Oral

Olanzapine is a second-generation antipsychotic that blocks 5-HT2 and dopamine receptors, making it useful for the prevention of chemotherapy-induced nausea and vomiting. It is considered a first line treatment in many protocols for nausea and vomiting prophylaxis in HSCT patients (Yuda et al. 2022). Olanzapine can cause a mild degree of prolonged QT interval on an ECG (Porta-Sánchez et al. 2017) and is associated with orthostatic hypotension accompanied by tachycardia and dizziness (Jana et al. 2015).

Mechanism: ***QT prolongation***: Olanzapine prolongs cardiac repolarization by blocking the rapid component of the delayed rectifier potassium current. ***Orthostatic hypotension***: The orthostatic hypotension associated with olanzapine is attributed to alpha-1 adrenergic receptor antagonism.

Onset: ***QT prolongation***: QT prolongation usually occurs rapidly. ***Orthostatic hypotension***: The onset of olanzapine-induced orthostatic hypotension is rapid and typically occurs during the initial dose titration.

Risk factors: ***QT prolongation***: Risk factors for olanzapine-induced QT prolongation are similar to other drugs with this adverse effect and include females, age > 65 years, structural heart disease (e.g., history of MI or heart failure with a reduced EF), history of drug-induced TdP, genetic defects of cardiac ion channels, congenital long QT syndrome, baseline QTc interval prolongation (e.g., > 500 ms) or lengthening of the QTc by $\geq$ 60 ms electrolyte disturbances (e.g., hypokalemia, hypocalcemia, hypo-magnesemia), bradycardia, hepatic or kidney impairment, substance use, and coad-ministration of multiple medications ($\geq$ 2) that prolong the QT interval or increase drug interactions that raise serum drug concentrations of QTc prolonging medica-tions (Li and Ramos 2017). ***Orthostatic hypotension***: Risk factors for the develop-ment of orthostatic hypotension include older adults, rapid dose titration, known cardiovascular disease (history of myocardial infarction, ischemic heart disease, heart failure, or conduction abnormalities), hypovolemia/dehydration, and concomi-tant medications that also cause or exacerbate orthostatic hypotension (e.g., tricyclic antidepressant).

Prevention: ***QT prolongation***: ECG monitoring is recommended in high-risk patients, especially when the injection form is administrated. Serum potassium and magnesium levels should also be monitored and corrected. ***Orthostatic hypoten-sion***: Blood pressure should be closely monitored in high-risk patients receiving olanzapine before any titrating dose.

Management: ***QT prolongation***: Olanzapine and any concomitant drug known to prolong the QT interval should be withheld if QTc > 450 mesc (in males) or > 460 mesc (in females). Any electrolyte abnormalities (e.g., hypocalcemia, hypokalemia, hypomagnesemia) should also be corrected (Khatib et al. 2021). ***Orthostatic hypotension***: Slowly tapering may be helpful in managing orthostatic hypotension.

References

Ahmad SR, Singer SJ, Leissa BG. Congestive heart failure associated with itraconazole. The Lancet. 2001;357(9270):1766–7.

Airey C, Dodwell D, Joffe J, Jones W. Etoposide-related myocardial infarction. Clin Oncol. 1995;7(2):135.

Alidina A, Lawrence D, Ford L, Baer MR, Bambach B, Bernstein SH, et al. Thiotepa-associated cardiomyopathy during blood or marrow transplantation: association with the female sex and cardiac risk factors. Biol Blood Marrow Transp. 1999;5(5):322–7.

Alkan Y, Haefeli W, Burhenne J, Stein J, Yaniv I, Shalit I. Voriconazole-induced QT interval prolongation and ventricular tachycardia: a non-concentration-dependent adverse effect. Clin Infect Dis. 2004;39(6):e49–52.

Angelucci E, Lucarelli G, Baronciani D, Durazzi S, Galimberti M, Giardini C, et al. Sudden cardiac tamponade after chemotherapy for marrow transplantation in thalassaemia. The Lancet. 1992;339(8788):287–9.

Appelbaum F, Strauchen J, Graw JRR, Savage D, Kent K, Ferrans V, et al. Acute lethal carditis caused by high-dose combination chemotherapy: a unique clinical and pathological entity. The Lancet. 1976;307(7950):58–62.

Ashley ED, Perfect J. Pharmacology of azoles. UpToDate. In: Kauffman CA, editor. UpToDate. Waltham, MA; 2017

Atkison P, Joubert G, Barron A, Grant D, Wall W, Rosenberg H, et al. Hypertrophic cardiomyopathy associated with tacrolimus in paediatric transplant patients. The Lancet. 1995;345(8954):894–6.

Ballantyne CM, Corsini A, Davidson MH, Holdaas H, Jacobson TA, Leitersdorf E, et al. Risk for myopathy with statin therapy in high-risk patients. Arch Intern Med. 2003;163(5):553–64.

Bearman S, Petersen F, Schor R, Denney J, Fisher L, Appelbaum F, et al. Radionuclide ejection fractions in the evaluation of patients being considered for bone marrow transplantation: risk for cardiac toxicity. Bone Marrow Transp. 1990;5(3):173–7.

Bidaki R, Kian M, Owliaey H, Zarch MB, Feysal M. Accidental chronic poisoning with methotrexate; report of two cases. Emergency. 2017;5(1)

Birchall I, Lalani Z, Venner P, Hugh J. Fatal haemorrhagic myocarditis secondary to cyclophosphamide therapy. Br J Radiol. 2000;73(874):1112–4.

Bleeker J, Gertz M, Pellikka P, Buadi F, Dingli D, Dispenzieri A, et al. Cardiomyopathy following high dose melphalan conditioning prior to autologous peripheral blood stem cell transplantation for multiple myeloma and primary amyloidosis. Biol Blood Marrow Transp. 2011;17(2):S203.

Brockstein B, Smiley C, Al-Sadir J, Williams S. Cardiac and pulmonary toxicity in patients undergoing high-dose chemotherapy for lymphoma and breast cancer: prognostic factors. Bone Marrow Transp. 2000;25(8):885–94.

Choi SW, Reddy P. Current and emerging strategies for the prevention of graft-versus-host disease. Nat Rev Clin Oncol. 2014;11(9):536–47.

Chou H-W, Wang J-L, Chang C-H, Lai C-L, Lai M-S, Chan KA. Risks of cardiac arrhythmia and mortality among patients using new-generation macrolides, fluoroquinolones, and β-lactam/β-lactamase inhibitors: a Taiwanese nationwide study. Clin Infect Dis. 2015;60(4):566–77.

Cleary JD, Stover KR, Farley J, Daley W, Kyle PB, Hosler J. Cardiac toxicity of azole antifungals; 2013

Dhesi S, Chu MP, Blevins G, Paterson I, Larratt L, Oudit GY, et al. Cyclophosphamide-induced cardiomyopathy: a case report, review, and recommendations for management. J Investigat Med High Impact Case Rep. 2013;1(1):2324709613480346.

Elazhary S, Alawyat HA. Bradycardia associated with antithymocyte globulin treatment of a pediatric patient with sickle cell disease: a case report and literature review. Hematol Transfus Cell Ther. 2022;44:284–7.

Elewa H, El-Mekaty E, El-Bardissy A, Ensom MH, Wilby KJ. Therapeutic drug monitoring of voriconazole in the management of invasive fungal infections: a critical review. Clin Pharmacokinet. 2015;54:1223–35.

Feliz V, Saiyad S, Ramarao SM, Khan H, Leonelli F, Guglin M. Melphalan-induced supraventricular tachycardia: Incidence and risk factors. Clin Cardiol. 2011;34(6):356–9.

Friedrich MG, Sechtem U, Schulz-Menger J, Holmvang G, Alakija P, Cooper LT, et al. Cardiovascular magnetic resonance in myocarditis: a JACC white paper. J Am Coll Cardiol. 2009;53(17):1475–87.

Gill D, Mann K, Kaur S, Ruiz V, Dean R. Rare cause of cardiotoxicity Arch Med. 2017;9:1.

Godown J, Deal AM, Riley K, Bailliard F, Blatt J. Worsening bradycardia following antithymocyte globulin treatment of severe aplastic anemia. J Pediat Pharm Therapeut. 2011;16(3):218–21.

Gorelik E, Masarwa R, Perlman A, Rotshild V, Abbasi M, Muszkat M, et al. Fluoroquinolones and cardiovascular risk: a systematic review, meta-analysis and network meta-analysis. Drug Saf. 2019;42:529–38.

Hernández Silva G, Puerto Chaparro RG, Martínez Melo JÁ, Porras Bueno CO, Martínez Rodríguez JE, González Trillos SJ. Hypertrophic cardiomyopathy secondary to tacrolimus therapy in a kidney transplant patient: a case report and focused review of the literature. Clinical Case Rep. 2022;10(11): e6539.

Hodak SP, Moubarak JB, Rodriguez I, Gelfand MC, Alijani MR, Tracy CM. QT prolongation and near fatal cardiac arrhythmia after intravenous tacrolimus administration: a case report. Transplant. 1998;66(4):535–7.

Hoffmann WJ, McHardy I, Thompson GR III. Itraconazole induced hypertension and hypokalemia: mechanistic evaluation. Mycoses. 2018;61(5):337–9.

Jana AK, Praharaj SK, Roy N. Olanzapine-induced orthostatic hypotension. Clin Psychopharmacol Neurosci. 2015;13(1):113.

Jones RB. Clinical pharmacology of melphalan and its implications for clinical resistance to anticancer agents. Clin. Relev. Resis. Cancer Chemo. 2002:305–22

Kamezaki K, Fukuda T, Makino S, Harada M. Cyclophosphamide-induced cardiomyopathy in a patient with seminoma and a history of mediastinal irradiation. Intern Med. 2005;44(2):120–3.

Kang J, Wang L, Chen X-L, Triggle DJ, Rampe D. Interactions of a series of fluoroquinolone antibacterial drugs with the human cardiac K+ channel HERG. Mol Pharmacol. 2001;59(1):122–6.

Kanj SS, Sharara AI, Shpall EJ, Jones RB, Peters WP. Myocardial ischemia associated with high-dose carmustine infusion. Cancer. 1991;68(9):1910–2.

Khatib R, Sabir FR, Omari C, Pepper C, Tayebjee MH. Managing drug-induced QT prolongation in clinical practice. Postgrad Med J. 2021;97(1149):452–8.

Kockx M, Kritharides L. Cyclosporin A-induced hyperlipidemia. Lipoproteins-Role in Health and Diseases. Rijeka: InTech Publishers; 2012. pp. 337–54

Kounis NG. Kounis syndrome (allergic angina and allergic myocardial infarction): a natural paradigm? Int J Cardiol. 2006;110(1):7–14.

Kuittinen T, Husso-Saastamoinen M, Sipola P, Vuolteenaho O, Ala-Kopsala M, Nousiainen T, et al. Very acute cardiac toxicity during BEAC chemotherapy in non-Hodgkin's lymphoma patients undergoing autologous stem cell transplantation. Bone Marrow Transp. 2005;36(12):1077–82.

Lee C, Harman G, Hohl R, Gingrich R. Fatal cyclophosphamide cardiomyopathy: its clinical course and treatment. Bone Marrow Transp. 1996;18(3):573–7.

Lee C-C, Lee M-G, Chen Y-S, Lee S-H, Chen Y-S, Chen S-C, et al. Risk of aortic dissection and aortic aneurysm in patients taking oral fluoroquinolone. JAMA internal medicine. 2015;175(11):1839–47

Li M, Ramos LG. Drug-induced QT prolongation and torsades de pointes. Pharm Therapeut. 2017;42(7):473.

Liu R, Li D, Sun F, Rampoldi A, Maxwell JT, Wu R, et al. Melphalan induces cardiotoxicity through oxidative stress in cardiomyocytes derived from human induced pluripotent stem cells. Stem Cell Res Ther. 2020;11:1–17.

Lorraine I, John A. Physiologic and pharmacologic effects of corticosteroids. Holland-Frei Canc Med. 2013:34–67

Mahal HS. Fluconazole-induced type 1 Kounis syndrome. Am J Ther. 2016;23(3):e961–2.

Mancano MA. ISMP adverse drug reactions: allergic angina caused by fluconazole rhabdomyolysis caused by risperidone high incidence of hyponatremia with high-dose trimethoprim-sulfamethoxazole lithium carbonate-induced hypersalivation persistent hemorrhage after idarucizumab administration. Hosp Pharm. 2017;52(7):455–8.

Marr KA, Leisenring W, Crippa F, Slattery JT, Corey L, Boeckh M, et al. Cyclophosphamide metabolism is affected by azole antifungals. Blood. 2004;103(4):1557–9.

McEvoy G, Miller J, Snow E. AHFS 2010 drug information. Bethesda, MD: American Society of Health-Systems Pharmacists Inc.; 2010.

Mileshkin L, Seymour J, Wolf M, Gates P, Januszewicz E, Joyce P, et al. Cardiovascular toxicity is increased, but manageable, during high-dose chemotherapy and autologous peripheral blood stem cell transplantation for patients aged 60 years and older. Leuk Lymphoma. 2005;46(11):1575–9.

Miller LW. Cardiovascular toxicities of immunosuppressive agents. Am J Transp. 2002;2(9):807–18.

Mohr J, Johnson M, Cooper T, Lewis JS, Ostrosky-Zeichner L. Current options in antifungal pharmacotherapy. Pharmacotherapy: J. Human Pharmacol. Drug Therapy. 2008;28(5):614–45

Morandi P, Ruffini P, Benvenuto G, La Vecchia L, Mezzena G, Raimondi R. Serum cardiac troponin I levels and ECG/Echo monitoring in breast cancer patients undergoing high-dose (7 g/m2) cyclophosphamide. Bone Marrow Trans. 2001;28(3):277–82.

Nakamae H, Tsumura K, Hino M, Hayashi T, Tatsumi N. QT dispersion as a predictor of acute heart failure after high-dose cyclophosphamide. The Lancet. 2000;355(9206):805–6.

Nakamae H, Hino M, Akahori M, Terada Y, Yamane T, Ohta K, et al. Predictive value of QT dispersion for acute heart failure after autologous and allogeneic hematopoietic stem cell transplantation. Am J Hematol. 2004;76(1):1–7.

Nakata Y, Yoshibayashi M, Yonemura T, Uemoto S, Inomata Y, Tanaka K, et al. Tacrolimus and myocardial hypertrophy. Transp. 2000;69(9):1960–2.

Negrin RS. Early complications of hematopoietic cell transplantation. UpToDate. 2020. https://www.uptodate.com/contents/earlycomplications-of…

Newbery G, de Alcantara LN, Gurgel LA, Driscoll R, Lima CC. Persistent heart failure following melphalan and fludarabine conditioning. J Cardiol Cases. 2019;20(3):88–91.

Ng M, Celermajer D. Glucocorticoid treatment and cardiovascular disease. BMJ Publishing Group Ltd; 2004. pp. 829–30

Nieto Y, Cagnoni PJ, Bearman SI, Shpall EJ, Matthes S, Jones RB. Cardiac toxicity following high-dose cyclophosphamide, cisplatin, and BCNU (STAMP-I) for breast cancer. Biol Blood Marrow Transp. 2000;6(2):198–203.

Nowis D, Mączewski M, Mackiewicz U, Kujawa M, Ratajska A, Wieckowski MR, et al. Cardiotoxicity of the anticancer therapeutic agent bortezomib. Am J Pathol. 2010;176(6):2658–68.

Ogino MH, Tadi P. Cyclophosphamide. 2020

Pannuti F, Barboni F. Histopathological study of the heart-damaging effects of thio-tepa in the chick embryo "in situ". 3. Quaderni Sclavo di diagnostica clinica e di laboratorio. 1972;8(2):669–78

Patel R, Busulfan TP. StatPearls [Internet]: StatPearls Publishing; 2021

Philips J, Marty F, Stone R, Koplan B, Katz J, Baden L. Torsades de pointes associated with voriconazole use. Transpl Infect Dis. 2007;9(1):33–6.

Phillips GL, Meisenberg B, Reece D, Adams V, Badros A, Brunner J, et al. Amifostine and autologous hematopoietic stem cell support of escalating-dose melphalan: a phase I study. Biol Blood Marrow Transp. 2004;10(7):473–83.

Porta-Sánchez A, Gilbert C, Spears D, Amir E, Chan J, Nanthakumar K, et al. Incidence, diagnosis, and management of QT prolongation induced by cancer therapies: a systematic review. J Am Heart Assoc. 2017;6(12): e007724.

Qu Y, Fang M, Gao B, Amouzadeh HR, Li N, Narayanan P, et al. Itraconazole decreases left ventricular contractility in isolated rabbit heart: mechanism of action. Toxicol Appl Pharmacol. 2013;268(2):113–22.

Raj SR, Stein CM, Saavedra PJ, Roden DM. Cardiovascular effects of noncardiovascular drugs. Circulation. 2009;120(12):1123–32.

Ram R, Storb R. Pharmacologic prophylaxis regimens for acute graft-versus-host disease: past, present and future. Leuk Lymphoma. 2013;54(8):1591–601.

Ritchie D, Seymour J, Roberts A, Szer J, Grigg A. Acute left ventricular failure following melphalan and fludarabine conditioning. Bone Marrow Transp. 2001;28(1):101–3.

Saag KG, Furst DE. Major side effects of systemic glucocorticoids. 2012

Sampat PJ, Riaz S, Martinez F, Aiello D. Bortezomib plus melphalan-induced cardiomyopathy presenting as sinus tachycardia and systolic heart failure. Cureus. 2020;12(7)

Shah S, Haeger-Overstreet K, Flynn B. Methotrexate-induced acute cardiotoxicity requiring veno-arterial extracorporeal membrane oxygenation support: a case report. J Med Case Rep. 2022;16(1):447.

Shanholtz C. Acute life-threatening toxicity of cancer treatment. Crit Care Clin. 2001;17(3):483–502.

Sharkey PK, Rinaldi MG, Dunn JF, Hardin TC, Fetchick RJ, Graybill JR. High-dose itraconazole in the treatment of severe mycoses. Antimicrob Agents Chemother. 1991;35(4):707–13.

Singh S, Nautiyal A. Aortic dissection and aortic aneurysms associated with fluoroquinolones: a systematic review and meta-analysis. Am. J. Med. 2017;130(12):1449–57. e9

Taler SJ, Textor SC, Canzanello VJ, Schwartz L. Cyclosporin-induced hypertension: incidence, pathogenesis and management. Drug Saf. 1999;20:437–49.

Teaford HR, Saleh OMA, Villarraga HR, Enzler MJ, Rivera CG. The many faces of itraconazole cardiac toxicity. Mayo Clin Proceed: Innovat, Qual Outcom. 2020;4(5):588–94.

Tricco AC, Soobiah C, Antony J, Hemmelgarn B, Moher D, Hutton B, et al. Safety of serotonin (5-HT3) receptor antagonists in patients undergoing surgery and chemotherapy: protocol for a systematic review and network meta-analysis. Syst Rev. 2013;2:1–5.

Tuzovic M, Mead M, Young PA, Schiller G, Yang EH. Cardiac complications in the adult bone marrow transplant patient. Curr Oncol Rep. 2019;21:1–16.

Weinberger A, Pinkhas J, Sandbank U, Shaklai M, de Vries A. Endocardial fibrosis following busulfan treatment. JAMA. 1975;231(5):495

Yanamandra U, Gupta S, Khadwal A, Malhotra P. Melphalan-induced cardiotoxicity: ventricular arrhythmias. Case Reports. 2016:bcr2016218652

Yeh ET, Bickford CL. Cardiovascular complications of cancer therapy: incidence, pathogenesis, diagnosis, and management. J Am Coll Cardiol. 2009;53(24):2231–47.

Yeh J, Whited L, Saliba RM, Rondon G, Banchs J, Shpall E, et al. Cardiac toxicity after matched allogeneic hematopoietic cell transplant in the posttransplant cyclophosphamide era. Blood Adv. 2021;5(24):5599–607.

Yuda S, Fuji S, Savani B, Gatwood KS. Antiemetic strategies in patients who undergo hematopoietic stem cell transplantation. Clin Hemat Inter. 2022;4(3):89–98.

Pre-HSCT Cardiovascular Evaluation

Mehrdad Payandeh, Mohammad Eslami Jouybari, Mohammad Dabiri, Elgar Enamzadeh, and Mina Mohseni

Abstract Curing multiple hematologic malignancies and occasionally non-malignant and benign conditions in a potential manner, the procedure of hematopoietic stem cell transplantation (HSCT) poses a significant threat to morbidity and mortality. An extensive pretransplant workup, aimed at minimizing risks, is part of the efforts to make transplantation safer. This includes the optimal selection of transplant candidates. HSCT offers a potential cure for multiple and different hematologic malignancies and, in rare cases, non-malignant conditions. However, it is important to note that this procedure carries a considerable risk of morbidity and mortality. An extensive pretransplant workup and optimal selection of transplant candidates comprise attempts to improve the safety of the procedure. Prior research has demonstrated an association between pre-transplant abnormalities and transplant outcomes. The assessment of individual parameters, such as pulmonary function tests (PFTs), findings in echocardiography tests, and the evaluation of the Charlson Comorbidity Index (CCI), like the Hematopoietic cell transplantation—specific comorbidity index (HCT-CI), are part of this process. These studies reveal correlations between pretransplant health status and eventual transplant success. Cardiac complications including heart failure, pulmonary edema, sudden cardiac death, arrhythmia, and pericarditis arising in the first 100 days following HSCT are categorized as acute cardiotoxicity related to the procedure. Prior anthracycline treatment is an established predisposing factor for cardiotoxicity. Notably, diminished cardiac systolic and diastolic function can present few months to years after initial anthracycline exposure. Consequently,

M. Payandeh
Kermanshah University of Medical Sciences, Kermanshah, Iran

M. E. Jouybari
Gastrointestinal Cancer Research Center, Mazandaran University of Medical Sciences, Sari, Iran

M. Dabiri
Hematology-Oncology Department, Cancer Institute, Imam Khomeini Hospital, School of medicine, Tehran University of Medical Sciences, Tehran, Iran

E. Enamzadeh (✉) · M. Mohseni
Cardio-Oncology Research Center, Rajaie Cardiovascular Medical and Research Center, Iran University of Medical Sciences, Tehran, Iran
e-mail: Elgar_enamzadeh@yahoo.com

comprehensive cardiac evaluation is imperative for all patients undergoing HSCT. This chapter will examine recommended assessments of cardiac status in HSCT candidates with a history of anthracycline therapy.

Keywords Hematopoietic stem cell transplantation · Pretransplant · Cardiovascular evaluation · Cardiac function · Heart failure · Echocardiogram · Cardio-toxicit

Abbreviations

AML	Acute myeloid leukemia
ADT	Androgen deprivation therapies
ACE	Angiotensin-converting enzyme
BNP	Brain natriuretic peptide
CV	Cardiovascular
CVD	Cardiovascular disease
CML	Chronic myeloid leukaemia
CGA	Complete geriatric assessment
CHF	Congestive heart failure
GLS	Global longitudinal strain
HF	Heart failure
HSCT	Hematopoietic stem cell transplantation
IMiDs	Immunomodulatory drugs
LVEF	Left ventricular ejection fraction
MRA	Magnetic resonance imaging
MUGA	Multigated acquisition
MI	Myocardial infarction
PIs	Proteasome inhibitors
PAH	Pulmonary arterial hypertension
PH	Pulmonary hypertension
RIC	Reduced-intensity conditioning
TKIs	Tyrosine kinase inhibitors
VEGF	Vascular endothelial growth factor

1 Introduction

1. Used as a potential cure for certain non-malignant diseases and various hematological cancers, HSCT has been employed as a life-saving therapy since its introduction in 1980. There has been a continuous rise in the quantity of allogeneic stem cell transplantations conducted since the initial implementation of

this treatment, such that by 2013, more than 320,000 people had received this type of transplant (Pasquini and Zhu 2014).

2. Allogeneic HSCT is widely accepted as the prevailing treatment for various hematologic disorders. However, this procedure carries a significant burden of potential complications and fatalities. In order to enhance the safety of transplantation, significant endeavors have been made to meticulously choose suitable candidates for the procedure and conduct thorough examinations prior to the actual transplant.

3. While guidelines regarding the pretransplant assessment have been issued, there is considerable heterogeneity in the scope and logistical aspects of the pre-transplant examination across different institutions (Pasquini and Zhu 2014).

4. Several investigations have indicated a link between pretransplant anomalies and the success of transplantation. This connection can be observed in various aspects, including specific metrics like pulmonary function tests and echocardiography, as well as through the application of comorbidity scores like the specialized index for hematopoietic cell transplantation-related comorbidities (Lehmann et al. 2000).

5. Nonetheless, limited research has been conducted to investigate whether an in-depth pre-transplant assessment effectively decreases transplant-associated mortality rates through the identification of hidden infections or undetectable organ impairments.

6. The preparative chemotherapy regimen and the patient's body immune response are the primary contributors to treatment-related toxicity in induvial received HSCT. In this procedure, myelosuppressive chemotherapy, with or without total body irradiation, is administered prior to the infusion of a hematopoietic stem cell graft. This graft induces an immunologic response against malignant cells.

7. Acute cardiotoxicity refers to the development of cardiac complications, such as heart failure, sudden cardiac death, pulmonary edema, arrhythmia, and pericarditis in the first 100 days of HSCT. Research indicates that post-transplant acute cardiac complications occur at varying incidences, ranging from 5 to 43%, and are associated with a death rate of 1.2% (Lehmann et al. 2000).

8. During HSCT, cardiovascular stress is a significant concern due to the need for cardiovascular reserve to manage various factors such as volume expansion from cytotoxic therapy, anemia, vasoparalysis, and sepsis-induced myocardial suppression. Additionally, patients undergoing HSCT are almost certain to experience sepsis, further adding to the cardiovascular burden. To achieve optimal outcomes, it is essential to address any decline in the function of the cardiovascular system caused by cytotoxicity, even if it is not permanent.

9. Identifying patients at risk for cardiotoxicity in the population undergoing HSCT, particularly in cases of AML and heavily pre-treated patients with lymphoma, is crucial due to the known association between previous anthracycline therapy and cardiac impairment. Significantly, the reduction in heart function may become evident several months to several years following a previous encounter. It is widely believed that monitoring the ejection fraction is adequate for identifying patients who may be susceptible to cardiotoxicity.

Hematopoietic stem cell transplantation related cardiovascular complications

Causes and risk factors	Diagnostics	Treatment and prevention
• Anthracyclines • Radiotherapy • Tyrosine kinase inhibitors and other molecular-targeted agents • Allogenic hematopoietic stem cell transplantation • High-risk conditioning regimen: cyclophosphamide • Graft versus host disease (GvHD)	• Electrocardiogram (ECG) • 2D-echocardiography • 3D-echocardiography • Global longitudinal strain analysis (GLS) • Cardiac magnetic resonance • Troponins • Natriuretic peptides	• Risk factor evaluation • Pre-transplantation cardiovascular assessment • Post-transplantation cardiovascular assessment • Angiotensin-converting enzyme (ACE) inhibitors • Angiotensin II receptor blockers (ARB) • Beta-blockers (BBs) • Antioxidants

2 Cardiovascular Complications Associated with HSCT

Cardiac complications, both in the acute and late stages, continue to pose substantial obstacles for individuals undergoing hematopoietic stem cell transplantation (HSCT). These complications have the potential to contribute to adverse health outcomes and increased mortality rates among HSCT survivors.

2.1 Short-Term Cardiotoxicity during HSCT

Within the initial 100 days following transplantation or during HSCT, individuals may experience life-threatening cardiovascular conditions such as arrhythmias, acute heart failure,pericardial effusion or tamponade, and cardiac arrest. Research has demonstrated that these cardiac complications, which have the potential to jeopardize life, can manifest during HSCT or within the first 100 days following transplantation (Tuzovic et al. 2019). Lethal arrhythmia within one year was frequently observed in HSCT patients, leading to extended hospital stays and a higher likelihood of admission to an intensive care unit. In adult patients, the incidence of arrhythmia was estimated to range from 2 to 10% (Rotz et al. 2021). This highlights the significant impact of arrhythmia in HSCT patients, as it not only increases the risk of death but also prolongs the overall hospitalization period (Tonorezos et al. 2015; Goldberg et al. 1998). Cases of severe pericardial effusion/tamponade and myocardial infarction have been reported in recipients of hematopoietic stem cell transplantation (HSCT), albeit at a low rate (Versluys et al. 2018; Ferreira et al. 2014; Peres et al. 2010). Various studies have reported the incidence of severe and acute cardiotoxicity in HSCT recipients, ranging from 0.9 to 8.9% (Gul et al. 2015; Mo et al. 2013; Murdych and Weisdorf 2001; Alblooshi et al. 2021). Furthermore, the estimated incidence of new-onset congestive heart failure (CHF) among these recipients falls between 0.4% and 2.2% (Gul et al. 2015; Mo et al. 2013; Murdych and Weisdorf 2001; Alblooshi et al. 2021), as stated in different studies.

2.2 Long-Term Cardiac Complications from HSCT

Reports have indicated the occurrence of long-term cardiovascular complications, encompassing conditions such as ischemic heart disease, cardiomyopathy, vascular disorders, stroke, and cardiac complications associated with comorbidities like hypertension and diabetes mellitus. Risk factors for heart failure in this context include age, high blood pressure, diabetes mellitus, and previous exposure to anthracycline treatment at doses equal to or exceeding 250 mg/m^2. Notably, there is a stronger emphasis on the investigation of long-term cardiovascular complications related to HSCT in comparison with acute complications.

Important questions

A. **What tests are done before stem cell transplant?**

Blood tests are conducted to assess the functioning of the liver and kidneys, as well as to determine the level of blood cells. In addition, an echocardiogram, which is a heart and blood vessel scan, is performed. To evaluate the condition of organs like the lungs and liver, an X-ray and/or computerized tomography (CT) scan is also utilized.

B. **How to prepare for the PRP/stem-cell therapy procedure?**

Prior to the procedure, it is important to refrain from taking any anti-inflammatory medications, including Aspirin, Motrin, Advil, Aleve, or Naprosyn, for a period of seven days. Additionally, inform your doctor if you are currently taking any blood thinning medications. In the 1–2 days leading up to the procedure, ensure that you consume 64 oz of water daily.

C. **What is the most important factor in donor selection for HSCT?**

Before initiating cancer treatments with cardiotoxic effects, it is advisable to conduct a baseline cardiovascular evaluation and risk stratification for cancer patients. In the context of allogeneic hematopoietic stem cell transplantation (allo-HSCT), the histocompatibility between the donor and recipient is of paramount importance, and there are two donor types available: matched related siblings and unrelated donors. However, it is worth noting that having an HLA-identical sibling donor, which is typically considered the optimal choice for allo-HSCT, is accessible to fewer than 33% of patients (Pesto et al. 2016).

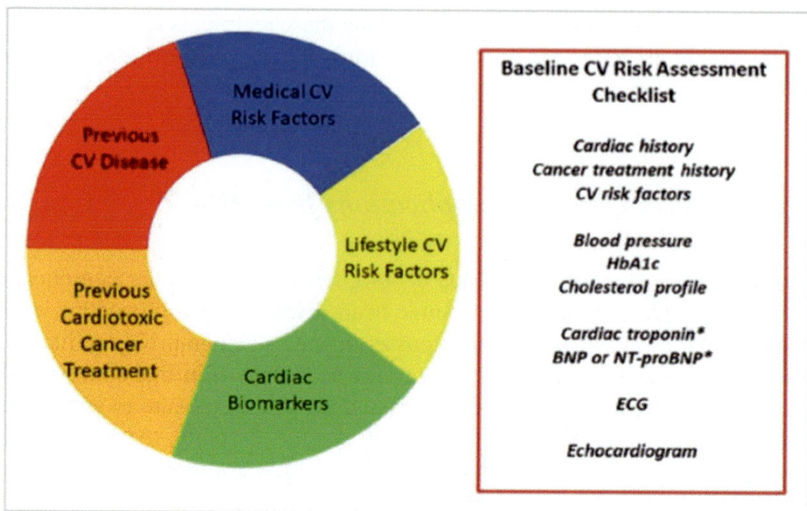

10. Before HSCT, heart failure (HF) is identified by either an ejection fraction (EF) below 50% or a 10% reduction in EF, signifying its significance among cardiac complications (Murdych and Weisdorf 2001; Singla et al. 2013). Various clinical manifestations can arise from HF, encompassing symptoms such as orthopnea, paroxysmal nocturnal dyspnea, exercise intolerance, night cough, wheezing, palpitations, peripheral edema, and syncope. Detecting HF at an early stage is crucial since there is currently no medical treatment available in routine HF management that can regenerate scar tissue, leading to heightened mortality and morbidity. As a result, patients are assessed for HF through cardiac ultrasound and physical examination (Peres et al. 2010).

11. An elevation of pulmonary artery pressure (PAP) exceeding 25 mmHg at rest defines pulmonary hypertension (PH) (Sureddi et al. 2012). In the assessment of PAP levels, echocardiography has emerged as a safe and accessible alternative to invasive catheterization, offering improved sensitivity (Goldberg et al. 1986). Notably, the EBMT guidelines have not reported an occurrence of elevated PAP among the mentioned cardiac complications. The emphasis in practice is usually on post-transplant PH and mortality, with symptoms such as tachypnea, hypoxia, and respiratory failure emerging after the transplant. The cause of mortality, leading to a mortality rate of 55%, was attributed to PH and its associated complications in 86% of cases.

12. Atrial flutter, Atrial fibrillation, and supraventricular tachycardia rank among the most prevalent arrhythmias, with incidence rates as low as 1%. The presence of arrhythmias is influenced by numerous risk factors, encompassing baseline renal impairment, prior history of arrhythmias, identification of premature ventricular complexes (PVCs) through long-term baseline electrocardiogram (ECG), advanced age, previous adjuvant administration of anthracycline, and

classification of baseline ejection fraction as reduced or preserved (Goldberg et al. 1986).

13. Comprehensive evaluation of cardiac risk factors, including hyperlipidemia, diabetes, hypertension, and holds paramount importance in the care of individuals who have undergone HSCT. Similarly critical is the prompt initiation of proactive and intensive interventions for managing left ventricular dysfunction. To effectively manage these patients, an optimal approach involves collaboration between the fields of hematology/oncology and cardiology, which can be facilitate through a cardio-oncology clinic.

    Figure 1 depicts the differentiation between primary and secondary prevention strategies in the assessment of patients before commencing cancer treatment. The assessment of individuals without evident CVD or prior cardiotoxicity falls under the scope of primary prevention. Conversely, interventions focused on evaluating patients with pre-existing CVD or previous cardiovascular toxicity evidences are classified as secondary prevention measures.

14. In recent years, reduced-intensity conditioning (RIC) regimens have emerged as an approach to mitigate the toxicity linked to myeloablative preparative regimens. A comprehensive comprehension of the pharmacokinetic tolerance and toxicity of conditioning drugs is imperative, particularly due to the heightened therapeutic efficacy in malignancy stemming from the immune response elicited

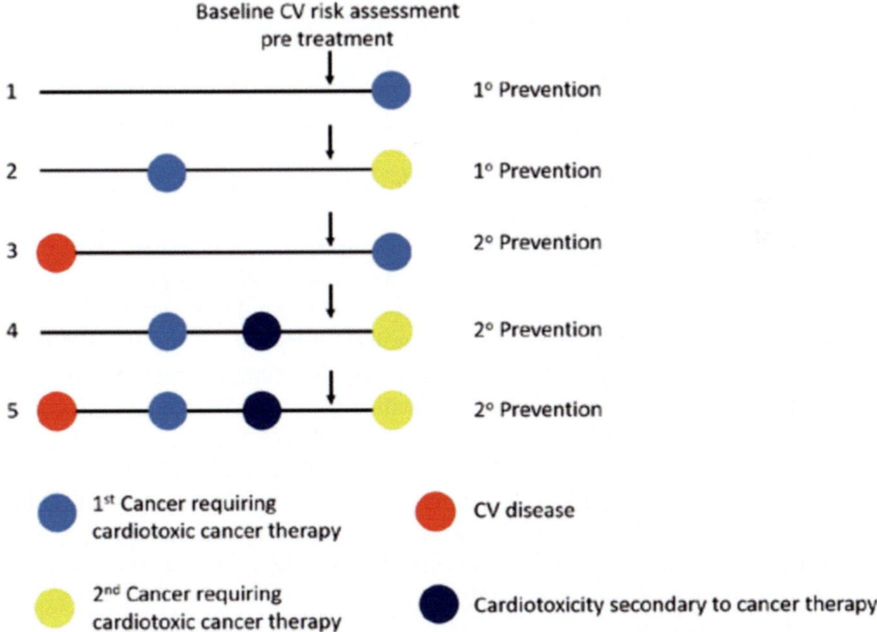

Fig. 1 Differentiation of prevention strategies in patient assessment pre-cancer treatment

by the graft in comparison to chemotherapy conditioning (Sengsayadeth et al. 2015).

15. Myocardial necrosis which clinically presents within ten days of drug administration with hypotension,dyspnea, pericardial effusion and tachycardia, decreased QRS voltage on electrocardiogram, can caused by High-dose CY. Heart failure occurred in as many as 28% of cases of HSCT using older regimens with higher dose CY, which is felt to be dose-dependent with higher rates of heart failure in patients receiving doses greater or equal to 170 mg/m^2. Consulting with a clinical pharmacist is especially important before practitioners select conditioning drugs in old patients and cases with comorbid diseases, as high-dose CY can induce myocardial necrosis leading to complications like heart failure (Goldberg et al. 1986; Hertenstein et al. 1994; Mills and Roberts 1979; Gottdiener et al. 1981; Appelbaum et al. 1976; Steinherz et al. 1981).

16. The potential for cardiovascular complications should be considered by patients undergoing RIC, as reported case series have shown a range of heart failure severity, from reversible systolic dysfunction to fatal outcomes, associated with the combination of fludarabine and melphalan.The use of melphalan in RIC or autologous HSCT regimens has been linked to atrial fibrillation, atrial flutter, and supraventricular tachycardia in up to 9% of cases. It is important for patients receiving melphalan and fludarabine in RIC to be aware of the cardiovascular risks associated with these agents (Peres et al. 2010; Olivieri et al. 1998; McMurray et al. 2012; Kim et al. 2015; Pesto et al. 2016; Parasuraman et al. 2016; Dandoy et al. 2013).

17. The advancement of RIC has expanded the treatment options for patients with comorbidities and advanced age, allowing them to consider HSCT as a viable approach. Recent data also suggests that individuals with LV systolic ejection fraction above 50% are still eligible candidates for HSCT. Furthermore, patients with marginal left ventricular systolic dysfunction can undergo HSCT safely, with no substantial impact on overall survival or treatment-related consequences (Dandoy et al. 2013). Performing a comprehensive geriatric assessment (CGA) could prove advantageous for older individuals, particularly those above 65 years of age, who are undergoing transplants. This is particularly relevant considering the aging population and the increasing number of transplants being performed on individuals with pre-existing cardiovascular conditions (Dandoy et al. 2013). In the HSCT population, the role of CGAs is an area that is rapidly evolving, although there is limited available data on this subject (Pesto et al. 2016). Evaluations in the CGA extend beyond comorbidities and include assessments for various conditions such as sarcopenia, falls, osteoporosis, incontinence, failure to thrive (FTT), delirium, polypharmacy, and dementia (Parasuraman et al. 2016).

18. In the context of the pre-transplantation assessment, it is advisable to include the assessment of cardiac systolic function through various methods such as cardiac magnetic resonance imaging (MRI), cardiac ultrasound also known as

Echocardiography, and multiple gated acquisition scan (MUGA) (Sengsayadeth et al. 2015).

3 Biochemical and Electrocardiographic Parameters

Previous cardiovascular disease		
Heart failure or cardiomyopathy	Very high	C
Prior proteasome inhibitor cardiotoxicity	Very high	C
Venous thrombosis (DVT or PE)	Very high	C
Cardiac amyloidosis	Very high	C
Arterial vascular disease (IHD, PCI, CABG, stable angina, TIA, stroke, PVD)	Very high	C
Prior immunomodulatory drug CV toxicity	High	B
Baseline LVEF < 50%	High	C
Borderline LVEF 50–54%	Medium	C
Arrhythmias	Medium	C
Left ventricular hypertrophy	Medium	C

Thoroughly examining the screening electrocardiogram for signs of premature supraventricular complexes and left ventricular systolic dysfunction or is crucial due to the increased susceptibility to arrhythmias during transplantation. In this particular population, the utilization of preventive therapy, such as angiotensin-converting enzyme (ACE) inhibitors and beta blockers, should be considered as a recommended approach (Armenian and Chow 2014). For individuals with prior cardiac disease, arrhythmias, or any cardiovascular risk factors, it is advisable to consider a referral to a cardio-oncology clinic (Sengsayadeth et al. 2015).

19. Patients with hematologic malignancies who undergo HSCT and high-dose chemotherapy exhibit elevated blood concentrations of NT-pro BNP. Furthermore, post-procedure electrocardiographic recordings revealed higher Tp-e interval and Tp-e/QT ratio values than pre-treatment results.

20. Patients with blood neoplasms who have undergone HSCT may encounter elevated levels of NT-proBNP, with the use of cyclophosphamide, advanced age, and higher cholesterol levels identified as independent risk factors for this increase. Moreover, the use of cyclophosphamide and fludarabine as cancer treatments has been shown to increase the Tp-e/QT interval ratio during electrocardiographic evaluation after HSCT.

4 Lifestyle Modifications

- Inducing cardiac disease, tobacco use is recognized as a significant risk factor. It is crucial to counsel cancer survivors, especially those in the younger age group, to quit smoking, considering that up to 30% of them engage in this habit. Furthermore, tobacco use is not only linked to the development of secondary cancers but also poses a threat to heart health.
- Their counseling should include additional lifestyle modifications such as regular exercise, maintaining a healthy weight, and a low-fat diet.
- It is recommended that Individuals undergoing HSCT engage in a minimum of 150 min of moderate exercise per week, regardless of their age or the treatment employed (Peres et al. 2010).

5 Electrocardiogram Screening

- Screening with an electrocardiogram should be considered for individuals with risk factors of cardiovascular disease, including preexisting cardiovascular disease, prior chest/mediastinal radiation therapy, and cardiac amyloidosis. However, it is important to note that such screening is not recommended for the general population, as per the US Preventive Services Task Force (UPSTF). The prevalence of cardiac involvement in patients with amyloidosis exceeds 50%, making this specific patient population highly vulnerable to arrhythmias. As a result, the monitoring of electrocardiograms plays a crucial role in providing valuable insights and aiding in the management of their condition (Sengsayadeth et al. 2015). Recipients of chest/mediastinal radiation therapy have a considerable cardiac disease rate of 17% by the age of 35 years, and screening electrocardiograms can offer valuable assistance in this regard. Among individuals who have undergone mediastinal radiation therapy, it is important to emphasize that cardiac deaths play a major role in overall mortality (Appelbaum et al. 1976).
- Two years following the HSCT chemotherapy treatment, it is advisable for childhood HSCT survivors to undergo a baseline electrocardiogram, as advised (Armenian and Chow 2014).

6 Cardiac Imaging in Survivors

Assessment of cardiac function before HSCT is generally advised to reduce the occurrence and seriousness of cardiotoxicities. If available, routine follow-up of cardiovascular function in HSCT survivors should be performed utilizing biomarker data analysis or imaging techniques. By employing imaging techniques, early detection of cardiac dysfunction and injury becomes possible in HSCT patients, allowing timely implementation of therapeutic interventions. Echocardiography, an extensively employed and firmly established modality, serves as the principal approach for assessing systolic cardiac function through the quantification of left ventricular ejection fraction (LVEF). Tissue Doppler, cardiac MRI, 2D/3D echocardiography, and speckle tracking (Strain rate imaging) are among the various imaging techniques available for assessing cardiac function.

When initiating the transplantation procedure, an LVEF exceeding 50% is typically regarded as the acceptable threshold (Ohmoto and Fuji 2021). In cases where the LVEF falls below 40%, it is advisable to contemplate cardioprotective treatment and the use of a less cardiotoxic regimen. For the early identification of potential cardiac and left ventricular dysfunction, the use of 3D echocardiography is strongly advised. If, during cancer treatment, the LVEF surpasses 50% in the first cardiac function screening test but later declines by more than 10% from the initial and baseline value after starting treatment, it is recommended to consider modifying or pausing the treatment protocol (Ohmoto and Fuji 2021). Alternatively, a feasible approach is to employ 2D echocardiography alongside contrast-enhanced imaging. By incorporating the assessment of global longitudinal strain (GLS), which measures longitudinal shortening, there exists an opportunity to identify dysfunction at an even earlier stage compared to conventional echocardiography measurements.

- Cardiac MRI,Stress echocardiography, and also carotid dopler study of intima-media thickness with ultrasound are not recommended for asymptomatic persons without any cardiac risk factors. However, individuals with a history of cardiac mediastinal radiation may find stress echocardiography to be a valuable method of diagnosis (Olivieri et al. 1998). Present research efforts are dedicated to investigating the viability of employing strain echocardiography and speckle tracking echocardiography as means of diagnosis (McMurray et al. 2012). Furthermore, ongoing studies are examining the efficacy of cardiac MRI in the detection of cardiac fibrosis and some left ventricular remodeling during screening. Although it may not be applicable to all HSCT survivors, it could be beneficial for specific patients who are detected to be high-risk [41, 61]. Meanwhile, due to the radiation exposure in MUGA scan tests, their administration is fully discouraged.

Carotid artery imaging and ankle-brachial index screening are not advised for the general population undergoing HSCT as a means of detecting arterial disease. However, it is suggested that childhood HSCT survivors and individuals who have received prior mantle radiation should undergo carotid artery ultrasound imaging around 10 years after HSCT (Kim et al. 2015). Various clinical studies have examined the utilization of serum cardiac biomarkers, including troponins and brain natriuretic peptides (BNP), as additional assessment tools alongside echocardiography. These biomarkers hold promise in offering additional insights and aiding in the identification of HSCT patients who might be susceptible to cardiovascular complications. However, the utility of cardiac biomarkers for monitoring cardiovascular side effects remains under investigation, as the relationship between biomarker elevation and clinically relevant outcomes such as left ventricular dysfunction is not yet well-established. Further research is needed to determine whether routine biomarker monitoring should be incorporated into standard practice for individuals receiving potentially cardiotoxic therapies.

7 Drugs with Cardio-Toxicity

QT prolongation, arrhythmias, left ventricular failure, and vascular problems such as hypertension and myocardial infarction (MI), are among the various cardiovascular complications associated with the drug class. The estimation of risk has been conducted for all of these complications (Steinherz et al. 1981).

Anthracycline chemotherapy

Atrial fibrillation and ventricular fibrillation, left ventricular failure, and heart failure, are the primary cardiovascular complications observed in anthracycline chemotherapy agents (Parasuraman et al. 2016, [69]).

Anti-HER2 therapies

Anti-HER2 therapies are primarily linked to cardiovascular complications such as heart failure, systemic hypertension, and left ventricular failure.

VEGF inhibitors

Cardiovascular complications of significant concern associated with VEGF inhibitors, recognized as VEGF tyrosine kinase inhibitors (TKIs) or angiogenesis inhibitors, encompass heart failure, left ventricular dysfunction, and systemic hypertension. Additionally, these agents have demonstrated an association with the prolongation of QTc intervals and the occurrence of venous thromboembolism (VTE) events, including MI (Sengsayadeth et al. 2015).

Tyrosine kinase inhibitors (TKIs) that target BCR-ABL are the standard multi-targeted tyrosine kinase inhibitors for treatment of chronic myeloid leukaemia (CML)

Multi-targeted tyrosine kinase inhibitors which are used for treatment of chronic myeloid leukemia (CML) targeting BCR-ABL are associated with several significant cardiovascular complications. These complications include venous thromboembolism (VTE), systemic hypertension, left ventricular dysfunction, heart failure (HF), accelerated atherosclerosis, QTc prolongation, and pulmonary hypertension. Specifically, these inhibitors have been linked to stroke, VTE leading to MI, LV dysfunction, systemic hypertension, peripheral arterial occlusive disease (PAOD), dasatinib-induced pulmonary arterial hypertension, nilotinib-induced QTc prolongation, and accelerated atherosclerosis induced by ponatinib and nilotinib (Olivieri et al. 1998).

Immunomodulatory imide drugs (IMiDs) and Proteasome inhibitors (PIs)

The combination of IMIDs and PIs is primarily associated with cardiovascular complications including left ventricular dysfunction and HF, myocardial infarction and myocardial ischemia, ventricular and atrial arrhythmias, as well as VTE events (Pesto et al. 2016).

RAF inhibitor/MEK inhibitor combinations

For all combinations of RAF and MEK inhibitors, the primary cardiovascular complications include LV dysfunction, HF, and systemic hypertension. Additionally, one Combination of cobimetinib and vemurafenib is associated with QTc prolongation (Pesto et al. 2016).

Androgen deprivation therapies (ADT)s

As explained below, the use of gonadotropin-releasing hormone (GnRH) agonists for prostate cancer treatment, specifically androgen deprivation therapy (ADT), is linked to an elevated likelihood of diabetes mellitus, hypertension, and atherosclerosis (Parasuraman et al. 2016).

Immune checkpoint inhibitors (ICIs)

Acute coronary syndromes (ACSs) characterized by vasculitis and atherosclerotic plaque rupture, sudden cardiac death, atrioventricular block, ventricular arrhythmias, non-inflammatory heart failure, and fulminant myocarditis (FM) are among the various manifestations of myocarditis.

8   Baseline Cardiovascular Risk Stratification Proforma

By employing these simple guidelines, a risk level can be ascertained when assessing the summary. Additionally, it is advisable to adhere to the following overarching principles until more comprehensive guidance becomes accessible.

Cardiac biomarkers (where available)		
Elevated baseline troponinc	Medium2	C
Elevated baseline BNP or NT-proBNpc	High	B
Demographic and cardiovascular risk factors		
Age ≥ 75 years	High	C
Age 65–74 years	Medium	C
Hypertensiond	Medium	C
Diabetes mellituse	Medium	C
Hyperlipidaemiaf	Medium	C
Chronic kidney disease	Medium	C
Family history of thrombophilia	Medium	C

Previous cardiotoxic cancer treatment		
Prior anthracycline exposure	High	C
Prior thoracic spine radiotherapy	Medium	C
Current myeloma treatment		
High-dose dexamethasone > 160 mg/month	Medium	C
Lifestyle risk factors		
Current smoker or significant smoking history	Medium	C
Obesity (BMI > 30 kg/m^2)	Medium	C
Risk level		

- Individuals with one or more than one risk factors are then categorized based on the most significant risk factor they possess.
- If patients have one or more very high-risk factors, their risk level is considered "very high."
- On the other hand, if patients have one or more than one high-risk feature risk factors, their level is classified as "high."
- Individuals who do not exhibit any other risk factors are stratified as "low risk."
- Medium risk factors are assigned point values known as medium1 or medium2 based on their significance.
- Patients who present multiple medium-risk factors resulting in an accumulation of 5 or more points are categorized as having a "high-risk" level.
- Patients who have solely one medium1 risk factor are categorized as having a "low-risk" level.
- During the course of treatment, cancer patients categorized as medium-risk necessitate heightened vigilance in monitoring their CV well-being. Alternatively, it is prudent to contemplate referring them for a comprehensive cardio-oncology visit.
- Patients who exhibit a solitary medium 2 risk factor or possess one or more intermediate 1 risk factors amounting to a cumulative score of 2–4 points are classified as having a "medium risk" level.
- In accordance with guidelines at local, national, and international levels, cancer patients classified as low-risk should persist with their treatment regimen while adhering to recommended CV surveillance protocols.
- Patients classified as high and very high-risk levels should be recommended for a thorough evaluation by cardio-oncology or cardiology specialists, with a preference for specialized cardio-oncology services if accessible. This approach guarantees the utmost effectiveness in managing their modifiable CV and pre-existing CVD risk factors, alongside the establishment of an individualized surveillance plan throughout the course of cancer treatment.

References

Alblooshi R, Kanfar S, Lord B, Atenafu EG, Michelis FV, Pasic I, Gerbitz A, Al-Shaibani Z, Viswabandya A, Kim DDH, et al. Clinical prevalence and outcome of cardiovascular events in the first 100 days postallogeneic hematopoietic stem cell transplant. Eur J Haematol. 2021;106:32–9.

Appelbaum F, Strauchen JA, Graw Jr RG, et al. Acute 23. lethal carditis caused by high-dose combination che- motherapy. A unique clinical and pathological entity. Lancet. 1976;1:58–62

Armenian SH, Chow EJ. Cardiovascular disease in sur- vivors of hematopoietic cell transplantation. Cancer. 25. 2014;120:469–79. This paper outlines a comparison of the United States Services Preventive Task Force cardiovascular screenings guidelines and those outlined by the Children's Oncology Group and the 26. Center for International Blood and Marrow Transplant

Dandoy CE, Hirsch R, Chima R, Davies SM, Jodele S. Pulmonary hypertension after hematopoietic stem cell transplantation. Biol Blood Marrow Transplant. 2013;19:1546–1556. [PubMed] [Google Scholar]

Ferreira DC, de Oliveira JS, Parisio K, Ramalho FM. Pericardial effusion and cardiac tamponade: clinical manifestation of chronic graft-versus-host disease after allogeneic hematopoietic stem cell transplantation. Rev Bras Hematol Hemoter. 2014;36:159–61.

Goldberg MA, Antin JH, Guinan EC, Rappeport JM. Cyclophosphamide cardiotoxicity: an analysis of dos- ing as a risk factor. Blood. 1986;68:1114–8.

Goldberg SL, Klumpp TR, Magdalinski AJ, Mangan KF. Value of the pretransplant evaluation in predicting toxic day-100 mortality among blood stem-cell and bone marrow transplant recipients. J Clin Oncol. 1998;16:3796–802.

Gottdiener JS, Appelbaum FR, Ferrans VJ, Deisseroth A, Ziegler J. Cardiotoxicity associated with high-dose cy- clophosphamide therapy. Arch Intern Med. 1981;141:758–63.

Gul Z, Bashir Q, Cremer M, Yusuf SW, Gunaydin H, Arora S, Slone S, Nieto Y, Sherwani N, Parmar S, et al. Short-term cardiac toxicity of autologous hematopoietic stem cell transplant for multiple myeloma. Leuk Lymphoma. 2015;56:533–5.

Hertenstein B, Stefanic M, Schmeiser T, et al. Cardiac 21. toxicity of bone marrow transplantation: predictive value of cardiologic evaluation before transplant. J Clin Oncol. 1994;12:998–1004. 22

Kim J, Shapiro L, Flynn A. The clinical application of mesenchymal stem cells and cardiac stem cells as a therapy for cardiovascular disease. Pharmacol Ther. 2015;151:8–15. [PubMed] [Google Scholar]

Lehmann S, Isberg B, Ljungman P, Paul C. Cardiac systolic function before and after hematopoi- etic stem cell transplantation. Bone Marrow Transplant. 2000;26:187–192. [PubMed] [Google Scholar]

McMurray JJ, Adamopoulos S, Anker SD, Auricchio A, Böhm M, Dickstein K, Falk V, Filippatos G, Fonseca C, Gomez-Sanchez MA, Jaarsma T, Køber L, Lip GY, Maggioni AP, Parkhomenko A, Pieske BM, Popescu BA, Rønnevik PK, Rutten FH, Schwitter J, Seferovic P, Stepinska J, Trindade PT, Voors AA, Zannad F, Zeiher A; Task Force for the Diagnosis and Treatment of Acute and Chronic Heart Failure 2012 of the European Society of Cardiology, Bax JJ, Baumgartner H, Ceconi C, Dean V, Deaton C, Fagard R, Funck- Brentano C, Hasdai D, Hoes A, Kirchhof P, Knuuti J, Kolh P, McDonagh T, Moulin C, Popescu BA, Reiner Z, Sechtem U, Sirnes PA, Tendera M, Torbicki A, Vahanian A, Windecker S, McDonagh T, Sechtem U, Bonet LA, Avraamides P, Ben Lamin HA, Brignole M, Coca A, Cowburn P, Dargie H, Elliott P, Flachskampf FA, Guida GF, Hardman S, Iung B, Merkely B, Mueller C, Nanas JN, Nielsen OW, Orn S, Parissis JT, Ponikowski P; ESC Committee for Practice Guidelines. ESC guidelines for the diagnosis and treatment of acute and chronic heart failure 2012: The Task Force for the Diagnosis and Treatment of Acute and Chronic Heart Failure 2012 of the European Society of Cardiology. Developed in collaboration with the Heart Failure Association (HFA) of the ESC. Eur J Heart Fail. 2012;14:803–869. [PubMed] [Google Scholar]

Mills BA, Roberts RW. Cyclophosphamide-induced cardiomyopathy: a report of two cases and review of the English literature. Cancer. 1979;43:2223–6.

Mo XD, Xu LP, Liu DH, Zhang XH, Chen H, Chen YH, Han W, Wang Y, Wang FR, Wang JZ, et al. Heart failure after allogeneic hematopoietic stem cell transplantation. Int J Cardiol. 2013;167:2502–6.

Murdych T, Weisdorf DJ. Serious cardiac complications during bone marrow transplantation at the University 17. of Minnesota, 1977–1997. Bone Marrow Transplant. 2001;28:283–7

Ohmoto A, Fuji S. Cardiac complications associated with hematopoietic stem-cell transplantation. Bone Marrow Transplant. 2021;56:2637–43.

Olivieri A, Corvatta L, Montanari M, Brunori M, Offidani M, Ferretti GF, Centanni M, Leoni P. Paroxysmal atrial fibrillation after high-dose melphalan in five patients autotransplanted with blood progenitor cells. Bone Marrow Transplant. 1998;21(10):1049–53.

Parasuraman S, Walker S, Loudon BL, Gollop ND, Wilson AM, Lowery C, Frenneaux MP. Assessment of pulmonary artery pressure by echocardiography-A comprehensive review. Int J Cardiol Heart Vasc. 2016;12:45–51. [PMC free article] [PubMed] [Google Scholar]

Pasquini MC, Zhu X. Current uses and outcomes of hematopoietic stem cell transplantation. 2014 CIBMTR Summary Slides. Available at: http://www.cibmtr.org. 2014

Peres E, Levine JE, Khaled YA, Ibrahim RB, Braun TM, Krijanovski OI, Mineishi S, Abidi MH. Cardiac complications in patients undergoing a reduced-intensity conditioning hematopoietic stem cell transplantation. Bone Marrow Transplant. 2010;45:149–52.

Peres E, Levine JE, Khaled YA, et al. Cardiac complications in patients undergoing a reduced-intensity conditioning hematopoietic stem cell 19. transplantation. Bone Marrow Transplant. 2010;45:149–52

Pesto S, Begic Z, Prevljak S, Pecar E, Kukavica N, Begic E. Pulmonary hypertension - new trends of diagnostic and therapy. Med Arch. 2016;70:303–307 [PMC free article] [PubMed] [Google Scholar]

Rotz SJ, Ryan TD, Hayek SS. Cardiovascular disease and its management in children and adults undergoing hematopoietic stem cell transplantation. J Thromb Thrombolysis. 2021;51:854–69.

Sengsayadeth S, Savani BN, Blaise D, Malard F, Nagler A, Mohty M. Reduced intensity conditioning allogeneic 27. hematopoietic cell transplantation for adult acute myeloid leukemia in complete remission—a review from the Acute Leukemia Working Party of the EBMT. Haematologica. 2015;100:859–69

Singla A, Hogan WJ, Ansell SM, et al. Incidence of supraventricular arrhythmias during autologous peripheral blood stem cell transplantation. Biol Blood Marrow Transplant. 2013;19:1233–7

Steinherz LJ, Steinherz PG, Mangiacasale D, et al. Car- 24. diac changes with cyclophosphamide. Med Pediatr Oncol. 1981;9:417–22

Sureddi RK, Amani F, Hebbar P, et al. Atrial fibrillation following autologous stem cell transplantation in patients with multiple myeloma: incidence and risk factors. Ther Adv Cardiovasc Dis. 2012;6(229–36):20.

Tonorezos ES, Stillwell EE, Calloway JJ, Glew T, Wessler JD, Rebolledo BJ, Pham A, Steingart RM, Lazarus H, Gale RP, et al. Arrhythmias in the setting of hematopoietic cell transplants. Bone Marrow Transplant. 2015;50:1212–6.

Tuzovic M, Mead M, Young PA, Schiller G, Yang EH. Cardiac complications in the adult bone marrow transplant patient. Curr Oncol Rep. 2019;21:28.

Versluys AB, Grotenhuis HB, Boelens MJJ, Mavinkurve-Groothuis AMC, Breur J. Predictors and outcome of pericardial effusion after hematopoietic stem cell transplantation in children. Pediatr Cardiol. 2018;39:236–44.

Long Term Cardiotoxicity Surveillance in HSCT

Seyed Reza Safaei Nodehi, Azin Alizadehasl, Kamran Roudini, Hossein Ranjbar, and Niloufar Akbari Parsa

Abstract Improved survival after HSCT is associated with late health issues. Among them, cardiovascular complications result in increased mortality rates and significantly affect survivors' quality of life. Therefore, it is crucial to reduce cardiovascular late effects in post-HSCT patients as a life-threatening issue.

Keywords Hematopoietic stem cell transplant · Long-term surveillance · Cardiovascular disease · Cardiotoxicity

Abbreviations

HSCT Hematopoietic Stem Cell Transplant
CVD   Cardiovascular Disease
GVHD Graft Versus Host Disease
CVA   Cerebro Vascular Accident
TBI   Total Body Irradiation

S. R. Safaei Nodehi
Hematology and Medical Tehran Oncology Ward, Department of Internal Medicine, Cancer Research Center, Cancer Institute, Imam Khomeini Hospital Complex, Tehran University of Medical Sciences, Tehran, Iran

A. Alizadehasl
Cardio-Oncology Research Center, Rajaie Cardiovascular Medical and Research Center, Iran University of Medical Sciences, Tehran, Iran

K. Roudini
Hematology Research Center, Imam Khomeini Hospital, Tehran University of Medical Sciences, Tehran, Iran

H. Ranjbar
Imam Khomeini Hospital, Tehran University of Medical Sciences, Tehran, Iran

N. Akbari Parsa (✉)
Dr. Heshmat Hospital, Guilan University of Medical Sciences, Rasht, Iran
e-mail: dr.niloufar.parsa@gmail.com

© The Author(s), under exclusive license to Springer Nature Switzerland AG 2024
A. Alizadehasl et al. (eds.), *Cardiovascular Considerations in Hematopoietic Stem Cell Transplantation*, https://doi.org/10.1007/978-3-031-53659-5_6

HF   Heart Failure
GLS   Global Longitudinal strain
cMRI   Cardiac Magnetic Resonance Imaging

1 Introduction

Advances in HSCT have resulted in longer survival of patients, (Roziakova et al. 2012) which is mainly because of more improved transplantation protocols as well as better post-transplantation support (Zhao et al. 2022). Almost 80% of patients who survive more than 2 years are anticipated to become long-term HSCT survivors who may be about half a million patients worldwide (Roziakova et al. 2012). The number is increasing constantly and is expected to grow fivefold before 2030 (Zhao et al. 2022). But, an increased survival rate is not necessarily equal to full restoration of health (Armenian and Chow 2014). In the early post-transplantation, keeping the patient alive is the main objective. But in long-term survivors, recovery of baseline health status, proper quality of life, and return to normal social integration at home and at work are expected. Long-term survivors are at increased risk for late complications (Tichelli et al. 2008a) such as diseases after the early phase (that means after the first year) and clinical consequences on long-term survivors (Tichelli et al. 2008b). 65% of post-HSCT patients have at least a single long-term late effect (Zhao et al. 2022) and 20% of HSCT survivors experience serious life-threatening complications over a lifetime (Roziakova et al. 2012) and compared with the general population, late mortality after HSCT is increased (Tichelli et al. 2008a). Patients who survive at least 5 years after HSCT will have a 4–9 fold increased mortality rate in the next 30-year period after HSCT, and the life expectancy is 30% less than healthy people (Zhao et al. 2022). Exposure to chemotherapy before and during transplantation, along with long-term immune suppression (Armenian and Chow 2014) can lead to damage to healthy tissue and premature onset of chronic disease (Tichelli et al. 2008b), compromise survivors' quality of life, (Majhail 2017) affecting this population with the risk of chronic health problems, such as cardiac disease, endocrine complications, and other malignancies, (Baker et al. 2010) cataracts, chronic kidney disease, osteonecrosis, impairment in functional status due to chronic GVHD, (Armenian and Chow 2014) sexual and reproductive system problems, financial toxicity, and integration back to society (Majhail 2017). Moreover, premature mortality compared with healthy individuals of the same sex and age, (Diesch-Furlanetto et al. 2021) remains of serious issue not only for healthcare professionals but also patients who have successfully survived the period immediately following HSCT (Baker et al. 2010) (Fig. 1). Due to the high morbidity burden among HSCT survivors and the delayed development of late complications, patients require life-long follow-up (Tichelli et al. 2008b). The recognition of late effects is essential for the optimal treatment of survivors, for finding

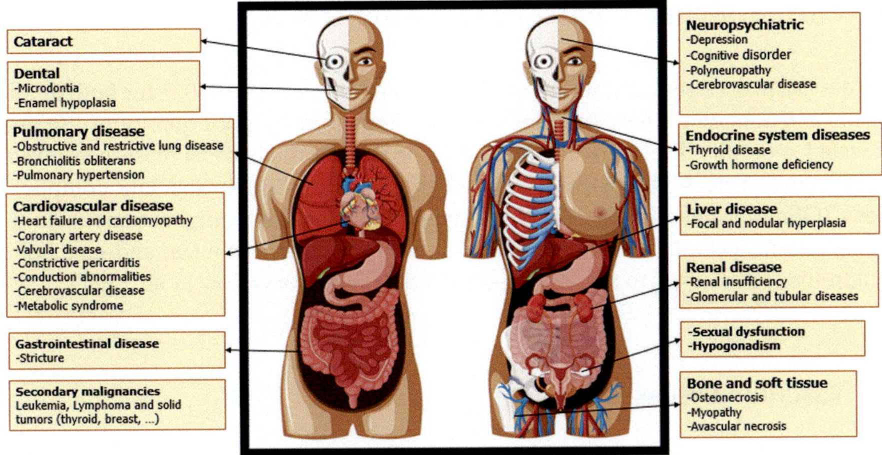

Fig. 1 Long-term consequences in survivors of hematopoietic stem cell transplantation

strategies to lessen the complications, for the organization of specific multidisciplinary follow-up care, and for more studies, which can affect future development potential (Lawitschka and Peters 2018).

This chapter focuses on the risk factors, pathophysiology, incidence, screening strategies, and clinical treatment of late cardiovascular consequences after HSCT.

2 Cardiotoxicity in Long-Term Survivors

Cardiovascular events were thought not to be late complications of HSCT because CVD is so frequent in the general population and with long-term intervals between the transplantation and an event, the correlation of the two conditions becomes less obvious (Rovó and Tichelli 2012). However, late cardiac toxicity after HSCT is now well-known. In a single-center retrospective cohort study of 1491 patients, HSCT patients developed a significantly higher incidence of cardiovascular mortality, ischemic heart disease, heart failure, stroke, vascular disease, arrhythmias, hypertension, diabetes, and dyslipidemia in comparison to controls in the general population (Chow et al. 2011). The median time for the development of CVD was 4 years after HSCT (1.1–19.4 years). At present, it is believed that one of the most serious consequences for long-term HSCT survivors is CVD (Rovó and Tichelli 2012).

3   Cardiovascular Consequences After HSCT

Cardiovascular complications are not really common (less than 10% for both autologous and allogenic transplantation) but are the reason for poor quality of life and elevated mortality in survivors. Therefore, it is important to decrease cardiovascular consequences after HSCT because they are life-threatening late complications (Ohmoto and Fuji 2021). Numerous cardiovascular complications such as Heart failure, valvular heart disease, vascular disease, arrhythmias, and metabolic syndrome are believed to affect long-term HSCT survivors (Rotz et al. 2021).

4   Vascular Diseases

The risk of arterial complications, like CAD and stroke, has been > 20% at 20 years after HSCT (Tuzovic et al. 2019). Arterial event is a major reason for nonrelapse mortality and morbidity in HSCT survivors, with a two-fold increased risk in comparison with the general population (Saunders et al. 2020).

Risk factors for these post-HSCT complications include radiation exposure, hypertension, diabetes, dyslipidemia, allogeneic HSCT, and modifiable risk factors like physical inactivity, smoking, and obesity; also, an interesting newly identified risk factor seems to be clonal hematopoiesis of indeterminate potential in adults treated with autologous HSCT, which is related to increased mortality from CVD (Ryan and Salim Hayek 2021).

5   Heart Failure

Armenian et al. assessed the incidence of HF a year after HSCT in a case–control study. They demonstrated that the HF incidence was 11.7% and the average time for diagnosis of HF was 3 years after HSCT. In this study, female lymphoma survivors were with the highest incidence (14.5% in 15 years), mostly because women are more susceptible to anthracycline-related cardiotoxicity (Armenian et al. 2008). The incidence of HF is reported to be about 5.6% to 10.8% in patients who survived at least 10 years after HSCT (Massey et al. 2020). Heart failure risk is elevated with the administration of prior anthracyclines and cyclophosphamide, TBI at the time of HSCT, chest radiation, and GVHD after HSCT. Furthermore, the severity of risk might be affected by the comorbid factors and host characteristics like the sex of the recipient and age at exposure (Baker et al. 2010), diabetes, arterial hypertension, dyslipidemia, smoking, physical inactivity, and obesity. Post-HSCT survivors with heart failure have poor outcomes, and less than 50% will survive after 5 years (Saunders et al. 2020). Moreover, HSCT survivors showed a higher probability of iron overload about 28 months after transplantation and they are expected to have a higher burden of iron deposition in the myocardium as well, which may lead to cardiomyopathy in combination with other cardiac risk factors (Rotz et al. 2017).

6 Metabolic Syndrome

Metabolic syndrome is reported to be about 17.1% in adult patients who underwent HSCT in childhood, and 32% in children and adolescents after HSCT. The incidence of metabolic syndrome in post-HSCT adults is twofold compared with the general adult population.

- *Dyslipidemia*—It has been delineated that dyslipidemia risk enhances from 28% before HSCT to 74% 3 months after it and to 89% about 25 years thereafter. Factors related to post-HSCT dyslipidemia include obesity, family history of dyslipidemia, acute and chronic GVHD, high-dose TBI, and use of immunosuppressants like corticosteroids, calcineurin inhibitors, and sirolimus.
- *Hypertension*—About 70% of long-term HSCT survivors will suffer hypertension in 2 years after transplantation. Administration of cyclosporine can be a predictive factor for the development of hypertension.
- *Diabetes*—Diabetes prevalence is reported about 30% within 2 years after HSCT. Risk factors for developing diabetes include previous TBI or radiation to the abdomen, which is able to destroy the pancreas and lead to insulin resistance. Also, corticosteroids may cause insulin resistance while cyclosporine and tacrolimus can compromise insulin production (Saunders et al. 2020).

7 Factors Impacting Incidence of Cardiotoxicity

- *Previous anthracycline therapy*

Most acute leukemia patients who are candidates for HSCT receive anthracyclines as one of the induction or/and consolidation parts of treatment (Tuzovic et al. 2019). A prominently increased rate of cardiovascular death has been due to anthracycline exposure pre-HSCT. The American Society of Clinical Oncology (ASCO) guidelines explain a cumulative dose of doxorubicin $\geq$ 250 mg/m^2 or epirubicin $\geq$ 600 mg/m^2 as a high risk for cardiotoxicity (Zhao et al. 2022). Anthracycline-related cardiomyopathy is defined as a progressive impairment in left ventricular systolic function which is dose-dependent. The incidence is 0.14% when the total dose is less than 400 mg/m^2, but reported to be 7% when the dose is 550 mg/m^2 and measured more than 18% at a dose of 700 mg/m^2 (Majhail and Rizzo 2012). The mechanism of anthracycline-related cardiotoxicity is an increment in intracellular reactive oxygen species which destruct mitochondrial structure and also disturb repair of DNA damage (Zhao et al. 2022). Cardiac dysfunction might manifest a few months or even years after exposure (Coghlan et al. 2007).

– Genetic susceptibility to anthracycline

Serious cardiac toxicity has been announced even at cumulative doses < 250 mg/ m^2; on the other hand, doses exceeding 1000 mg/m^2 may have no long-term effect. As mentioned, 40% of clinical HF cases had a cumulative dose of < 250 mg/m^2. Genetic susceptibility can explain this heterogeneity. Studies suggest that genetic polymorphism in the metabolism of anthracycline and transport might affect the risk of CVD in survivors (Baker et al. 2010).

– *Molecular-Targeted Agents and Tyrosine Kinase Inhibitors (TKIs)*

Pretreatment with TKIs is another risk factor associated with late cardiac sequelae in long-term survivors. Sorafenib and midostaurin are reported to diminish the relapse rate with improved survival; but, sorafenib might impair left ventricular systolic function (Zhao et al. 2022).

– *Radiotherapy*

Chest radiation enhances the incidence of cardiomyopathy in HSCT survivors and a dose of **more than** 30 Gy can induce cardiotoxicity. In a group of adult people who were Hodgkin lymphoma survivors with a history of radiotherapy (≥ 30 Gy in half of them) before HSCT, CT angiography showed CAD in 39% of the patients (Zhao et al. 2022). Also, mediastinal radiation can produce inflammation and fibrosis of all cardiac structures. Arrhythmia due to conduction pathways fibrosis, and restrictive cardiomyopathy, with left-sided valves thickening and regurgitations, are some of the deleterious long-term HSCT consequences (Majhail and Rizzo 2012).

Graft-Versus-Host Disease

GvHD is known to be the commonest lethal consequence after HSCT. This complication happens when donor T cells assume the recipient is a foreign body and may be acute or chronic. A study demonstrated that recipients who suffered grade 2–4 acute GvHD had cardiovascular risk factors like as diabetes mellitus, hypertension, and hypersensitivity lung disease (Zhao et al. 2022). These comorbidities might be related to the adverse consequences of calcineurin inhibitors and steroids that are the most common drugs for the management of GVHD. Finally, GVHD can cause injury and inflammation to vessels and lead to arterial events (Tuzovic et al. 2019).

– *Conventional cardiovascular risk factors*

Conventional cardiovascular risk factors such as hypertension, diabetes, dyslipidemia, and obesity are important considerations for cardiovascular disease (Armenian and Chow 2014). Metabolic syndrome is diagnosed in 39% of HSCT survivors compared with 8% of patients with leukemia managed only with conventional chemotherapy (Diesch-Furlanetto et al. 2021). For some patients, these risk factors might be transient, and resolve just after quitting immunosuppressive drugs. But in most patients, they persist long thereafter, thus it is important to manage risk factors as early as we can to reduce long-term CVD risk. The latency period between transplantation and hospitalization for CVD or cardiac death can be as soon as 6 years. So we have a limited window of opportunity (Armenian and Chow 2014).

8 Endothelial Injury to Cardiovascular Disease

Hypothyroidism, growth hormone incompetence, and especially gonadal dysfunction can be responsible for the problem. In individuals with a growth hormone insufficiency, Insulin sensitivity is impaired leading to hypertriglyceridemia, hypertension, and low HDL cholesterol. Subclinical Thyroid dysfunction and also overt hypothyroidism can induce metabolic syndrome post-HSCT as well. There is a powerful association between metabolic syndrome and hypogonadism in HSCT survivors. Also, low testosterone levels in men can be a frequent reason for metabolic syndrome, and these patients demonstrate better insulin sensitivity after testosterone replacement (Hamidi Madani et al. 2012). On the other hand in women, blood pressure, lipid metabolism, body composition, and also vascular tone are influenced by estrogen levels. Non-endocrinal risk factors may also be related to metabolic syndrome post-HSCT. Chronic magnesium deficiency is a usual detection in survivors. It may happen due to less dietary intake or gastrointestinal GVHD and intestinal loss and also calcineurin inhibitors can have deleterious effects by suppressing magnesium reabsorption. This hypomagnesemia can be related to insulin resistance and also metabolic syndrome. Endothelial injury which happens after HSCT can lead to premature atherosclerosis as the primary incident and then cardiovascular events in long-term HSCT survivors. Besides, vascular endothelium is known to be a TBI target. Radiation damages the endothelium of the microvasculature system and then results in vascular stiffening, which leads to CVD (Rovó and Tichelli 2012).

9 Screening

9.1 Screening of the Modifiable Cardiovascular Risk Factors

Some CVD risk factors can be modified post-HSCT. Regular screening requires blood pressure measurement regularly, control of waist circumference and body weight, and some tests for evaluation of dyslipidemia or abnormal glucose levels. In accordance with recommendations and prophylactic strategies for long-term post-HSCT patients, measurement of blood pressure must be done at least every 2 years. There is controversy about the best method of screening for diagnosis of type 2 diabetes as soon as possible. Fasting blood glucose, HbA1C, and a 2-h post-challenge glucose might be performed (Rovó and Tichelli 2012).

- If blood pressure is > 135/80 mmHg, the recommendation is to measure fasting blood glucose each 3 years.
- In childhood survivors, the Children's Oncology Group states to measure fasting blood glucose every 2 years post-HSCT, especially in patients who had cranial radiation or TBI in a preparative regimen.
- Measurement of fasting lipids is recommended every 5 years after 45 years for women and after 35 years for men but should begin at the age of 20 years when other cardiovascular risk factors such as smoking, obesity, hypertension, or diabetes are present.
- In childhood survivors, lipids level should be measured every 2 years which begins 2 years post-HSCT, especially in patients who had cranial radiation or TBI (Blaes et al. 2016). When dyslipidemia medication is prescribed, lipid values should be tested every 6–8 weeks until control of the serum level. In patients who are in the low-risk group because of a normal lipid profile after 1 year, the post-HSCT screening recommendations are like the general population (Rovó and Tichelli 2012).

10 Detecting Cardiac Injury

- **Electrocardiography**

Resting electrocardiogram and 24-h monitoring are the cornerstones of cardiovascular screening. 75% of Hodgkin lymphoma survivors with a history of mediastinal radiotherapy had conduction pathways abnormality on their electrocardiogram. Electrocardiogram should be repeated annually for patients with structural heart disease like impaired systolic function, or those with a history of thorax radiation (Rotz et al. 2017).

–In childhood survivors, baseline electrocardiography should be done 2 years after HSCT (Blaes et al. 2016).

- **Echocardiography**

Early diagnosis of cardiovascular diseases may enable us to make timely interventions, and echocardiography has an undeniable role. In children with a history of anthracyclines administration, ventricular wall thickness alterations might predict a significant decrease in ejection fraction or shortening fraction. 3D echocardiography or 2D echocardiography along with contrast-enhanced imaging are strongly recommended for the assessment of left ventricular systolic function. Also, GLS analysis enables us to estimate the longitudinal shortening of the left ventricle, so we can detect left ventricular dysfunction much sooner in comparison with conventional echocardiography measurements (Zhao et al. 2022). We can benefit from exercise stress echocardiography to find occult anthracycline cardiomyopathy with more sensitivity as well (Rotz et al. 2017). Echocardiographic screening intervals can vary from every 1 to every 5 years, which depends on the cumulative dose of anthracycline in the regimen, the patient's age at exposure, and history of mediastinal radiotherapy (Bhatia 2011).

– **Magnetic resonance imaging**

cMRI can help in patients receiving cardiotoxic drugs like anthracyclines by providing a characterization of tissue and compared with echocardiography, it is likely more precise at finding smaller alterations in LV ejection fraction. It determines premature changes in myocardial contrast, before affecting LV walls motion which might lead to cardiac morbidity in long-term follow-up, although more pieces of literature about its feasibility for screening cardiac health of HSCT survivors is needed (Rotz et al. 2017).

– **Biomarkers**

Markers like NT-pro BNP are exciting for stratification of cardiotoxicity risk, further investigation is needed to evaluate if biomarkers can estimate which patients might benefit from preventive or therapeutic interventions (Rotz et al. 2017). Also, defining an established cutoff for these patients is challenging, which has limited its utility in routine screening of HSCT survivors to date (Armenian and Chow 2014). Cardiac troponin is a sensitive and specific marker for assessment of myocytes injury which is often used in acute coronary syndrome. But, we do not have strong documents demonstrating its association with chronic LV dysfunction after HSCT. Therefore, we have currently no indication of its utility for the evaluation of LV dysfunction in survivors. Other markers showing injured endothelium like vWF and hs- CRP might help in the assessment of vascular injury but further studies are required (Armenian and Chow 2014).

11 Cancer Survivorship Clinics

– The increased possibility of cardiovascular disease in survivors necessitate recommendations for routine surveillance program and organizing a cancer survivorship clinic for routine follow-ups and consultations to promote survivors' knowledge about potential risks of cardiac complications and also screening recommendations (Blaes et al. 2016) This surveillance program should start less than 2 years after the end of cardiotoxic treatment (Tuzovic et al. 2019).
– Cooperation between oncologists and cardiologists is crucial to provide the best evidence-based supervision for long-term HSCT survivors (Blaes et al. 2016) (Table 1).

Table 1 Recommended screening and management protocols for long-term HSCT survivors

	Screening		Recommendation and management
	Tools and laboratory marker	Frequency	
Clinical assessment	– Evaluation of cardiovascular symptoms – Weight, height, waist circumference, and BMI assessment at every clinic visit	Annually	– Weight control – Counseling a healthy heart lifestyle with appropriate diet and physical activity – Smoking cessation
Risk factor assessment	Diabetes – Fasting glucose – HbA1C – 2-h post-challenge glucose	– After 45 years: Every 3 years (earlier if BP is more than 135/80 mmHg) – In childhood survivors a fasting plasma glucose every 2 years post-HSCT	– Appropriate management of CVD risk factors such as HTN, diabetes, and hyperlipidemia for HSCT survivors – In patients administered pharmacologic treatment for dyslipidemia, it is recommended to plan regular follow-ups which is the assessment of fasting lipid profile each 6–8 weeks until the lipid target is achieved
	Hypertension – Blood pressure measurement in > 18 years old	Every 1–2 years	
	Hyperlipidemia – Fasting lipids assessment	– Each 5 years from age 45 years for women and age 35 years for men – For higher-risk individuals (smoking, diabetes, obesity, hypertension, or a positive family history of CVD), screening after the age of 20 years is recommended – In childhood survivors, screening each 2 years which begins 2 years post HSCT is suggested	

(continued)

Table 1 (continued)

| | Screening | | Recommendation and management |
	Tools and laboratory marker	Frequency	
Electrocardiogram	For measurement of ischemia and arrhythmia	– Not routinely indicated for low-risk patients – Annually for high-risk patients (structural heart disease like baseline HF, history of radiotherapy in thorax region, or using arrhythmia-provoking drugs) – For childhood survivors, a baseline ECG 2 years post-HSCT is recommended	– Management of ischemia or arrhythmia
Echocardiogram	– 2D and 3D echocardiography – Strain imaging	Intervals vary from annually to each 5 years which depend on the patients' history (drugs and dose, age at exposure, history of radiation, baseline cardiac structural diseases, …)	– Standard management for heart failure, valvular diseases, pericardial diseases, and …
Cardiac MRI	Accurate at the detection of minimal decline in LVEF	Not routinely indicated for screening	
Biomarkers	Troponin, NT-ProBNP	If cardiomyopathy is suspected with HF symptoms, assessment of biomarkers may be reasonable In patients with borderline baseline cardiac function biomarkers might be useful	

12 Management

12.1 Management of Cardiovascular Risk Factors

Two principal strategies may be approached: baseline endocrine disorder treatment which might be the etiology for the risk factors for cardiac events, or management of the risk factors themselves. Substitution for hormone deficiency can have a favorable result which might be added to the standard management of body fat composition, glucose tolerance, and lipid profile. However, sole treatment of endocrine deficiency would not be sufficient to control cardiac risk factors (Rovó and Tichelli 2012).

- *Lifestyle modifications*—Regular exercise meaning at least 150 min of exercise with moderate intensity, a low-fat dietary regimen, and keeping a normal weight are recommended. Patients should be counseled to quit smoking which could lead to malignancies and also cardiac disease (Zhao et al. 2022).
- *Diabetes*—Substitution of corticosteroid by other immunosuppressants or decreasing its dose must be in mind. If these measures combined with exercise do not control blood sugar, pharmacologic treatment is recommended to enhance glucose metabolism. About two-thirds of HSCT survivors are receiving oral hypoglycemic drugs and about one-third are treated with insulin (Rovó and Tichelli 2012).
- *Dyslipidemia*—The mainstay of dyslipidemia management is normalizing LDL levels. At first, we recommend changing to a healthy lifestyle and diet which consists of limiting saturated fat and trans-fatty acid ingestion and also planning for regular exercise. Minimizing immunosuppressive drug doses and also replacing them with a drug with fewer side effects on lipid profiles is suggested. Finally, if the above measures did not control the LDL level, lipid-lowering agents should be prescribed. Atorvastatin and rosuvastatin are both high-intensity statins that also have favorable effects on triglyceride levels. If the patient is receiving cyclosporine, myopathy might happen more frequently because of interaction with CYP3A4 inhibitors and safer drugs like sirolimus can be an alternative. It is recommended to assess creatinine kinase and liver function, especially when cyclosporine or other CYP3A4 inhibitors are in the regimen as well (Rovó and Tichelli 2012).
- *Hypertension*—Again, modification of lifestyle is the first step in managing high blood pressure particularly mild HTN, which includes weight loss, DASH and low-sodium diet, reduction of alcohol, and regular exercise. If hypertension is severe or resistant, and if chronic kidney disease is present, antihypertensive agents are indicated based on hypertension guidelines recommendations (Rovó and Tichelli 2012)

12.2 Cardiomyopathy

It has been proved in many clinical trials that beta blockers and ACE inhibitors have favorable effects of protection from HF induced by anthracyclines. For example, enalapril could prevent LVEF decline caused by doxorubicin when the cumulative dose was more than 200 mg/m^2 in non-HSCT patients. Also in the setting of HSCT, BBs and ACE inhibitors were prescribed for 6 months after transplantation as a combination therapy in 90 patients in the OVERCOME trial. In follow-up, no significant reduction in LVEF was found, although a definite decline in the placebo group was demonstrated (Bosch et al. 2013; Gupta et al. 2018).

Statins had some preventive effect on anthracycline-induced cardiomyopathy in a retrospective study on women with breast cancer. But if the effect is the direct cardio-protective effect from anthracycline is questioned and the effect was not evaluated in HSCT patients (Rotz et al. 2017).

– *Antioxidants as Cardio-Protective Treatment*

Cardioprotective drug mechanisms include the activation of endogenous antioxidants, restriction of ROS genesis, and the repair of damages induced by ROS. The most famous chelating drug is dexrazoxane and it inhibits anthracycline-iron complexes formation and diminishes ROS cardiac myocytes, so it can decrease cardiotoxicity in breast cancer patients receiving anthracyclines (Zhao et al. 2022). The other one is tannic acid which also can reduce cell death induced by doxorubicin by restricting activation of PARP-1. Erythropoietin may also attenuate the anti-migratory and antiproliferative role of doxorubicin in animal models. Moreover, vitamin D might have a beneficial effect on preventing cardiotoxicity which demands further investigation (Rotz et al. 2017) (Table 1).

13 Conclusion and Future Perspective

The constant increase in late CVD in post-HSCT survivors is expected in the future decades and CVD is considered to be one of the most dreadful nonmalignant late complications post-HSCT. However, this event might probably be postponed by regular screening and treatment of the modifiable risk factors. Organized screening for hypertension, diabetes, and hyperlipidemia and adequate preventive therapy is mandatory as well as early management of left ventricular systolic dysfunction. Cooperation between oncologists and cardiologists as cardio-oncology clinic team members would be the ideal approach to survivors management.

As preclinical and clinical studies and research are insufficient, cardioprotective treatments recommended in HSCT survivors are mostly based on chemotherapy-induced cardiac toxicity guidelines. Therefore, more clinical data is needed to establish documented transplantation, screening, preventive, and management protocols that are specific to transplantation-associated cardiovascular complications. By reducing cardiac events, the quality of life and clinical outcomes of post-HSCT survivors will be improved significantly.

References

Armenian SH, Chow EJ. Cardiovascular disease in survivors of hematopoietic cell transplantation. Cancer. 2014;120(4):469–79. https://doi.org/10.1002/cncr.28444.

Armenian SH, Sun CL, Francisco L, et al. Late congestive heart failure after hematopoietic cell transplantation. J Clin Oncol. 2008;26(34):5537–43. https://doi.org/10.1200/JCO.2008.17.7428.

Baker KS, Armenian S, Bhatia S. Long-term consequences of hematopoietic stem cell transplantation: current state of the science. Biol Blood Marrow Transplant. 2010;16(1 Suppl):S90–6. https://doi.org/10.1016/j.bbmt.2009.09.017.

Bhatia S. Long-term health impacts of hematopoietic stem cell transplantation inform recommendations for follow-up. Expert Rev Hematol. 2011;4(4):437–52; quiz 453–4. https://doi.org/10.1586/ehm.11.39.

Blaes A, Konety S, Hurley P. Cardiovascular complications of hematopoietic stem cell transplantation. Curr Treat Options Cardiovasc Med. 2016;18(4):25. https://doi.org/10.1007/s11936-016-0447-9.

Bosch X, Rovira M, Sitges M, et al. Enalapril and carvedilol for preventing chemotherapy-induced left ventricular systolic dysfunction in patients with malignant hemopathies: the OVERCOME trial (preventi on of left Ventricular dysfunction with Enalapril and caRvedilol in patients submitted to intensive ChemOtherapy for the treatment of Malignant hEmopathies). J Am Coll Cardiol. 2013;61:2355–62.

Chow EJ, Mueller BA, Baker KS, et al. Cardiovascular hospitalizations and mortality among recipients of hematopoietic stem cell transplantation. Ann Intern Med. 2011;155(1):21–32. https://doi.org/10.7326/0003-4819-155-1-201107050-00004.

Coghlan JG, Handler CE, Kottaridis PD. Cardiac assessment of patients for haematopoietic stem cell transplantation. Best Pract Res Clin Haematol. 2007;20(2):247–63. https://doi.org/10.1016/j.beha.2006.09.005.

Diesch-Furlanetto T, Gabriel M, Zajac-Spychala O, et al. Late effects after haematopoietic stem cell transplantation in ALL, long-term follow-up and transition: a step into adult life. Front Pediatr. 2021;9: 773895. https://doi.org/10.3389/fped.2021.773895.

Gupta V, Kumar Singh S, Agrawal V, Bali Singh T. Role of ACE inhibitors in anthracycline-induced cardiotoxicity: a randomized, double-blind, placebo-controlled trial. Pediatr Blood Cancer. 2018;65:e27308.

Hamidi Madani A, Heidarzadeh A, Akbari Parsa N, et al. A survey on relative frequency of metabolic syndrome and testosterone deficiency in men with erectile dysfunction. Int Urol Nephrol. 2012;44(3):667–72. https://doi.org/10.1007/s11255-011-0086-8.

Lawitschka A, Peters C. Long-term effects of myeloablative allogeneic hematopoietic stem cell transplantation in pediatric patients with acute lymphoblastic leukemia. Curr Oncol Rep. 2018;20(9):74. https://doi.org/10.1007/s11912-018-0719-5.

Majhail NS. Long-term complications after hematopoietic cell transplantation. Hematol Oncol Stem Cell Ther. 2017;10:220–7. https://doi.org/10.1016/j.hemonc.2017.05.009.

Majhail NS, Rizzo JD, et al. Recommended screening and preventive practices for long-term survivors after hematopoietic cell transplantation. Bone Marrow Transplant. 2012;47(3):337–41. https://doi.org/10.1038/bmt.2012.5.

Massey RJ, Diep PP, Ruud E, Burman MM, Kvaslerud AB, Brinch L, Aakhus S, Gullestad LL, Beitnes JO. Left ventricular systolic function in long-term survivors of allogeneic hematopoietic stem cell transplantation. JACC CardioOncol. 2020;2(3):460–71. https://doi.org/10.1016/j.jac cao.2020.06.011.

Ohmoto A, Fuji S. Cardiac complications associated with hematopoietic stem-cell transplantation. Bone Marrow Transplant. 2021;56(11):2637–43. https://doi.org/10.1038/s41409-021-01427-2.

Rotz SJ, Ryan TD, Hayek SS. Cardiovascular disease and its management in children and adults undergoing hematopoietic stem cell transplantation. J Thromb Thrombolysis. 2021;51(4):854–69. https://doi.org/10.1007/s11239-020-02344-9.

Rotz SJ, Ryan TD, Hlavaty J, George SA, El-Bietar J, Dandoy CE. Cardiotoxicity and cardiomyopathy in children and young adult survivors of hematopoietic stem cell transplant. Pediatr Blood Cancer. 2017;64(11). https://doi.org/10.1002/pbc.26600.

Rovó A, Tichelli A. Late effects working party of the european group for blood and marrow transplantation. Cardiovascular complications in long-term survivors after allogeneic hematopoietic stem cell transplantation. Semin Hematol. 2012;49(1):25–34. https://doi.org/10.1053/j.seminh ematol.2011.10.001.

Roziakova L, Bojtarova E, Mistrik M, et al. Serial measurements of cardiac biomarkers in patients after allogeneic hematopoietic stem cell transplantation. J Exp Clin Cancer Res. 2012;31(1):13. https://doi.org/10.1186/1756-9966-31-13.

Ryan TD, Salim Hayek S, Rotz S. Review of LATE CV effects after hematopoietic stem cell transplantation. American College of Cardiology; 2021.

Saunders IM, Tan M, Koura D, Young R. Long-term follow-up of hematopoietic stem cell transplant survivors: a focus on screening, monitoring, and therapeutics. Pharmacotherapy. 2020;40(8):808–41. https://doi.org/10.1002/phar.2443. PMID: 32652612.

Tichelli A, Bhatia S, Socié G. Cardiac and cardiovascular consequences after haematopoietic stem cell transplantation. Br J Haematol. 2008a;142(1):11–26. https://doi.org/10.1111/j.1365-2141.2008.07165.x.

Tichelli A, Rovó A, Gratwohl A. Late pulmonary, cardiovascular, and renal complications after hematopoietic stem cell transplantation and recommended screening practices. Hematology Am Soc Hematol Educ Program. 2008:125–33. https://doi.org/10.1182/asheducation-2008.1.125.

Tuzovic M, Mead M, Young PA, Schiller G, Yang EH. Cardiac complications in the adult bone marrow transplant patient. Curr Oncol Rep. 2019;21(3):28. https://doi.org/10.1007/s11912-019-0774-6.

Zhao Y, He R, Oerther S, Zhou W, et al. Cardiovascular complications in hematopoietic stem cell transplanted patients. J Pers Med. 2022;12(11):1797. https://doi.org/10.3390/jpm12111797.

Cardiac Manifestations
of Graft-Versus-Host Disease

Sina Salari, Kamran Roudini, and Hanieh Hajiali Fini

Abstract Cancer is the second most common cause of death globally. Hematological malignancy is an important contributor to the global cancer burden and hematopoietic stem cell transplantation (HSCT) is a curative therapeutic method in the advanced stages of such malignancies. HSCT is associated with severe complications, such as graft versus host disease (GVHD). The cardiovascular system is rarely reported as a target organ of GVHD, although this will affect survivor mortality and morbidity. The prevention, early diagnosis and adequate treatment of cardiovascular GVHD are important in post-HSCT patients.

Keywords Cancer · Hematological malignancy · Bone marrow transplantation · Hematopoietic stem cell transplantation (HSCT) · Graft versus host disease (GVHD) · Cardiac complication

Abbreviations

APC   Antigen-presenting cell
BMT   Bone marrow transplantation

S. Salari
Cardio-Oncology Research Center, Rajaie Cardiovascular Medical and Research Center, Iran University of Medical Sciences, Tehran, Iran

Medical Oncology-Hematology at Shahid Beheshti University of Medical Sciences, Taleghani Hospital, Tehran, Iran

S. Salari
e-mail: s.salari@sbmu.ac.ir

K. Roudini
Hematology Research Center, Tehran, Iran

H. Hajiali Fini (✉)
Echocardiography Research Center, Rajaie Cardiovascular Medical and Research Center, Iran University of Medical Sciences, Tehran, Iran
e-mail: haniyehajiali97@gmail.com

© The Author(s), under exclusive license to Springer Nature Switzerland AG 2024
A. Alizadehasl et al. (eds.), *Cardiovascular Considerations in Hematopoietic Stem Cell Transplantation*, https://doi.org/10.1007/978-3-031-53659-5_7

CAD   Coronary artery disease
CRP   C-reactive protein
CV   Cardiovascular
DLP   Dyslipidemia
DM   Diabetes mellitus
ECG   Electrocardiogram
ESR   Erythrocyte sedimentation rate
GVHD   Graft versus host disease
HLA   Human leukocyte antigen
HSCT   Hematopoietic stem cell transplantation
HTN   Hypertension
IL   Interleukin
IMT   Intima-media thickness
LV   Left ventricle
TNF   Tumor necrosis factor
TTE   Trans thoracic echocardiography

1 Introduction

Cancer is the second leading cause of death globally and is anticipated to take first place by 2060 (Mattiuzzi and Lippi 2019). Statistical review indicates that there were about 19.3 million new cancer patients and 10 million cancer-related deaths worldwide in 2020 (Sung et al. 2021). Although cancer patient outcomes have improved with the advent of advanced early diagnostic modalities and appropriate therapeutic approaches, these novel therapeutic methods and increased patient survival can result in the development of other disorders (Paterson et al. 2022).

The uncontrolled growth of hematopoietic and lymphoid elements can lead to hematological malignancy (Tietsche de Moraes Hungria et al. 2019). These types of malignant disorder make them important contributors to the global cancer burden as it is estimated that they form 6.5% of all cancers worldwide (Tietsche de Moraes Hungria et al. 2019; Keykhaei et al. 2021). The incidence of hematological malignancy depends on factors such as the subtype as well as the patient's age, gender and socioeconomic level. It accounted for about 407,000 leukemia patients and 309,000 deaths in 2018 (Keykhaei et al. 2021).

Hematopoietic stem cell transplantation (HSCT) and bone marrow transplantation (BMT) are therapeutic methods in advanced-stage hematological malignancies where chemoradiotherapy has not been sufficient to cure the disease and long-term survival is the goal. The procedure consists of bone marrow ablation, high-dose chemotherapy or radiation (conditioning regimen) followed by administration of autologous or allogenic hematopoietic stem cells (Zhao et al. 2022; Tuzovic et al. 2019). The use of HSCT has increased in recent years at a rate of over 7% per year (about 90,000 per year). It accounted for 1.5 million patients received transpalntation in 2019 (Zhao et al. 2022).

Improvements in transplantation strategies and post-transplantation supportive care have increased the survival rates of HSCT patients. A five-fold increase in the long-term survival rate after HSCT is expected by 2030 (Zhao et al. 2022). Although there is remarkable increase in patient outcomes, HSCT also has been related to complications that include relapse, failure to engraft, cardiovascular events and graft versus host disease (GVHD), which can affect survivor mortality and morbidity. About two-thirds of transplantation survivors experience one or more of these complications. One post-transplantation follow up project reported 93.2% incidence of long term complications within an average time of 7.2 years (Zhao et al. 2022). Severe or life-threatening complications also have been showed in 18% of HSCT survivors (Zhao et al. 2022).

2 GVHD

GVHD is an important complication of allogenic HSCT and occurs in 30–70% of patients (Jagasia et al. 2014). It is caused by an immunologic response to host tissues by donor lymphocytes and is associated with mortality and morbidity in these patients (Dogan et al. 2013). Traditionally, 100 days post-transplantation has been used as the length of time in which to classify acute and chronic disease. Acute GVHD generally occurs within the first three months, but chronic GVHD manifests beyond this point after bone marrow transplantation (Dogan et al. 2013; Nazari et al. 2013). A "normal" immune system can reject donor T-cells. However, in post HSCT situation, the recipient immune system will have been compromised by the use of chemoradiotherapy and differences between donor and recipient tissue antigens can lead to the activation of allogenic T-cells and the onset of GVHD (Ghimire et al. 2017).

Acute GVHD consists of three phases. Firstly, there is conditioning mediated tissue damage. Tissue damage will occur before donor-cell infusion from factors such as the underlying disease, infection, therapeutic methods and/or the conditioning regimen. The damaged tissue releases pro-inflammatory cytokines that include tumor necrosis factor (TNF), interleukin-1 (IL-1) and lead to activate host antigen-presenting cells (APCs). In the afferent phase, host APCs activate donor T-cells. The response will depend on the difference between the donor and recipient human leuko-cyte antigens (HLAs). Low intensity or non-myeloablative conditioning can decrease the toxicity and severity of GVHD. Several studies have shown that delaying the donor cell transfer after conditioning can reduce the risk of GVHD. Finally, T-cell induced inflammation can result in tissue damage and target organ dysfunction as an efferent phase (Ghimire et al. 2017) (Fig. 1). Immunosuppressive methods for prevention and treatment tend to focus on donor T-cell activation. Chronic GVHD is an important reason of non-relapsing mortality after HSCT (Ghimire et al. 2017). Inflammation, cell-mediated and humoral immunity as well as fibrosis can contribute to the pathophysiology of chronic GVHD (Jagasia et al. 2014). Acute GVHD, where the immune cells and organs are the primary targets, is a main risk factor for chronic GVHD. Thymus destruction and insufficient selection of donor T-cells by the thymus can cause alloimmunity and autoimmunity, which are associated with chronic GVHD. Sclerotic lesions in different organs are known hallmarks of chronic GVHD (Ghimire et al. 2017). The most common risk factors for GVHD include donor/recipient HLA mismatch, donor/recipient age (older host, especially with a younger donor), donor/recipient apposite gender and T-lymphocytes number in the inoculum (Dogan et al. 2013; Ferreira et al. 2014).

The National Institute of Health in the USA has revised classification criteria for acute versus chronic GVHD. The latter is now categorized by a combination of clinical symptoms rather than the time of onset. Acute GVHD is associated with epithelial injury and commonly manifests as skin, gastrointestinal tract and liver involvement. Chronic GVHD is associated with sclerotic and fibrotic changes in one or more organs and mesenchymal tissue is mostly involved. Chronic GVHD commonly affects the skin, mouth, eyes, nails, genitals, gastrointestinal tract, liver, lung, kidney and heart (Dogan et al. 2013; Ghimire et al. 2017; Jamil and Mineishi 2015). Overall, liver, lung, skin and gasterointestinal tract are organs that most commonly affected by GVHD (Dogan et al. 2013).

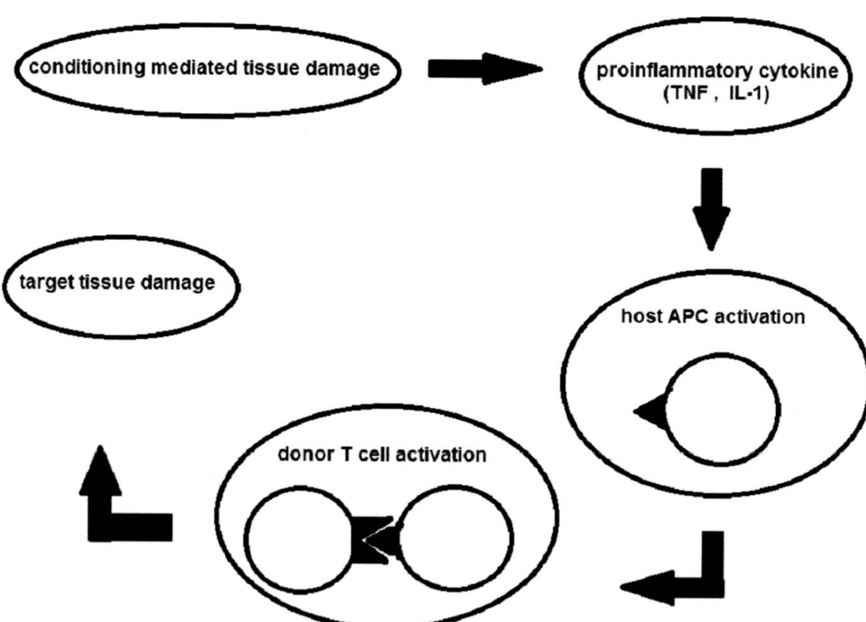

Fig. 1 Acute GVHD pathophysiology- damaged tissue releases pro-inflammatory cytokines and activates host antigen-presenting cells. Host APCs activate donor T cells and then T cell induced inflammation results in target tissue damage and organ dysfunction. APCs, antigen-presenting cells. IL-1,interleukine -1.TNF, tumor necrosis factor

Cardiovascular complications are not common after HSCT, although it is associated with high mortality rates and a reduced quality of life (Zhao et al. 2022). Clinical manifestations of chronic GVHD often present whitin the first year after HSCT, but some patients can develop the condition many years after transplantation (Jagasia et al. 2014).

3 Cardiac GVHD

3.1 Cardiac GVHD Epidemiology

The cardiovascular system is rarely reported as a target organ in GVHD, but difficulties in histological diagnostic methods could be the reason for the underreporting. Cardiac involvement often occurs in the early phases after HSCT or in later flare-ups in chronic GVHD (Tichelli et al. 2008).

3.2 Cardiac GVHD Mechanisms

Inflammation, cardiac infiltration of donor T-cells and some immune-suppressive responses are common precursors of cardiac GVHD. Studies have revealed increases in C-reactive protein (CRP) and erythrocyte sedimentation rate (ESR) plasma levels are indicators of systemic inflammation in post-HSCT patients who develop GVHD. These plasma levels are higher in patients with more severe symptoms (Dogan et al. 2013). The grade of acute GVHD is significantly related to early-onset cardiac dysfunction. GVHD has a direct cardiotoxicity effect via donor T-cell infiltration in the myocardium and indirect effect through cytokine release such as TNF factor α (TNF-α) and IL-2. TNF can affect muscle electrical activity and reduce myocardial contractility. IL-2 is associated with arrhythmia (tachyarrhythmia, bradyarrhythmia and third-degree atrioventricular block). In acute GVHD, suppression of regulatory T-cells can cause activation of responder T-cell mediated immune reactions and result in inflammation of the myocardium, thereby reducing cardiac function (Moriyama et al. 2022; Rackley et al. 2005).

Various immunosuppressive agents which can have cardiotoxic effects often are used for acute or chronic GVHD treatment (Tuzovic et al. 2019). Immunosuppressive prophylaxis for GVHD in the early post-HSCT phase can also increase cardiovascular risk factors. Post-HSCT survivors who develop grades II–IV acute GVHD have a nine-fold increased risk of hypertension, six-fold increased risk of diabetes and three-fold increased risk of dyslipidemia compared to autologous HSCT survivors, Steroids and calcineurin inhibitors such as cyclosporine are used in post-HSCT patients who present with high-grade GVHD, although cardiovascular risk factors such as diabetes, hypertension and hyperlipidemia can develop from these agents. On the other hand, cessation of these drugs may not resolve the cardiovascular risk factors, necessarily. Ruxolitinib, an off-label treatment which has been used for steroid-refractory GVHD also is known to affect the lipid profile (Tuzovic et al. 2019; Armenian et al. 2012).

Arterial wall inflammation, lipid storage in the endothelium and further damage can contribute to atherosclerosis. As chronic GVHD stems from immune mediated inflammation, allogenic reactions and vascular endothelial injury can result in vascular events. The loss of thrombomodulin, a natural anticoagulant, has been observed in biopsies of GVHD patients. Endothelial damage also can lead to steroid resistance and failure of GVHD recovery (Tuzovic et al. 2019; Nazari et al. 2013; Ghimire et al. 2017).

3.3 Cardiac GVHD Manifestation

Cardiac GVHD patients develop cardiomyopathy, coronary artery disease (CAD), arrhythmia and pericardial disease. Pericardial effusion is known as a sign of polyserositis related to chronic GVHD, not as a cardiac presentation. Patient manifestations can vary from asymptomatic to fatal. The number of studies which focus on prophylaxis and therapeutic methods for HSCT-induced cardiovascular complications including cardiac GVHD are limited (Zhao et al. 2022; Tichelli et al. 2008; Rackley et al. 2005).

Cardiomyopathy

Investigative research has focused on the effect of GVHD on left ventricle (LV) systolic function. Most studies, primarily have focused on animals, relate to the effect of chronic GVHD on LV diastolic function. A prospective study by Dogan et al. evaluated cardiac involvement in patients who had undergone bone marrow transplantation and compared the groups who did and did not develop GVHD. They reported a significant increase in LV mass and wall thickness accompanied by a decrease in the E/A ratio in patients who developed chronic GVHD compared to the non-GVHD group. The LV diameter and LV ejection fraction showed no significant difference between groups. The studies indicated that GVHD patients who receive immune suppressive agents such as cyclosporine showed significantly increased LV thickness and mass compared to patients did not. This could be explained by the effect of cyclosporine on the neurohormonal system, which can cause renal and systemic vasoconstriction and activate rennin-angiotensin aldosterone and the sympathetic nervous systems (Dogan et al. 2013).

Durley and colleague assessed the incidence and features of cardiac events in posttransplantation patients who received cyclophosphamide as a prophylactic agent for GVHD. They reported that the use of post-transplantation cyclophosphamide lead to decrease the incidence of GVHD, although it is related to LV systolic dysfunction and cardiac events within first 100 days behind the transplantation (Duléry et al. 2021).

Coronary Artery Disease and Thromboembolism Events

Traditional risk factors, inflammatory responses and endothelial damage can contribute to atherosclerosis in GVHD patients. Atherosclerosis and CAD are rare and life-threatening late complications in these patients. The pathogenesis and morphology of the lesions are not well known; however, atherosclerosis has been reported in young post-HSCT patients, Unexpectedly and significant stenosis has been observed in young women (Nazari et al. 2013; Smukowska-Gorynia et al. 2019).

Intima-media thickness (IMT) is a known marker of atherosclerosis in patients without BMT. Nazari et al. performed a cross-sectional study to assess IMT as an early predictor of atherosclerosis in GVHD patients and reported higher IMT in ultrasonography of post BMT patients with chronic GVHD (Nazari et al. 2013). Guideline based diagnosis and management are used in GVHD patients with CAD based on limited previous case reports (Smukowska-Gorynia et al. 2019; Liu et al. 2022).

Venous and arterial thromboembolism is often related to inflammation. Endothelial dysfunction and reduced thrombomodulin-dependent generation of activated protein C are involved in GVHD pathogenesis and lead to a procoagulant state. Some studies have accounted thromboembolism incidence in allogenic HSCT recipients with a higher risk in GVHD patients (Jurdi et al. 2021).

Arrhythmia

GVHD can cause arrhythmia, particularly bradyarrhythmia (Ferreira et al. 2014). IL-2 can induce tachyarrhythmia, bradyarrhythmia and third-degree atrioventricular block. The association between bradyarrhythmia and GVHD is aided by its resemblance to cardiac graft rejection. Rouah et al. reported on a post-BMT patient who developed acute GVHD followed after chronic GVHD and needed a pacemaker because of a complete heart block. Lymphohistiocytic infiltration, necrosis region along with scarring of the atrium, ventricle myocardium, atrioventricular node, and the bundle branches were detected in the patient autopsy one month later (Rackley et al. 2005).

Bradycardia in GVHD patients can develop to sinus node dysfunction or third degree heart block; thus, such patients should be strictly monitored. Patients without typical manifestations of GVHD should be assessed by electrophysiological study and endomyocardial biopsy to seek lymphocytic infiltration. Although bradycardia-associated GVHD often ameliorates by increasing immunosuppressive agents (Rackley et al. 2005), a differential diagnosis for bradyarrhythmia in these patients is drug toxicity, especially in those who received rapid infusion of high dose steroids. In fact, high doses of methylprednisolone ($\geq$ 4 mg/kg/day) can lead to lymphocyte death and a sudden release of cytokines. Anti-thymocyte globulin infusion sometimes results in the same effect. A high serum level of cyclosporine is related to bradyarrhythmia (Rackley et al. 2005). Ibrutinib is used for the treatment of chronic GVHD and is related to atrial fibrillation (Tuzovic et al. 2019). Overall, in post-HSCT patients who have developed GVHD and present with unexplained dysrhythmia or coronary arteries disease, cardiac GVHD should be considered (Tichelli et al. 2008).

Pericardial Disease

Constrictive pericarditis related to GVHD is a rare but potentially reversible condition. Donor immune cells can contribute to the inflammation of the pericardium and such patients usually present with dyspnea and peripheral edema. This has been resolved by systemic immunosuppressive therapy in the early stage of the disease and before permanent and irreversible thickening of pericardium, although surgical pericardiectomy has been reported to be a viable option to delivering acceptable functional recovery (Morley-Smith et al. 2013; Pieri et al. 2020). Pieri et al. reported a post-HSCT patient with frequent admission due to progressive exertional dyspnea and peripheral edema. Cardiac imaging showed thickening of pericardium and a constrictive physiology. Surgical partial pericardiectomy and immunomodulatory therapy with ruxolitinib improved patient symptoms, without the need for hospital admission. Histological analysis confirmed GVHD features (Pieri et al. 2020) (Table 1).

4 Cardiovascular Disease Screening

Based on the latest guidelines, Cardiovascular evaluation is recommended for patients who Candidate for HSCT,which includes basic cardiovascular (CV) assessment, natriuretic peptides (NPs), electrocardiogram (ECG), and transthoracic echocardiography (TTE). In low-risk patients, basic CV evaluation and ECG are recommended 3 and 12 months after HSCT and then yearly. If a patient presents with new cardiac symptoms, TTE and NP are recommended as additional evaluations. In high-risk patients, a whole set of CV assessment, including basic CV evaluation, ECG, NP and TTE, is recommended for post HSCT patients after 3 and 12 months, as early surveillance. The more comprehensive factors related to GVHD should be attented and evaluated in post-HSCT patients with long-term cardiovascular disease (Zhao et al. 2022).

Table 1 Cardiac manifestations of GVHD

Cardiac manifestation	Pathophysiology	Presentation	Diagnostic tool
Cardiomyopathy	Donor T-cell infiltration Inflammatory cytokines (TNF, IL-2)	Ventricle systolic dysfunction Ventricle diastolic dysfunction Increased LV wall thickness and mass	Echocardiograph
Coronary artery disease	Traditional risk factors (HTN, DLP, DM) Arterial wall inflammation Endothelial dysfunction Loss of thrombomodulin	Coronary artery disease and pre-mature atherosclerosis	Coronary angiography ECG Cardiac troponin Echocardiography
Arrhythmia	Inflammatory cytokine (IL-2) Lymphocyte infiltration Drug	Tachyarrhythmia Bradycardia Complete heart block	ECG, EPS
Pericardial disease	Chronic inflammation of pericardium	Pericardial thickening Constrictive physiology	Cardiac imaging (echocardiography, CMR)

CMR, cardiac magnetic resonance; DLP, dyslipidemia; DM, diabetes; ECG, electrocardiogram; EPS, electrophysiological study; IL-2, interleukine-2; TNF, tumor necrosis factor

5 Conclusion and Future Horizons

Regardless of malignancies in post-HSCT patients, the average survival rate is 24–76%. Primary and secondary complications can increase the mortality rate and affect the quality of life in these patients and cardiovascular complications can present in the both the short- and long-term.

Alizadehasl et al. (2023) performed a systematic review and meta-analysis in 2022 and reported a 16.84% prevalence of cardiovascular disease in post-HSCT patients. Cardiovascular complications are more common in allogenic HSCT. immunosuppressive drugs have not been used, so there was not an increased risk of GVHD in autologous transplantation. Despite the effect of cardiac GVHD on post-HSCT outcomes, limited studies have focused on early diagnostic methods or preventive and therapeutic strategies for cardiac GVHD. The evaluation of molecular markers and risk stratification tools accompanied by assessment of adequate treatment for such patients can help develop early diagnostic, preventive and therapeutic approaches that could improve patient survival and quality of life.

References

Alizadehasl A, Ghadimi N, Hosseinifard H, Roudini K, Hossein Emami A, Ghavamzadeh A, et al. Cardiovascular diseases in patients after hematopoietic stem cell transplantation: systematic review and meta-analysis. Curr Res Transl Med. 2023;71(1):103363. https://doi.org/10.1016/j.retram.2022.103363

Armenian SH, Sun CL, Vase T, Ness KK, Blum E, Francisco L, Venkataraman K, Samoa R, Wong FL, Forman SJ, Bhatia S. Cardiovascular risk factors in hematopoietic cell transplantation survivors: role in development of subsequent cardiovascular disease. Blood. 2012;120(23):4505–12. https://doi.org/10.1182/blood-2012-06-437178. Epub 2012 Oct 3.

Dogan A, Dogdu O, Ozdogru I, Yarlioglues M, Kalay N, Inanc MT, Ardic I, Celik A, Kaynar L, Kurnaz F, Eryol NK, Kaya MG. Cardiac effects of chronic graft-versus-host disease after stem cell transplantation. Tex Heart Inst J. 2013;40(4):428–34. PMID: 24082373; PMCID: PMC3783135.

Duléry R, Mohty R, Labopin M, et al. Early cardiac toxicity associated with post-transplant cyclophosphamide in allogeneic stem cell transplantation. J Am Coll Cardiol CardioOnc. 2021;3(2):250–9. https://doi.org/10.1016/j.jaccao.2021.02.011.

El Jurdi N, Elhusseini H, Beckman J, et al. High incidence of thromboembolism in patients with chronic GVHD: association with severity of GVHD and donor-recipient ABO blood group. Blood Cancer J. 2021;11:96. https://doi.org/10.1038/s41408-021-00488-2.

Ferreira DC, de Oliveira JS, Parísio K, Ramalho FM. Pericardial effusion and cardiac tamponade: clinical manifestation of chronic graft-versus-host disease after allogeneic hematopoietic stem cell transplantation. Rev Bras Hematol Hemoter. 2014;36(2):159–61. https://doi.org/10.5581/1516-8484.20140034.

Ghimire S, Weber D, Mavin E, Wang X, Dickinson AM, Holler E. Pathophysiology of GvHD and Other HSCT-related major complications. Front Immunol. 2017;8:79. https://doi.org/10.3389/fimmu.2017.00079.

Jagasia MH, Greinix HT, Arora M, Williams KM, Wolff D, Cowen EW, et al. National institutes of health consensus development project on criteria for clinical trials in chronic graft-versus-host disease: I. The 2014 diagnosis and staging working group report. Biol Blood Marrow Transplant. 2015;21(3):389–401.e1. https://doi.org/10.1016/j.bbmt.2014.12.001. Epub 2014 Dec 18.

Jamil MO, Mineishi S. State-of-the-art acute and chronic GVHD treatment. Int J Hematol. 2015;101:452–66. https://doi.org/10.1007/s12185-015-1785-1.

Keykhaei M, Masinaei M, Mohammadi E, Azadnajafabad S, Rezaei N, et al. A global, regional, and national survey on burden and Quality of Care Index (QCI) of hematologic malignancies; global burden of disease systematic analysis 1990–2017. Exp Hematol Oncol. 2021;10(1):11. https://doi.org/10.1186/s40164-021-00198-2.

Liu Y, Cheng C, Xu R, Xie Y. Acute myocardial infarction in a 13-year-old adolescent with chronic graft-versus-host disease: a case report. Anatol J Cardiol. 2022;26(11):849–51. https://doi.org/10.5152/AnatolJCardiol.2022.1760.

Mattiuzzi C, Lippi G. Current Cancer Epidemiology. J Epidemiol Glob Health. 2019;9(4):217–22. https://doi.org/10.2991/jegh.k.191008.001.

Moriyama S, Fukata M, Hieda M, et al. Early-onset cardiac dysfunction following allogeneic haematopoietic stem cell transplantation. Open Heart. 2022;9: e002007. https://doi.org/10.1136/openhrt-2022-002007.

Morley-Smith AC, Cowie MR, Vazir A. Pericardial constriction attributable to graft-versus-host disease: importance of early immunosuppression. Circ Heart Fail. 2013;6(5):e59-61. https://doi.org/10.1161/CIRCHEARTFAILURE.113.000462.

Nazari M, Morteza A, Irvani M, Ghavamzadeh A, Al-e-Esmael A, Khalilzade·Hadi Rokni-Yazdi O. Evaluation of intima-media thickness of carotid artery in patients with chronic graft-versus-host disease using ultrasound. Res Article—Imaging Med. 2013; 5(1):19–23.

Paterson DI, Wiebe N, Cheung WY, Mackey JR, Pituskin E, Reiman A, Tonelli M. Incident cardiovascular disease among adults with cancer: a population-based cohort study. JACC CardioOncol. 2022;4(1):85–94. https://doi.org/10.1016/j.jaccao.2022.01.100.

Pieri CA, Roberts N, Gribben J, Manisty C. Graft-versus-host disease: a case report of a rare but reversible cause of constrictive pericarditis. Eur Heart J Case Rep. 2020;4(2):1–5. https://doi.org/10.1093/ehjcr/ytaa009.

Rackley C, Schultz KR, Goldman FD, Chan KW, Serrano A, Hulse JE, Gilman AL. Cardiac manifestations of graft-versus-host disease. Biol Blood Marrow Transplant. 2005;11(10):773–80. https://doi.org/10.1016/j.bbmt.2005.07.002.

Smukowska-Gorynia A, Iwańczyk S, Janus M, Lesiak M, Magoń J, Mularek-Kubzdela T, Lesiak M. Severe coronary artery disease secondary to graft-versus-host disease after bone marrow transplantation in a 24-year-old woman. Kardiol Pol. 2019;77(12):1202–3. https://doi.org/10.33963/KP.15062. Epub 2019 Nov 18.

Sung H, Ferlay J, Siegel RL, Laversanne M, Soerjomataram I, Jemal A, Bray F. Global cancer statistics 2020: GLOBOCAN estimates of incidence and mortality worldwide for 36 cancers in 185 countries. CA Cancer J Clin. 2021;71:209–49. https://doi.org/10.3322/caac.21660.

Tichelli A, Bhatia S, Socié G. Cardiac and cardiovascular consequences after haematopoietic stem cell transplantation. Br J Haematol. 2008;142(1):11–26. https://doi.org/10.1111/j.1365-2141.2008.07165.x.

de Moraes Hungria VT, Chiattone C, Pavlovsky M, Abenoza LM, Agreda GP, et al. Epidemiology of hematologic malignancies in real-world settings: findings from the hemato-oncology Latin America observational registry study. J Glob Oncol. 2019;5:1–19. https://doi.org/10.1200/JGO.19.00025

Tuzovic M, Mead M, Young PA, Schiller G, Yang EH. Cardiac complications in the adult bone marrow transplant patient. Curr Oncol Rep. 2019;21(3):28. https://doi.org/10.1007/s11912-019-0774-6.

Zhao Y, He R, Oerther S, Zhou W, Vosough M, Hassan M. Cardiovascular complications in hematopoietic stem cell transplanted patients. J Pers Med. 2022;12:1797. https://doi.org/10.3390/jpm12111797.

HSCT in Low EF Patients

Nasim Naderi, Mehdi Dehghani, and Seyed Hossein Mirpour Hassankiadeh

Abstract Hematopoietic stem cell transplantation (HSCT) has been used as a treatment method for decades. The most common diseases in autologous hematopoietic stem cell transplantation are multiple myeloma and Hodgkin's and non-Hodgkin lymphoma. The success rate of HSCT depends on two factors including: the underlying disease and the other is the stage of the disease. Chemotherapy is a common method of cancer treatment and is effective in cancer treatment, but adverse effects such as chemotherapy-induced cardiotoxicity (CIC) are known, especially by anthracyclines and trastuzumab in HER2-positive breast cancer. Radiation therapy to the mediastinum may aggravate atherosclerosis and lead to early onset of coronary artery disease as well as valvular disease that usually affects the left side valves, as well as acute and chronic pericardial involvement may be seen after radiation therapy.

Keywords Hematopoietic stem cells transplantation · Chemotherapy-induced cardiotoxicity · Radiation-induced cardiovascular disease

Abbreviations

HSCT   Hematopoietic stem cells transplantation
CIC   Chemotherapy-induced cardiotoxicity

N. Naderi
Internal Medicine, Hematology, Medical Oncology and Stem Cell Transplantation, Hematology Research Center, Shiraz University of Medical Sciences, Shiraz, Iran

Rajaie Cardiovascular Medical and Research Center, Iran University of Medical Sciences, Tehran, Iran

M. Dehghani (✉)
Hematology Research Center, Shiraz University of Medical Sciences, Shiraz, Iran
e-mail: mehdi_dehghani6@yahoo.com

S. H. Mirpour Hassankiadeh
Department of Internal Medicine, School of Medicine, Guilan University of Medical Sciences, Rasht, Iran

A. Alizadehasl et al. (eds.), *Cardiovascular Considerations in Hematopoietic Stem Cell Transplantation*, https://doi.org/10.1007/978-3-031-53659-5_8

RICVD	Radiation-induced cardiovascular disease
MRI	Magnetic resonance imaging
LV	Left ventricle
ACE	Angiotensin converting enzyme
Gy	Gray
LVEF	Left ventricular ejection fraction
GLS	Global longitudinal strain
HF	Heart failure
HCT	Hematopoietic cell transplantation
GDMT	Guideline directed medical therapies
MDT	Multidisciplinary team
ACEI	Angiotensin-converting enzyme inhibitors
ARB	Angiotensin II receptor blockers

1 Hematopoietic Stem Cells Transplantation, Chemotherapy-Induced Cardiotoxicity and Radiation-Induced Cardiovascular Disease

Hematopoietic stem cells transplantation (HSCT) might be performing from bone marrow or peripheral blood and has been used as a treatment method for about four decades. In order to recommend a bone marrow transplant, first of all, it is necessary to weigh the risk of the disease against the risk of transplantation, which depends on various factors including: underlying disease, age of the patient, and time interval between diagnosis of the disease and stem cell transplant. The type and gender of the donor (siblings or unrelated individuals), and individual characteristics also need to be considered (Hołowiecki 2008).

Generally, the most-common indications for autologous stem cell transplant are myeloma, malignant lymphoma and acute myeloblastic leukemia while the main indication for allogenic stem cell transplantation is acute myeloblastic leukemia, lymphoblastic leukemia, myelodysplastic syndrome, Chronic myeloblastic leukemia in the blastic phase or in case of resistant to tyrosine kinase inhibitor treatment or non-malignant disorders (bone marrow aplasia, severe immunodeficiency, paroxysmal nocturnal hemoglobinuria as well as non-hematological disorders, such as genetic disorders and autoimmune diseases (Hołowiecki 2008; Duarte et al. 2019).

The transplantation is used in different stages of the disease, including in the initial phase, when the cancer is first diagnosed, and in the advanced stage, when other treatments have failed. Apart from hematological disorders, HSCT is used to treat sickle cell anemia, thalassemia, genetic disorders and autoimmune disease. In autoimmune diseases, HSCT is used to reset the immune system and stop the disease progression. HSCT is used to treat diseases such as multiple sclerosis, systemic sclerosis, and autoimmune hemolytic anemia (Duarte et al. 2019).

HSCT is a versatile and effective treatment option for various hematological and non-hematological disorders. The transplantation is used to replace the faulty stem cells with healthy ones and provides long-term remission and even cure in some cases (Duarte et al. 2019).

The success of HSCT depends on various factors, such as the type and stage of the disease, the age and health of the patient, the donor's tissue compatibility, and the timing of the procedure. Complications including: infections, graft-versus-host disease and organ damage may occur (Snowden et al. 2022).

Despite the risks, HSCT remains a promising treatment option for many patients with hematological and immune disorders. With advances in technology and medical research, HSCT continues to evolve, offering hope for improved outcomes and quality of life for patients (Tayuwijaya et al. 2023).

Chemotherapy and radiation therapy as a treatment method in cancer patients can make hematopoietic stem cell transplantation difficult, and due to cardiac side effects that may occur during treatment, this treatment method may be difficult and complicated. Chemotherapy is a common method of cancer treatment that involves the use of drugs to treat and control the disease. However, while chemotherapy can be effective in treating cancer, it can also have adverse effects on the body that vary depending on the type of drug. One of the most important side effects of chemotherapy is cardiotoxicity, which can damage the heart and increase the risk of heart disease. Chemotherapy-induced cardiotoxicity (CIC) can manifest as several different types of heart damage, including arrhythmias, cardiomyopathy, heart failure, and ischemic heart disease. The risk of CIC is influenced by factors such as the type of chemotherapy drugs used, the dose and duration of treatment, and the patient's pre-existing cardiovascular health (Lipshultz et al. 2013; Magdy et al. 2016).

Several chemotherapy drugs have been identified as having a high risk of causing CIC, including anthracyclines, such as doxorubicin and epirubicin, and trastuzumab, a targeted therapy used to treat HER2-positive breast cancer. Cardiotoxicity were reported by paclitaxel and tyrosine kinase inhibitors can causes QT prolongation and arrhythmia. These drugs can cause damage to the heart muscle, leading to reduced heart function and an increased risk of heart failure(Lipshultz et al. 2013; Magdy et al. 2016).

To prevent CIC, it is essential to identify patients at risk and closely monitor their cardiovascular health during treatment. Imaging tests, such as echocardiography and cardiac magnetic resonance imaging study can be used to evaluate heart function before and during treatment. In some cases, the use of cardioprotective drugs, such as beta-blockers and ACE inhibitors, aldosterone antagonists, and statins, can significantly improve LV systolic function and It may also be helpful (Li et al. 2020).

Radiation induce cardiovascular damage, is a well-known complication of radiation therapy, especially in cancer patients. This disease is caused by damage to the heart and blood vessels due to exposure to ionizing radiation, and leading to an increase in the risk of cardiovascular events. Ionizing radiation through damage to endothelial cells can accelerate atherosclerosis and also cause pro-thrombotic changes in the coagulation pathway and also inflammation and fibrosis of myocardial, pericardial, valves and conductive system. Radiation induce cardiovascular damage has become a significant health concern due to the increasing number of cancer patients receiving radiation therapy (Koutroumpakis et al. 2022).

Prevention of cardiovascular damage caused by radiation can be minimized. In this way, if the heart is subjected to radiation treatment, it is possible to minimize cardiac damage by treating the underlying cardiovascular risk factors and also by preventive drug therapy after exposure to radiation(Koutroumpakis et al. 2022).

The mechanism of cell damage when they are exposed to ionizing radiation is that in less than one microsecond after the radiation, the cells undergo a stress response. This response begins with the impact of ionizing radiation on biological materials and acts directly or indirectly through the formation of reactive oxygen species with cellular biomolecules such as proteins, lipids and DNA, which can lead to cell damage. This reaction affects and interferes with all cellular organs and has the ability to influence the molecular mechanisms of cells (Baselet et al. 2019).

Radiation therapy to the mediastinal area accelerates the process of atherosclerosis and causes premature coronary artery disease. The effect of radiation therapy on the heart valves is observed mostly on the left side, therefore, the mitral valve and the aorta are more affected, among which aortic insufficiency is the most common. Of course, it may rarely lead to aortic stenosis that requires surgical interventions. Another heart problem that is more often seen with radiation therapy is pericardial involvement, which includes acute and chronic pericarditis and pericardial effusion. Radiation therapy by itself causes the dysfunction of vascular endothelial cells, which can cause cardiovascular events for patients that show themselves years after the completion of radiation therapy (Raghunathan et al. 2017).

Several studies have been conducted to assess the risk of radiation induce cardiovascular damage and the results have shown that the incidence of radiation induce cardiovascular damage varies depending on the dose and site of radiation, as well as other factors. One study found that the risk of radiation induce cardiovascular damage increased by 7.4% for each Gy of radiation dose to the heart. Another study reported that the risk of radiation induce cardiovascular damage was higher in patients with left-sided breast cancer who received radiation therapy compared to those with right-sided breast cancer.

Radiation induce cardiovascular damage is a potential complication of radiation therapy that can lead to significant treatment related morbidity and mortality in patients with cancer. Clinicians must consider the potential risk of RICVD when administering radiation therapy to cancer patients, particularly those with pre-existing cardiovascular conditions. Close monitoring of patients who receive radiation therapy is crucial to identify and manage radiation induce cardiovascular damage promptly.

2 Clinical Management of Patients with Reduced Left Ventricular Ejection Fraction Undergoing Hematopoietic Cell Transplantation

Hematopoietic cell transplantation (HCT) also called bone marrow transplantation (BMT) is a potential cure and probably the only treatment for certain advanced hematologic malignancies. Recently, advancements in transplantation protocols and post transplantation management strategies lead to remarkable improvement in survival in these patients. However, despite the significant improvements in outcomes, these patients encounter acute, sub-acute and chronic complications including cardiovascular complications (Lehmann et al. 2000; Gavriilaki et al. 2019; Zhao et al. 2022).

Although among all BMT-related complications, cardiovascular complications are uncommon (< 10% for both autologous and allogeneic transplantation) but the presence of LV systolic dysfunction prior to HCT as well as acute and late cardiovascular complications are still critical limitations for these patients and remarkably affect the morbidity and mortality of HCT survivors. Myocardial dysfunction and heart failure are among the main HCT-related cardiovascular complications (Lehmann et al. 2000; Gavriilaki et al. 2019; Zhao et al. 2022; Qazilbash et al. 2009; Piranfar et al. 2012; Mo et al. 2013; Hurley et al. 2015; Moriyama et al. 2022).

The most important concern regarding the HCT candidate is the presence of LV systolic dysfunction or reduced left ventricular ejection fraction (LVEF). Despite the remarkable progresses in the field of cancer therapy related cardiac dysfunction (CTRCD) and careful surveillance of cardiotoxicities, some of the HCT candidates will have LV systolic dysfunction prior to the HCT. Furthermore, the other etiologies for cardiac dysfunction can be considered in these patients including myocarditis, different types of cardiomyopathies or ischemic heart diseases.

Development of LV systolic dysfunction can also be seen both as short term cardiotoxicity during or early after HCT and as a long term cardiovascular complication (Lehmann et al. 2000; Gavriilaki et al. 2019; Zhao et al. 2022; Qazilbash et al. 2009; Piranfar et al. 2012; Mo et al. 2013; Hurley et al. 2015; Moriyama et al. 2022; Lyon et al. 2022).

3 HCT in Patients Who Have Already Reduced LVEF Prior to the HCT

The clinical studies regarding HCT in those with reduced LVEF are limited. Several years ago, patients with LVEF < 50% were considered high risk of non-relapse mortality and ineligible for HCT. However, the validity of this kind of risk assessment was unclear and a potentially curative treatment to patients with no alternate therapeutic options might be denied. So, many investigators tried to schedule these patients for HCT and have reported their experience. The cardiac complication rate following HCT in patients with reduced LVEF (cutoffs varying from < 55% to < 30%) has been reported between 6.7 and 42.9% in different studies (Lehmann et al. 2000; Gavriilaki et al. 2019; Zhao et al. 2022; Qazilbash et al. 2009; Piranfar et al. 2012; Mo et al. 2013; Hurley et al. 2015; Moriyama et al. 2022; Lyon et al. 2022).

There is no doubt that the mortality and morbidity is significantly higher in those with reduced LVEF and these patients may develop severe and sometimes refractory congestive heart failure following the HCT. However, it seems that many of patients with reduced LVEF can undergo HCT without increased risk of mortality or complication (Lehmann et al. 2000; Gavriilaki et al. 2019; Zhao et al. 2022; Qazilbash et al. 2009; Piranfar et al. 2012; Mo et al. 2013; Hurley et al. 2015; Moriyama et al. 2022; Lyon et al. 2022).

3.1 Practical Points (Lehmann et al. 2000; Gavriilaki et al. 2019; Zhao et al. 2022; Qazilbash et al. 2009; Piranfar et al. 2012; Mo et al. 2013; Hurley et al. 2015; Moriyama et al. 2022; Lyon et al. 2022)

1. A multi-disciplinary team is recommended to discuss about the best therapeutic strategies in those who will be potential candidates for HCT.
2. Careful selection of chemotherapy agents in terms of cardiotoxicity in those who are potential candidate for HCT after cancer therapies would be helpful.
3. Careful cardiovascular risk stratification and follow up as well as on time using the preventive and therapeutic measures according to the CTRCD guidelines (Lyon et al. 2022) would also be helpful in reducing the incidence of cardiac dysfunction after cancer therapies in potential candidates for HCT.
4. In those undergoing HCT, the presence of congestion (dyspnea, peripheral edema or ascites), arrhythmias (such as atrial fibrillation) and the other heart failure related signs and symptoms are better predictors of heart failure related morbidity and mortality than LVEF. It has been shown that asymptomatic patients with borderline LV systolic dysfunction can undergo HCT without increased risk of complication or mortality (Piranfar et al. 2012; Hurley et al. 2015).
5. The factors other than LVEF and congestion that may predispose a patient to post HCT cardiac complications, including hypertension, smoking, hyperlipidemia, coronary artery disease (CAD), arrhythmia and prior infarction. So, optimal risk factor modification as well as the prevention and management of CAD, ischemic heart syndromes and arrhythmia CVD in these patients is recommended and should generally follow guidelines for specific cardiovascular diseases.
6. Although very scant preclinical and clinical studies have been performed with a focus on the preventive and therapeutic strategies for HCT candidate particularly those with LV dysfunction, using the heart failure guidelines (McDonagh et al. 2021; Heidenreich et al. 2022) directed medical therapies (GDMT) including

angiotensin-converting enzyme (ACE) inhibitors (ACE-I) or angiotensin II receptor blockers (ARB) or angiotensin receptor–neprilysin inhibitor (ARNI), a selected beta-blockers (BBs), a mineral receptor antagonists (MRA) and a sodium glucose transporter 2 inhibitors (SGLT2I) in those HCT candidate with reduced EF would be rational even in asymptomatic ones unless the drugs are not tolerated or contraindicated.

7. It has been shown that aerobic exercise has beneficial effects in patients with reduced LVEF including the patients with CTRCD and would be recommended as cardiac rehabilitation programs for HCT candidates.

8. The HCT should be postponed in patients with symptomatic heart failure. Besides medical therapies using the above mentioned drugs, diuretics may be needed for relieving the congestion related signs and symptoms.

9. A multi-disciplinary team is recommended to discuss about the best strategies for heart failure therapy and related clinical and para clinical follow ups as well as decision making regarding the best time for HCT.

10. A multi-disciplinary team is recommended to discuss regarding the providing the advanced cardiovascular care facilities for patients with reduced LVEF undergoing HCT. These patients may develop severe decompensated heart failure following acute or sub-acute cardiotoxicity and may need advanced heart failure care including circulatory supports.

11. Cardiac monitoring and early follow ups would be helpful in on time diagnosis of cardiotoxicity related complications and prompt appropriate measures.

3.2 Management of LV Dysfunction During or After HCT(Lehmann et al. 2000; Gavriilaki et al. 2019; Zhao et al. 2022; Qazilbash et al. 2009; Mo et al. 2013; Hurley et al. 2015; Moriyama et al. 2022)

Although development of heart failure may be more common in those who have a baseline reduced LVEF, cardiotoxicity and heart failure can also be seen in patients without baseline LV systolic dysfunction (LVEF $\geq$ 50–55%) during, early and/or late after HCT.

Risk prediction has an outstanding role in preventing the cardiotoxicity related morbidities and mortalities. A number of risk prediction models for this purpose have been suggested aiming for stratification of patients into low and high cardiovascular risk in different clinical settings which is helpful in choosing the optimal treatment protocol, prophylactic treatment, conditioning regimens, and necessary cardiac function monitoring for HCT patients at different stages of transplantation and after that (Table 1).

A comprehensive cardiovascular assessment is crucial before HCT in all candidates who have normal cardiac function to detect un-diagnosed cardiac problems, risk stratification for chemotherapy related cardiotoxicity and decision making about the preventive measures.

Table 1 Shows risk factors of cardiotoxicity in patients undergoing HCT

Age
Preexisting cardiovascular disease
Infection
Traditional cardiovascular risk factors (HTN, DM, metabolic syndrome, smoking, …)
Acute or chronic Graft versus host disease (GVHD)
Pre-HCT exposure to anthracycline
Total body irradiation
Chest radiation
High-dose cyclophosphamide in the conditioning regimen or for GvHD
Corticosteroids
Tyrosine kinase inhibitors

This assessment should include physical examination, electrocardiogram (ECG), transthoracic echocardiography TTE), natriuretic peptides (NPs) and the other laboratory tests to assess diabetes mellitus and lipid profile and cardiopulmonary exercise tests (CPET).

Furthermore, a longitudinal follow up is recommended in HCT survivors using ECG, TTE, NPs and CPET. The long term surveillance should be at third months and 12 months after the HCT or any time if a new cardiac or heart failure related symptoms occur.

4 Recommendations

- It is recommended to start an ACE-I or ARB plus a beta-blockers (BB) for high risk patients undergoing HCT particularly those who have reduced GLS in echocardiography despite normal LVEF.
- If cardiac systolic dysfunction occurs during the surveillance, the patient should be treated and followed up according to the heart failure guideline recommendations (McDonagh et al. 2021; Heidenreich et al. 2022).
- As discussed earlier, the occurrence of acute or sub-acute cardiotoxicities and severe heart failure syndromes needs managing approach by a multidisciplinary team and providing advance heart failure care facilities.

5 Conclusion

HCT as a curative therapy for many patients can be performed with acceptable risk of complications in those who have LV systolic dysfunction. Starting GDMT for patients with LV systolic dysfunction before the HCT and during the procedure is

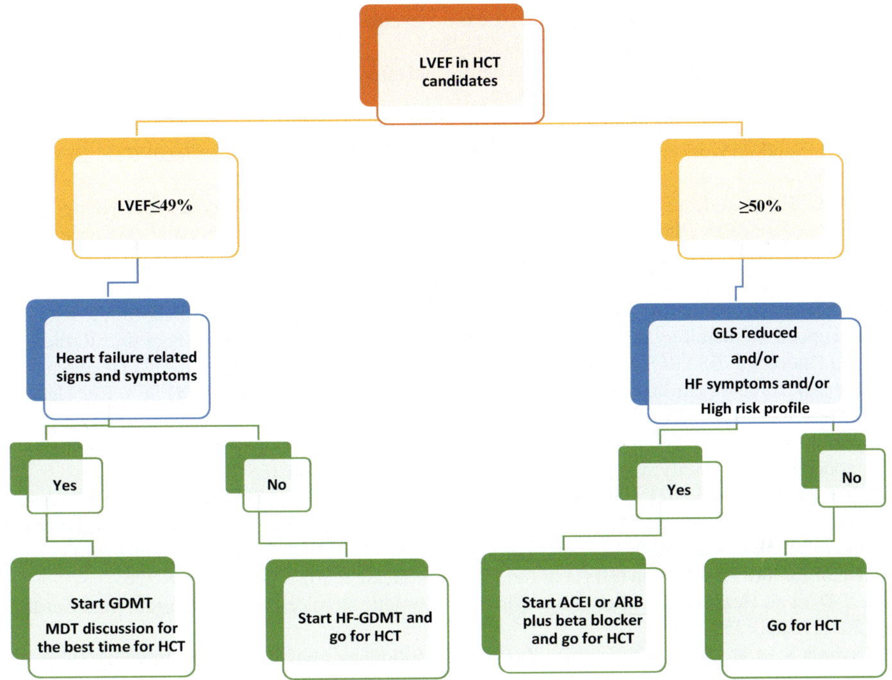

Fig. 1 Algorithm of decision making for hematopoietic cell transplantation consideringleft ventricular ejection fraction

recommended and crucial. Preventive strategies according to the CTRCD guidelines are recommended for those HCT candidates who have high risk profile for cardiotoxicities (Fig. 1).

References

Baselet B, et al. Pathological effects of ionizing radiation: endothelial activation and dysfunction. Cell Mol Life Sci. 2019;76:699–728.

Duarte RF, et al. Indications for haematopoietic stem cell transplantation for haematological diseases, solid tumours and immune disorders: current practice in Europe, 2019. Bone Marrow Transplant. 2019;54(10):1525–52.

Gavriilaki E, et al. Early prediction of cardiovascular risk after hematopoietic cell transplantation: are we there yet? Biol Blood Marrow Transplant. 2019;25(10):e310–6.

Heidenreich PA, et al. 2022 ACC/AHA/HFSA guideline for the management of heart failure: executive summary. J Cardiac Fail. 2022;28(5):810–30.

Hołowiecki, J., *Indications for hematopoietic stem cell transplantation*. Polskie archiwum medycyny wewnetrznej 2008;118(11):658–663.

Hurley P, et al. Hematopoietic stem cell transplantation in patients with systolic dysfunction: can it be done? Biol Blood Marrow Transplant. 2015;21(2):300–4.

Koutroumpakis E, et al. Radiation-induced cardiovascular disease: mechanisms, prevention, and treatment. Curr Oncol Rep. 2022;24(5):543–53.

Lehmann S, et al. Cardiac systolic function before and after hematopoietic stem cell transplantation. Bone Marrow Transplant. 2000;26(2):187–92.

Li X, et al. Role of cardioprotective agents on chemotherapy-induced heart failure: a systematic review and network meta-analysis of randomized controlled trials. Pharmacol Res. 2020;151: 104577.

Lipshultz SE, et al. Long-term cardiovascular toxicity in children, adolescents, and young adults who receive cancer therapy: pathophysiology, course, monitoring, management, prevention, and research directions: a scientific statement from the American Heart Association. Circulation. 2013;128(17):1927–95.

Lyon AR, et al. 2022 ESC Guidelines on cardio-oncology developed in collaboration with the European Hematology Association (EHA), the European Society for Therapeutic Radiology and Oncology (ESTRO) and the International Cardio-Oncology Society (IC-OS) Developed by the task force on cardio-oncology of the European Society of Cardiology (ESC). Eur Heart J. 2022;43(41):4229–361.

Magdy T, Burmeister BT, Burridge PW. Validating the pharmacogenomics of chemotherapy-induced cardiotoxicity: what is missing? Pharmacol Ther. 2016;168:113–25.

McDonagh TA, et al. 2021 ESC guidelines for the diagnosis and treatment of acute and chronic heart failure: developed by the task force for the diagnosis and treatment of acute and chronic heart failure of the European Society of Cardiology (ESC) With the special contribution of the Heart Failure Association (HFA) of the ESC. Eur Heart J. 2021;42(36):3599–726.

Mo X-D, et al. Heart failure after allogeneic hematopoietic stem cell transplantation. Int J Cardiol. 2013;167(6):2502–6.

Moriyama S, et al. Early-onset cardiac dysfunction following allogeneic haematopoietic stem cell transplantation. Open Heart. 2022;9(1): e002007.

Piranfar MA, et al. Bone marrow transplantation may augment cardiac systolic function in patients with a reduced left ventricular ejection fraction. J Cardiovasc Dis Res. 2012;3(4):310–4.

Qazilbash MH, et al. Outcome of allogeneic hematopoietic stem cell transplantation in patients with low left ventricular ejection fraction. Biol Blood Marrow Transplant. 2009;15(10):1265–70.

Raghunathan D et al (2017) Radiation-induced cardiovascular disease. Curr Atheroscler Rep. 19:1–8.

Snowden JA, et al. Indications for haematopoietic cell transplantation for haematological diseases, solid tumours and immune disorders: current practice in Europe, 2022. Bone Marrow Transplant. 2022;57(8):1217–39.

Tayuwijaya K, et al. Prognostic factors contributing to the survival of hematopoietic stem cell transplantation in the general population with leukemia: a systematic review. Regen Eng Transl Med. 2023;9(1):42–51.

Zhao Y, et al. Cardiovascular complications in hematopoietic stem cell transplanted patients. J Pers Med. 2022;12(11):1797.

Hypertension in HSCT

Seifollah Abdi, Ehsan Khalilipur, Amir Abdi, Sahar Tavakoli,
and Elgar Enamzadeh

Abstract Hypertension (HTN) is a prevalent and significant concern among patients who have undergone Hematopoietic Stem Cell Transplant (HSCT), contributing to increased morbidity and mortality in this population. The development of HTN in HSCT patients can be attributed to various factors, including chemotherapeutic and immunosuppressive medications, total body irradiation, chemotherapy, and graft-versus-host disease (GVHD). The management of HTN in HSCT recipients is crucial to minimize complications and improve outcomes. Understanding the risk factors associated with HTN in HSCT is essential for effective prevention and management. Medications used during HSCT, such as chemotherapeutic agents and calcineurin inhibitors, can contribute to elevated blood pressure levels. Radiation therapy, particularly total body and abdominal radiations has also been identified as a risk factor for HTN. Additionally, the presence of GVHD, both acute and chronic, increases the likelihood of HTN development due to the use of immunosuppressive agents and the impact of GVHD on vascular endothelium. HTN in HSCT patients is associated with various complications, including congestive heart failure, thrombotic microangiopathy, and posterior reversible encephalopathy syndrome (PRES). Screening and early detection of HTN in this population are crucial for prompt management of these complications. Regular blood pressure assessments, medication review, and

S. Abdi · E. Khalilipur (✉)
Cardiovascular Intervention Research Center, Rajaie Cardiovascular Medical and Research
Center, Iran University of Medical Science, Tehran, Iran
e-mail: Ehsankhalilipur@gmail.com

S. Abdi
e-mail: abdi@rhc.ac.ir

A. Abdi
Faculty of Medicine, Tehran Medical Sciences, Islamic Azad University, Tehran, Iran

S. Tavakoli
Hematology, Oncology and Bone Marrow Transplantation, Hematology and Oncology Research
Center Shariati Hospital, Tehran, Iran

E. Enamzadeh
Cardio-Oncology Research Center, Rajaie Cardiovascular Medical and Research Center, Iran
University of Medical Sciences, Tehran, Iran

monitoring of kidney function should be incorporated into routine care for HSCT survivors. Lifestyle modifications, including regular physical activity, maintaining a healthy weight, adopting a low-sodium diet, and limiting alcohol consumption, should be the first-line approach for managing mild HTN. Pharmacological treatment, with a focus on angiotensin-converting enzyme inhibitors (ACE-I), angiotensin receptor blockers (ARB), and calcium channel blockers (CCBs), is recommended for patients with higher blood pressure levels.

Keywords Hematopoietic stem cell transplant · Hypertension · Cardiovascular disease · Cardio-oncology

Abbreviations

HSCT	Hematopoietic Stem Cell Transplant
HTN	Hypertension
CVD	Cardiovascular Disease
GVHD	Graft Versus Host Disease
BP	Blood Pressure
SBP	Systolic Blood Pressure
DBP	Diastolic Blood Pressure
ACE-I	Angiotensin-converting Enzyme Inhibitor
ARB	Angiotensin Receptor Blocker
CCB	Calcium Channel Blockers

1   Introduction

Hypertension (HTN) is a frequent early or long-term occurrence following hematopoietic stem cell transplantation (HSCT) among both adult and pediatric populations, with studies reporting a widely differing range of HTN incidence as high as 70% following HSCT (range of 15–70%) (1, 2). This rise in blood pressure levels can be attributable to chemotherapy, radiation therapy, nephrectomy, and the use of corticosteroids and calcineurin inhibitors. Moreover, hypertension has been recognized as a late complication of stem cell transplant, with incidence linked to the development of chronic kidney disease (CKD). Optimal management of HTN in this population is paramount since HTN significantly contributes to the complications and mortality among these patients.

2 Risk Factors for the Development of Hypertension in HSCT

Medications and chemotherapeutic agents

Table 1 contains a comprehensive compilation of medications and agents that have been potentially linked to increased blood pressure in patients undergoing HSCT. Chemotherapeutic agents administered during chemotherapy and cancer treatments have been observed to be correlated with increases in the blood pressure levels of individuals diagnosed with cancer (3). Table 1 presents the agents that are frequently linked to hypertension, along with their potential mechanisms. These agents include VEGF inhibitors (VEGFi), alkylating agents, and erythropoietin. These agents induce elevations in blood pressure through various mechanisms, including increased vascular tone and peripheral resistance, impairment of endothelial function, and damage to the renal vasculature (4). VEGFi, along with more recent iterations of ABL-BCR tyrosine-kinase inhibitors like ibrutinib, as well as pharmaceuticals such as fluoropyrimidines, cisplatin, abiraterone, bicalutamide, and enzalutamide, constitute a range of pharmacological agents employed in the management of cancer (5). Furthermore, there is also ongoing consideration of non-cancer medications. When evaluating a new case of hypertension in cancer patients, it is crucial to consider multiple factors, including the utilisation of corticosteroids and non-steroidal anti-inflammatory drugs, the presence of stress and pain, excessive alcohol consumption, renal impairment, untreated sleep apnea, obesity, and reduced exercise (6). The rectification of these factors should be given precedence prior to contemplating the cessation of cancer treatment. Hypertension commonly arises as a result of the administration of Calcineurin inhibitors in individuals who have undergone HSCT or solid organ transplantation. Several potential underlying causes of hypertension have been identified in individuals using Calcineurin inhibitors. These include renal vasoconstriction, activation of the renin-angiotensin system, heightened sympathetic nervous tone, and elevated systemic vascular resistance (7). Furthermore, it has been observed that the exposure to cyclosporine is a significant predictor for the occurrence of new-onset hypertension within a period of 2 years following HSCT. According to the research conducted by Majhail et al., the administration of cyclosporine to individuals undergoing HSCT and experiencing graft-versus-host disease (GVHD) was found to be correlated with a heightened likelihood of developing hypertension after the transplantation procedure (Majhail et al., 8).

Radiation

The association between total body and abdominal radiations and hypertension has been consistently documented in the literature. Potential mechanisms underlying radiation-induced hypertension include endothelial damage, thrombotic microangiopathy, and the development of chronic kidney disease.

Table 1 Medications and factors commonly associated with hypertension in HSCT patients

	Mechanisms
Calcineurin inhibitors	• Vasoconstriction of the renal arterial system
Vascular endothelial growth factor (VEGF) inhibitors	• Vasoconstriction (by decreasing the production of nitric oxide)
Corticosteroids	• Sodium retention • Increased vascular tone • Increased blood volume
Alkylating agents	• Endothelial disruption/damage • Chronic kidney disease
Erythrocyte stimulating agents	• Direct vasopressor action of erythropoietin • Increase in blood viscosity • Endothelial disruption
Radiation	• Chronic kidney disease • Endothelial damage • Thrombotic microangiopathy

Graft-Versus-Host Disease (GVHD)

Graft-Versus-Host Disease (GVHD) is a prevalent and potentially fatal complication that frequently occurs following HSCT. GVHD arises when T cells with immune competence from the donor (referred to as the graft) perceive the recipient (known as the host) as an alien entity. Graft-versus-host disease GVHD manifests in two primary clinical forms, namely acute and chronic GVHD. Multiple hypotheses exist concerning the impact of GVHD on the cardiovascular function of patients undergoing HSCT. According to the research conducted by Armenian et al. (9), there was a significant correlation between the occurrence of high-grade GVHD and a 50% increased likelihood of developing hypertension within a decade after HSCT. Furthermore, a separate study revealed that individuals with a history of chronic GVHD exhibited a heightened susceptibility to hypertension, with a risk ratio of 3.2. The heightened susceptibility to hypertension in the aforementioned conditions primarily stems from the administration of commonly prescribed medications for managing or averting GVHD, such as immunosuppressive agents, steroids, and calcineurin inhibitors (3). Furthermore, the pathogenesis of GVHD has a direct effect on the vascular endothelium, resulting in damage and inflammation that may increase the susceptibility to hypertension (10). According to a study that included data on both autogenic and allogenic HSCT recipients, it was found that allogenic HSCT recipients had a significantly elevated risk of developing hypertension when compared to autogenic HSCT recipients (11).

3   Complications of HSCT Associated with HTN

According to available data, a significant proportion of individuals, specifically 70%, who have undergone HSCT, experience the development of hypertension within a period of two years following the transplantation procedure. HTN presents a substantial risk for the development of acute and long-term complications in HSCT patients. The prevention and management of complications associated with hypertension in these individuals necessitate regular monitoring of blood pressure and the implementation of screening measures for those with an elevated risk of developing hypertension and its subsequent complications. According to a study, the occurrence of hypertension among individuals who have undergone HSCT and have been exposed to high doses of anthracyclines is associated with a 35-fold higher likelihood of developing congestive heart failure (12).

Acute complications commonly observed in the context of hypertension in HSCT encompass thrombotic microangiopathy (TMA) and posterior reversible encephalopathy syndrome (PRES). These conditions, which pose a significant risk to life, necessitate prompt and efficient management (13). TMA is distinguished by the occurrence of microvascular injury and the formation of blood clots within the narrow blood vessels. The administration of transmembrane activator and calcium-modulating cyclophilin ligand interactor has been associated with the occurrence of organ damage, specifically kidney injury and dysfunction. Hypertension-induced posterior reversible encephalopathy syndrome (PRES) is a neurological condition that is distinguished by symptoms such as headaches, seizures, visual impairments, and changes in cognitive function. This condition possesses the potential to endanger one's life and necessitates immediate intervention in order to mitigate the risk of enduring irreversible harm to the brain. Furthermore, it has been observed that hypertension can play a role in the progression or exacerbation of CKD in HSCT recipients (14). CKD has the potential to progress to renal dysfunction, necessitating the use of renal replacement therapies like dialysis or kidney transplantation (14). Moreover, the lack of control over hypertension can exacerbate the prognosis and heighten the likelihood of complications, thereby impacting the long-term results and quality of life for individuals who have undergone HSCT.

4   Screening, Evaluation, and Treatment of HTN in HSCT

At present, there is a lack of specific guidelines pertaining to the definition and management of hypertension in individuals who have undergone HSCT. Nevertheless, a plethora of suggestions have been put forth regarding the management of modifiable risk factors in individuals who have survived cancer, including those who have undergone HSCT. According to established consensus guidelines, it is recommended that regular blood pressure evaluations be conducted for individuals undergoing cancer treatment during each medical appointment, with a minimum frequency of once per year (6). Furthermore, it is imperative to conduct a comprehensive assessment of the medications being administered to patients who have undergone HSCT in relation to the evaluation of elevated blood pressure. Finally, it is imperative to conduct meticulous and thorough surveillance of kidney function in this particular group, which entails evaluating serum creatinine levels as well as micro- and macro-albuminuria.

5 Lifestyle Modifications

Lifestyle modification, encompassing regular physical activity, attainment and maintenance of a healthy weight, adoption of a diet low in sodium (such as the DASH diet), and restriction or cessation of alcohol consumption, should be considered as the principal approach and may be adequate for achieving improvements in cases of mild hypertension (4, 6).

6 Pharmacological Treatment

The determination of when to commence blood pressure (BP) treatment and the specific BP goals to be pursued in the management of hypertension induced by cancer drugs are contingent upon the particular circumstances of the cancer and the predicted outcome. The suggested blood pressure target for individuals undergoing cancer therapy is 140 mmHg for systolic blood pressure and 90 mmHg for diastolic blood pressure. Therefore, in a general sense, the accepted criterion for commencing antihypertensive treatment in cancer patients is when their blood pressure exceeds 140/90 mmHg. Additionally, it may be deemed appropriate to consider a target of 130 mmHg systolic and 80 mmHg diastolic blood pressure levels during cancer therapy, assuming that the treatment is well tolerated (4, 6).

Angiotensin-converting enzyme inhibitors (ACE-I) and angiotensin receptor blockers (ARB) are the preferred initial pharmacological interventions for controlling blood pressure in individuals diagnosed with cancer. Additionally, it has been suggested that dihydropyridine calcium channel blockers (CCBs) may serve as a feasible alternative for secondary treatment in managing hypertension among cancer patients with inadequately controlled blood pressure. It is advisable for individuals diagnosed with cancer and presenting with systolic blood pressure (SBP) levels of 160 mmHg or higher, as well as diastolic blood pressure (DBP) levels of 100 mmHg or higher, to receive a combined treatment approach involving the administration of an angiotensin-converting enzyme inhibitor (ACE-I) or angiotensin receptor blocker (ARB) alongside a dihydropyridine calcium channel blocker (CCB). The administration of Diltiazem and Verapamil as a therapeutic approach for arterial hypertension in individuals with cancer is not recommended, primarily due to the potential for drug-drug interactions and the adverse inotropic effects associated with these pharmaceutical agents. In cases where blood pressure remains uncontrolled despite the administration of a combination of two medications, the addition of thiazide or thiazide-like diuretics and beta blockers with alpha blocking activity (such as carvedilol or labetalol) may be considered. Furthermore, in the event that the patient has been diagnosed with systemic hypertension and is currently undergoing medication, it is generally recommended to maintain the administration of all prescribed drugs during the course of treatment, unless there is a severe drug-drug interaction or if the antihypertensive medication significantly alters the metabolism of the drugs used in HSCT.

7　Conclusion and Future Perspective

In conclusion, hypertension (HTN) is a prevalent and consequential issue among individuals who have undergone hematopoietic stem cell transplantation (HSCT), as depicted in Fig. 1. The occurrence of HTN in this particular group can be ascribed to a range of factors, encompassing the use of chemotherapeutic and immunosuppressive drugs, total body irradiation, chemotherapy, and the presence of graft-versus-host disease (GVHD). HTN not only elevates the likelihood of cardiovascular complications but also plays a significant role in the morbidity and mortality rates observed among HSCT survivors. The importance of screening, early detection, and efficient management of HTN and its associated complications cannot be overstated in this particular group of patients. It is recommended to conduct routine assessment of blood pressure, evaluate the patient's medication history, and monitor kidney function. The initial strategy for managing mild hypertension should prioritise lifestyle modifications, including engaging in regular physical activity, managing weight, restricting sodium intake, and discontinuing alcohol consumption. Patients with elevated blood pressure levels may require pharmacological intervention, such as the administration of angiotensin-converting enzyme inhibitors (ACE-I), angiotensin receptor blockers (ARB), and calcium channel blockers (CCBs). Nevertheless, it is imperative to develop personalised treatment strategies that are tailored to the unique circumstances and cancer prognosis of each patient. It would be advantageous to establish and implement precise protocols for the characterization and control of hypertension in recipients of HSCT in the forthcoming years. Future research should prioritise the optimisation of strategies for managing HTN, the examination of long-term cardiovascular outcomes related to HTN in HSCT survivors, and the exploration of innovative therapeutic interventions aimed at addressing HTN in this specific population. Furthermore, the implementation of personalised strategies that take into account specific risk factors and patient characteristics has the potential to augment the efficacy of hypertension management in HSCT recipients, ultimately leading to enhanced cardiovascular well-being and improved quality of life.

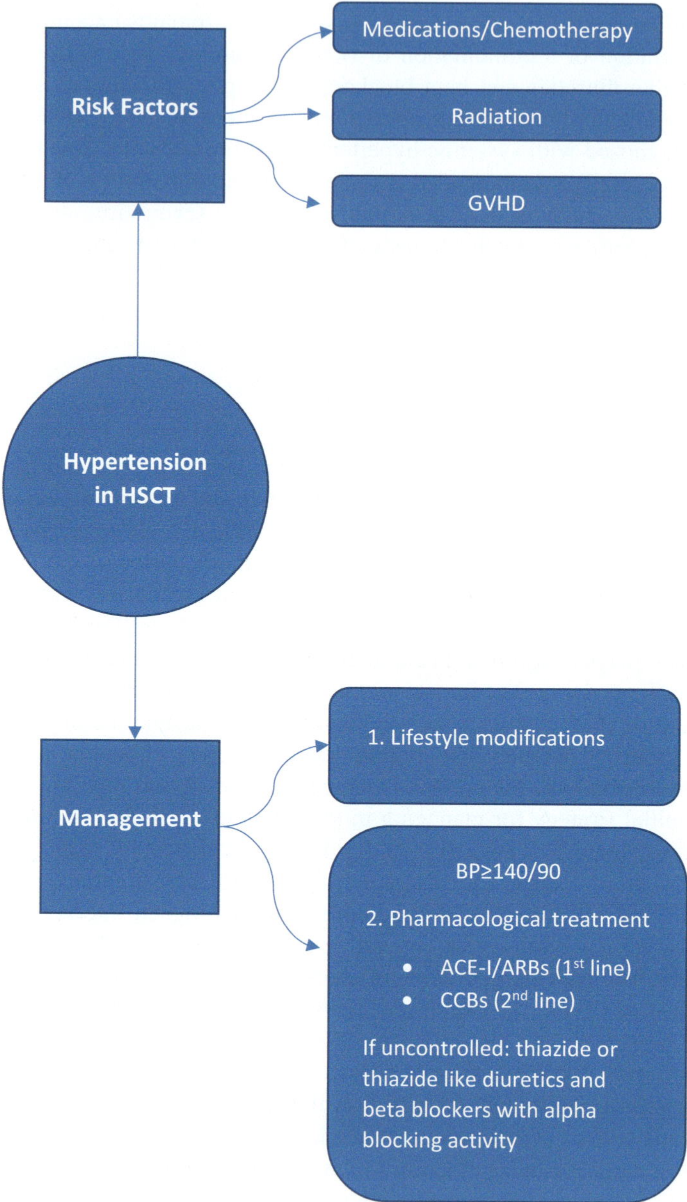

Fig. 1 Approach to hypertension in hematopoitic stem cell transplantation (HSCT), BP: blood pressure, GVHD: graft versus host disease

References

Armenian SH, Chow EJ. Cardiovascular disease in survivors of hematopoietic cell transplantation. Cancer 2014;120(4):469–79.

Armenian SH, Sun CL, Vase T, Ness KK, Blum E, Francisco L, et al. Cardiovascular risk factors in hematopoietic cell transplantation survivors: role in development of subsequent cardiovascular disease. Blood 2012;120(23):4505–12.

Armenian SH, Chemaitilly W, Chen M, Chow EJ, Duncan CN, Jones LW, et al. National institutes of health hematopoietic cell transplantation late effects initiative: the cardiovascular disease and associated risk factors working group report. Biol Blood Marrow Transplant. 2017;23(2):201–10.

Baker KS, Ness KK, Steinberger J, Carter A, Francisco L, Burns LJ, et al. Diabetes, hypertension, and cardiovascular events in survivors of hematopoietic cell transplantation: a report from the bone marrow transplantation survivor study. Blood 2007;109(4):1765–72.

Chalela CM, Uribe JC, Luna-Gonzalez M, Peña AM, Jimenez SI, Salazar LA, et al. Prevalence and associated factors for arterial hypertension in adults following hematopoietic stem cell transplantation. Blood 2019;134:5689.

Cohen JB, Brown NJ, Brown SA, Dent S, van Dorst DCH, Herrmann SM, et al. Cancer therapy-related hypertension: a scientific statement from the american heart association. Hypertension 2023;80(3):e46–57.

Hoorn EJ, Walsh SB, McCormick JA, Zietse R, Unwin RJ, Ellison DH. Pathogenesis of calcineurin inhibitor-induced hypertension. J Nephrol. 2012;25(3):269–75.

Kim CS, Han KD, Choi HS, Bae EH, Ma SK, Kim SW. Association of hypertension and blood pressure with kidney cancer risk: a nationwide population-based cohort study. Hypertension 2020;75(6):1439–46.

Laskin BL, Hingorani SR. Hypertension in oncology and stem-cell transplant patients. In: Flynn JT, Ingelfinger JR, Redwine KM, editors. Pediatric hypertension. Cham: Springer International Publishing; 2018. pp. 629–50.

Lyon AR, Lopez-Fernandez T, Couch LS, Asteggiano R, Aznar MC, Bergler-Klein J, et al. 2022 ESC guidelines on cardio-oncology developed in collaboration with the European Hematology Association (EHA), the European Society for Therapeutic Radiology and Oncology (ESTRO) and the International Cardio-Oncology Society (IC-OS). Eur Heart J Cardiovasc Imaging 2022;23(10):e333–465.

Majhail NS, Challa TR, Mulrooney DA, Baker KS, Burns LJ. Hypertension and diabetes mellitus in adult and pediatric survivors of allogeneic hematopoietic cell transplantation. Biol Blood Marrow Transplant. 2009;15(9):1100–7.

Mohammed T, Singh M, Tiu JG, Kim AS. Etiology and management of hypertension in patients with cancer. Cardio Oncol. 2021;7(1):14.

Palomo M, Diaz-Ricart M, Carreras E. Endothelial dysfunction in hematopoietic cell transplantation. Clin Hematol Int. 2019;1(1):45–51.

Souza VB, Silva EN, Ribeiro ML, Martins WA. Hypertension in patients with cancer. Arq Bras Cardiol. 2015;104(3):246–52.

Diabetes Mellitus in HSCT

Alireza Rezvani, Sara Adimi, Zahra Ghaemmaghami, and Amir Azimi

Abstract Diabetes mellitus (DM) is a chronic metabolic condition characterized byhyperglycemia resulting from defects in insulin production or function. Hematopoietic stem cell transplantation (HSCT) can impact glycemic control and diabetes risk both positively and negatively. This chapter discusses the relationship between DM and HSCT. HSCT recipients are at high risk of hyperglycemia and developing post-transplant diabetes mellitus (PTDM). Hyperglycemia is linked to worse transplant outcomes including increased infection risk, graft-versus-host disease (GVHD), hospital length of stay, and mortality. The development of PTDM is multifactorial, influenced by patient characteristics, immunosuppressive medications like corticosteroids, and GVHD. Screening and tight glycemic control are important for optimizing outcomes after HSCT. HSCT represents a promising treatment approach for certain types of diabetes. For type 1 diabetes mellitus (T1DM), HSCT aims to induce immune tolerance and regeneration of pancreatic beta cells through immunomodulation and myeloablation. Several clinical trials demonstrate HSCT can achieve insulin independence in T1DM patients for years with an acceptable safety profile. However, larger and longer-term studies are still needed. The potential mechanisms by which HSCT may treat type 2 diabetes mellitus (T2DM) include immune reset, beta cell regeneration, and anti-inflammatory effects, but efficacy data specifically for HSCT in T2DM is still limited. In conclusion, DM profoundly impacts HSCT outcomes and remains an important consideration in long-term survivorship care. Further, HSCT shows potential as a curative approach for some forms of diabetes but requires additional research to optimize outcomes and safety.

A. Rezvani
Hematology and Oncology Department, Firoozgar Hospital, Iran University of Medical Sciences, Tehran, Iran

S. Adimi · A. Azimi (✉)
Cardio-Oncology Research Center, Rajaie Cardiovascular Medical and Research Center, Iran University of Medical Sciences, Tehran, Iran
e-mail: Azimi7619@gmail.com

Z. Ghaemmaghami
Rajaie Cardiovascular Medical and Research Center, Iran University of Medical Sciences, Tehran, Iran

A. Alizadehasl et al. (eds.), *Cardiovascular Considerations in Hematopoietic Stem Cell Transplantation*, https://doi.org/10.1007/978-3-031-53659-5_10

Keywords Diabetes mellitus · Hematopoietic stem cell transplantation ·
Hyperglycemia · Post-transplant diabetes mellitus · Graft-versus-host disease

Abbreviations

BM-MSCs	Bone marrow mesenchymal stromal cells
BM-MNCs	Bone marrow mononuclear cells
CXCL12/CXCR4	Chemokine ligand 12/C-X-C chemokine receptor type 4
DKA	Diabetic ketoacidosis
DM	Diabetes mellitus
GVHD	Graft-versus-host disease
HSCT	Hematopoietic stem cell transplantation
IFN	Interferon
IL	Interleukin
LOS	Length of stay
NRM	Non-relapse mortality
OS	Overall survival
PTDM	Post-transplant diabetes mellitus
ROS	Reactive oxygen species
T1DM	Type 1 diabetes mellitus
T2DM	Type 2 diabetes mellitus
TGF	Transforming growth factor
UCB	Umbilical cord blood stem cells
WJ-MSCs	Wharton's jelly-derived mesenchymal stem cells

1 Introduction

Diabetes mellitus (DM) describes a group of metabolic conditions in which the body
is unable to properly control blood glucose levels due to defects in insulin production,
insulin function, or both (Association and Diagnosis and Classification of Diabetes
Mellitus 2014). Millions of people worldwide currently live with diabetes or are at
greater risk of DM developing in the future (Kahn et al. 2014). Long-term elevated
blood sugar associated with diabetes can damage and impair the function of multiple
organs over time, in particular the heart, blood vessels, kidneys, nerves, and eyes.

There are several mechanisms involved in developing diabetes, ranging from the
immune system mistakenly attacking the pancreatic beta cells, resulting in insufficient insulin production, to issues that cause resistance to the effects of insulin in the
body (Association and Diagnosis and Classification of Diabetes Mellitus 2014).

This chapter explores the relationship between diabetes mellitus (DM) and
hematopoietic stem cell transplantation (HSCT). HSCT has the potential to impact

glucose metabolism and diabetes risk. We will first discuss how hyperglycemia and DM may develop following HSCT procedures. We will then examine how HSCT may present an option for managing certain types of pre-existing DM. The overarching goal is to provide an overview of DM as both a potential complication and treatment target in the context of HSCT.

2   Impact of Hyperglycemia in HSCT Recipients

Although potentially curative for hematologic illnesses, HSCT is also linked with a high rate of morbidity and mortality. Hyperglycemia is one modifiable factor that can cause unfavorable outcomes. The reported prevalence of hyperglycemia among HSCT recipients ranges from 71 to 93% (Olausson et al. 2014). Diabetes complications are now understood to impact the cellular components and functions of the bone marrow, making it a target organ.

Several pathological manifestations of diabetes impact bone marrow function:

1. Mobilopathy: The ineffective release of hematopoietic stem cells from the bone marrow due to a changed gradient of chemokines is known as mobilopathy. Dysregulation of the CXCL12/CXCR4 axis contributes to this functional impairment.
2. Inflammation: Increased proinflammatory cytokines and downregulation of certain genes and proteins are linked to reduced HSC repopulation of the bone marrow.
3. Neuropathy: Diabetic autonomic neuropathy can impact the bone marrow niche's ability to function and the movement of stem cells.
4. Microangiopathy: Diabetes alters the small blood arteries, which affects bone marrow function. In particular, diabetic bone marrow endothelial cells exhibit higher levels of stress and damage, decreased mobility and network formation, and increased leakage. Blood flow is diminished as a result of this endothelial cell dysfunction in the bone marrow, and HSCs are lost.
5. Excessive ROS (Reactive oxygen species) production: Excessive ROS generation under pathological situations like diabetes can result in DNA damage and senescence, which may threaten stem cell viability. This can result in increasing adipogenesis while reducing osteoblastogenesis (Vinci et al. 2020).

Consequently, diabetes can have several negative effects, including:

– Infection: Through impaired immune cell communication, hyperglycemia increases the risk of infection and can also be driven by infections, creating a vicious cycle.

- Graft-versus-host-disease (GVHD): According to studies, hyperglycemia during neutropenia increases the likelihood that grade II–IV acute GVHD would manifest. Patients with normal or overweight blood sugar levels had severe hyperglycemia and a doubled risk of developing GVHD, although patients with obesity did not have an elevated risk.
- Higher length of stay in hospital (LOS): There are conflicting findings from research on the relationship between hyperglycemia and LOS, although some studies indicated an association between hyperglycemia and longer overall LOS.
- Overall survival (OS) and non-relapse mortality (NRM): According to several research, hyperglycemia is linked to lowered OS and NRM.
- Organ toxicity: Patients with hyperglycemia had higher rates of organ toxicities, including complications with the heart and kidneys (Olausson et al. 2014).

3 Post-transplant Diabetes Mellitus (PTDM) in the Context of HSCT

3.1 Definition and Prevalence

Post-transplant diabetes mellitus (PTDM) is a condition where diabetes develops after HSCT. While HSCT has improved outcomes and long-term survival rates, it is also associated with various late complications, including endocrine issues. Metabolic syndrome, insulin resistance, and glucose intolerance are commonly observed in HSCT recipients. Approximately 35% of them experience insulin resistance, and 25% have impaired fasting glucose levels (Fuji et al. 2016; Griffith et al. 2010; Dalla Via et al. 2020).

PTDM develops in 10–30% of patients after allogeneic HSCT (allo-HCSTD). New-onset PTDM after HSCT increases the risk of mortality threefold, and for half of the patients, hyperglycemia can reduce their chances of survival (Engelhardt et al. 2012; Hammer et al. 2009; Griffith et al. 2011). This prevalence is similar among both adults and children, with 15% of patients requiring medication for diabetes one year after HSCT (Lee et al. 2009).

Furthermore, patients who develop PTDM often exhibit other cardiovascular risk factors such as dyslipidemia and hypertension. Diabetes may negatively influence cancer recurrence or contribute to secondary cancer development after transplantation through epigenetic pathways. Hyperglycemia inhibits the phosphorylation of AMP-activated kinase and stabilizes the tumor suppressor TET2 protein, making individuals more susceptible to cancer (Engelhardt et al. 2019).

3.2   Pathogenesis

While PTDM is an important issue, its exact pathological mechanisms are still poorly understood (Engelhardt et al. 2012, 2019). Some known risk factors for developing PTDM include older age, non-white ethnicity, corticosteroid use, parenteral nutrition, and traumatic brain injury. Insulin resistance from previous irradiation, cytokine activity during infections, or graft-versus-host disease also contribute to the development of PTDM. In one long-term study, 6.5% of patients who received HSCT went on to develop PTDM over 10 years. Steroid therapy increases the risk for PTDM, and 64.1% of patients in the study developed diabetes during or shortly after initiating steroid treatment. In the other 35.9% of patients, diabetes was not related to steroid use. A daily prednisone dosage of 0.25 mg/kg or more has been associated with increased risk of developing DM (Majhail et al. 2009a).

GVHD is another risk factor for PTDM. Patients who experience grades 2–4 acute GVHD have a 2.2 times higher risk of developing PTDM. Chronic GVHD has also been associated with insulin resistance (Taskinen et al. 2000). The relationship between acute GVHD and the development of metabolic syndrome, which includes diabetes, is not fully understood. In one study of 86 adults after allogeneic hematopoietic stem cell transplantation, GVHD status was not associated with metabolic syndrome (Engelhardt et al. 2019; Majhail et al. 2009a; Taskinen et al. 2000; Subramanian and Trence 2007). Since patients suffering from acute GVHD often require high doses of corticosteroids, this may be a risk factor for developing PTDM instead of GVHD itself.

Patients may have subtle insulin resistance prior to transplantation that then progresses to more significant insulin resistance in the peripheral muscles and skeleton after HSCT (Engelhardt et al. 2019). Insulin resistance is associated with low-grade inflammation, which could have negative impacts during the HSCT process (Winer et al. 2009).

3.3   Prevention and Treatment and Long-Term Follow-Up of PTMD

Early intervention during the asymptomatic stage of hyperglycemia has been shown to improve long-term outcomes in the general population. It is advised that patients at high-risk for hyperglycemia and PTDM undergo screening for these conditions in the early period after undergoing HSCT and throughout long-term follow-up. Consensus guidelines for solid organ transplant recipients suggest monitoring fasting blood glucose levels at minimum on a weekly basis for the first month post-transplantation. Fasting blood glucose should then be checked at the 3, 6, and 12 month marks after transplantation. Annual monitoring of fasting blood glucose is also recommended according to the guidelines (Davidson et al. 2003).

While there are no specific guidelines for monitoring blood glucose levels in HSCT patients, regular testing is important given the risk of PTDM. Fasting blood glucose should be checked frequently during the initial 100 days post-transplant due to the high prevalence of PTDM during this period. After the first 100 days, periodic monitoring of fasting glucose levels is crucial, especially when immunosuppressive therapies known to increase diabetes risk are started. Random blood glucose readings of 200 mg/dL or higher should also prompt follow-up testing with a fasting glucose measurement and/or oral glucose tolerance test to check for diabetes. Patients should also be assessed for signs and symptoms of diabetes if an elevated casual blood sugar is found (Griffith et al. 2010).

Pharmacological treatments or lifestyle modifications aimed at reducing insulin resistance in skeletal muscle should be considered for both preventing and managing new cases of PTDM. Studies have shown that metformin or lifestyle changes can lower the risk of developing PTDM when compared to a placebo. Metformin and lifestyle interventions that target skeletal muscle insulin resistance may help prevent and treat new onset PTDM in transplant patients (Diabetes Prevention Program Research Group 2015). The suggested treatment approach begins with non-pharmacologic interventions such as diet and exercise. However, patients who have experienced weight loss due to previous therapies or have gastrointestinal GVHD may find these interventions challenging. Modification of immunosuppressive regimens, particularly reducing corticosteroid use, can improve glycemic control. If lifestyle modifications and adjustments to immunosuppressive therapy are insufficient, consideration may be given to orally administered drugs, taking into account side effects and drug interactions (Griffith et al. 2010).

Evidence is limited regarding the optimal pharmacological treatments for managing diabetes that develops after transplantation. Therefore, medication choices should be based on balancing safety, effectiveness, tolerability as well as considering each patient's specific transplant medications and existing medical conditions. With a lack of strong data to guide decisions, clinicians will need to weigh these factors carefully when selecting hypoglycemic agents for individual post-transplant diabetes patients. The treatment regimen must account for potential drug interactions and risks given the patient's transplant and other therapies (Griffith et al. 2010). In cases of high-dose corticosteroid use or severe hyperglycemia, insulin is often the most effective and safe treatment option. Sulfonylureas and meglitinide analogues may be considered, but the risk of hypoglycemia should be assessed, especially in patients with GVHD or variable oral intake. Cyclosporine and metformin can interact, increasing the risk of lactic acidosis. Hypoglycemic medications should be selected based on hepatic and renal clearance when treating patients at risk for liver disease or renal failure. Thiazolidinediones are not recommended, and caution should be exercised when using medications that affect gastrointestinal emptying. Monitoring of immunosuppressive medication levels may be necessary when adding certain agents (Griffith et al. 2010; Davidson et al. 2003).

HSCT survivors warrant systematic DM screening during follow-up, as carbohydrate metabolism declines over time post-transplantation (Wędrychowicz et al. 2013).

In conclusion, PTDM is a significant complication that can arise after HSCT. Timely screening and treatment optimize outcomes after transplantation. A multidisciplinary approach and individualized treatment plans are necessary to address the complex interactions between immunosuppressive therapies, glucose metabolism, and the unique challenges faced by HSCT recipients.

3.4 The Role of HCST in Treatment of Type 1 Diabetes Mellitus

The immune system of the body destroys beta cells in the pancreas, causing type 1 diabetes mellitus (T1DM), an autoimmune disease. This leads to an absolute deficiency of insulin and elevated blood glucose levels. Over 70% of beta cell mass is typically lost by the time T1DM is clinically diagnosed. In later stages of the disease, beta cell mass continues to decline, as indicated by low or undetectable plasma C-peptide levels. The complex interaction of genetic susceptibility and environmental factors leads to autoimmune diabetes, involving a variety of cells and cytokines. In type 1 diabetes, the autoimmune process affects both regulatory cells, which are linked to immunological tolerance, and effector cells, which are in charge of destroying beta cells. Among the effector cells and cytokines implicated are pathogenic B cells, CD4 and CD8 cells, activated dendritic cells, interleukin-12, and inducible nitric oxide synthase. On the other hand, immunological tolerance is linked to regulatory B and T cells that are influenced by interleukin-10, indoleamine-2,3-dioxygenase, transforming growth factor-beta, and other substances. Insulitis, characterized by inflammation and mononuclear cell invasion of the islets, is a hallmark of autoimmune diabetes. Beta cell death in insulitis occurs through direct contact with activated macrophages and T-cells or exposure to soluble mediators like cytokines, nitric oxide, and oxygen free radicals. However, recent research suggests that beta cells die through apoptosis, a process involving the up and down-regulation of hundreds of genes, in early T1DM. Despite significant progress in insulin therapy and management, T1DM remains challenging to treat. Treatment involves frequent administration of insulin injections, monitoring of glycaemic levels, and meticulous scheduling of meal times and composition, significantly affecting the quality of life of individuals with T1DM. Hyperglycemia contributes to the occurrence of complications, including microvascular complications such as retinopathy, nephropathy, and neuropathy, and macrovascular complications like cardiovascular disease, cerebrovascular accidents, and peripheral vascular disease. Since 2000 The transplantation of islets has demonstrated efficacy in the treatment of T1DM through the replacement of beta cells. Transplant recipients require long-term immunosuppression to balance autoimmune and alloimmune responses. The limitations posed by the side effects of drugs have impeded the broad implementation of pancreatic or islet transplantation therapies in the treatment of T1DM. Furthermore, the limited number of donors and the requirement for lifetime immunosuppression serve to further limit its utilization. Therefore, there is a need for innovative therapeutic interventions that restore beta-cell function and prevent the previously mentioned complications. Such interventions should ensure sustained recovery while improving the quality of life of individuals with T1DM (Carulli et al. 2023; Li et al. 2012; Couri et al. 2012).

Mechanism of Action

The utilization of HSCT is a developing therapeutic strategy for addressing T1DM. HSCT involves the administration of hematopoietic stem cells, either autologous or allogeneic, which are responsible for the generation of all blood cells, including those of the immune system. The multifaceted pathways involved in the mechanism of action of HSCT in diabetes have been extensively studied. First, HSCT eliminates autoreactive immune cells that are responsible for the destruction of beta cells via myeloablation, which includes high-dose chemotherapy and/or radiation therapy. Myeloablation results in the reduction of immune cells, specifically T and B cells, which play a crucial role in the immune response directed towards beta cells. As a consequence, there is a temporary phase of compromised immune function, wherein the body's immune system undergoes a reset. Secondly, HSCT induces immune tolerance towards beta cells. Upon re-infusion, the hematopoietic stem cells undergo differentiation into immune cells that exhibit a state of tolerance towards beta cells, thereby reducing the possibility of recurrent autoimmune attacks. Furthermore, HSCT facilitates the proliferation of regulatory T cells, a type of immune cell that inhibits autoimmunity and promotes immune tolerance. Thirdly, HSCT stimulates the regeneration of beta cells. This effect may be attributed to various mechanisms, including, the production of growth factors by hematopoietic stem cells, the recruitment of precursor cells from the bone marrow, and the activation of beta cell regeneration pathways. However, data from allogeneic bone marrow HSCT in non-obese diabetic mice indicate that immunomodulatory rather than regenerative effects are the primary mechanism of action of HSCT (Carulli et al. 2023; Couri et al. 2012).

Safety and Efficacy

A number of clinical trial studies have examined the safety and effectiveness of HSCT in the management of type 1 diabetes. The study conducted by Voltarelli et al. in Brazil included 15 individuals with T1DM and achieved sustained recovery of insulin independence in all individuals except one patient who had diabetic ketoacidosis. After a follow-up of up to 35 months, 13 individuals became continuously insulin-independent, and one patient resumed insulin treatment after 1 year. Adverse effects observed were mild and manageable, except for a case of bilateral pneumonia. In an extended follow-up with 23 patients, 12 became continuously insulin-independent, and eight were transiently insulin-independent, with four resuming insulin after an upper respiratory tract infection. With the addition of sitagliptin, an inhibitor of dipeptidyl peptidase 4, two individuals were able to regain insulin independence. The authors concluded that HSCT could reverse T1DM in humans for up to 4 years with tolerable negative side effects (Couri 2009; Voltarelli et al. 2007).

A small study in Poland of 8 patients with type 1 diabetes found that all became insulin independent after HSCT. HbA1c levels decreased significantly and C-peptide levels rose, with only one patient resuming insulin later. Twenty of the 24 individuals in the extended research remained insulin-free for a minimum of 9.5 months, and four remained insulin-free at follow-up. However, some experienced mild side effects like nausea and fever. A few developed serious antithymocyte globulin-related skin reactions/vasculitis. Following the implantation of a central line, one person developed pulmonary emphysema. Regrettably, a fatality occurred as a result of sepsis caused by Pseudomonas aeruginosa (Snarski et al. 2011).

Li et al. conducted a study on 13 Chinese patients with T1DM Only three patients developed insulin independence, while eight required lower insulin doses for proper glycemic control, and two were non-responders. The patients who participated in the study exhibited notably elevated post-prandial C-peptide and serum fasting levels, as well as dramatically reduced plasma HbA1c levels. The observed adverse effects were of a modest nature, with a singular instance of autoimmune thyroiditis occurring six months post-treatment. Additionally, the research revealed that CD4 + T cells exhibited a sustained decrease following immunological reconstitution, whereas levels of certain serum cytokines were notably diminished subsequent to HSCT.

Gu et al. conducted a two-phase study in China to investigate the effects of HSCT in patients with T1DM. The first phase of study included 28 patients with a recent T1DM diagnosis, and the results showed that patients with a history of DKA had a lower response rate to HSCT and a greater need for insulin after a year compared to those without DKA. In the second phase of study, 20 patients were treated with HSCT, and 14 patients achieved insulin independence, whereas just one patient in the group receiving insulin therapy temporarily became insulin independent before relapsing. The adverse event rate was comparable in both groups, and there was no significant difference in the development of endocrinopathies between the two groups (Li et al. 2012; Gu et al. 2012).

Cantú-Rodríguez et al. conducted a research on sixteen individuals who were diagnosed with T1DM less than three months prior. Results showed seven patients achieving complete insulin independence and six achieving only partial insulin independence. Over half of patients had side effects, mostly nausea, vomiting, fever and hair loss, but no major adverse events were reported. Complications of HSCT need to be weighed against the short and long-term complications of T1DM, which can be life-threatening and significantly reduce quality of life. The burden of T1DM is caused by long-term complications, which are not well-controlled with current treatment options. However, immunotherapy has unavoidable side effects, and the negative consequences of conventional T1D therapies are not negligible (Cantú-Rodríguez et al. 2016).

Future Directions and Conclusion

Despite promising clinical trial findings, several issues must be answered before HSCT can become a conventional therapy for T1DM. The optimization of the conditioning regimen, which is used to eliminate the existing immune cells and make space for the new transplanted stem cells, is one crucial area of research. A further research area is the timing of HSCT. Most studies to date have focused on patients with newly diagnosed T1DM, although it is uncertain if HSCT can be efficient for individuals with long-standing disease. Furthermore, the long-term safety and efficacy of HSCT remain uncertain, and more follow-up research is needed to determine the durability of remission and the risk of late sequelae. But to sum up, HSCT represents a promising avenue for the treatment of T1DM, and it has the potential to transform the lives of millions of people worldwide.

3.5   *The Role of HCST in Treatment of Type 2 Diabetes Mellitus*

Type 2 diabetes mellitus (T2DM) is a prevalent condition characterized by dysfunctional beta cell secretion and insulin resistance. Key contributing factors include obesity and aging. Recent research suggests that the immune system and inflammatory mediators play a role in T2DM's pathophysiology. Untreated T2DM can lead to serious long-term complications, including diabetic retinopathy, ischemic heart disease, stroke, and chronic kidney disease. While the initial events causing beta cell damage differ between T1DM and T2DM, the ultimate pathways to pancreatic cell death exhibit similarities, hinting at potential immunological and regenerative treatments. Conventional therapies like insulin therapy, dietary adjustments, and oral hypoglycemic medications may produce side effects and do not offer a permanent cure. Pancreas transplantation, the primary late-stage T2DM treatment, is expensive with recurrence risks, addressing symptoms rather than pancreatic cell function deterioration. Alternative approaches are essential, and stem cell-based therapies emerge as innovative options. Stem cells possess the ability to renew, regenerate, and release vital substances for maintaining various cell types. The three primary stem cell types used for T2DM treatment are Mesenchymal stem cells (MSCs), Hematopoietic stem cells (HSCs), and Induced Pluripotent Stem Cells (iPSCs). MSCs, derived from autologous or allogeneic tissues like bone marrow, adipose tissue, or blood, are most frequently utilized in clinical trials due to their versatility and low immunogenicity. Promising outcomes suggest that stem cell-based immunological and regenerative treatments may offer effective solutions for T2DM (Vinci et al. 2020; Peng et al. 2018; Voltarelli et al. 2011; Mathur et al. 2023).

Mechanism of Action

The precise mechanisms underlying the effectiveness of HSCT in treating individuals with T2DM are not fully understood. However, it is believed to operate through several pathways, including immune system reset, beta cell regeneration, and anti-inflammatory effects (Vinci et al. 2020).

Efficacy and Safety

Numerous clinical trials have explored the safety and efficacy of stem cell transplantation in the treatment of T2DM. However, there is a lack of specific clinical trials investigating the efficacy of HSCT in T2DM.

In a 2009 trial conducted in India, ten insulin-dependent T2DM patients who had failed triple oral anti-diabetic medication therapy underwent autologous bone marrow mononuclear cell (BM-MNC) transplantation. Over a six-month follow-up period, these patients experienced a significant reduction in their insulin requirements, with three of them eventually becoming insulin-independent. Additionally, the treatment resulted in substantial weight loss, increased C-peptide levels, and a decrease in HbA1c levels. The study's findings indicated that the treatment was both safe and effective, enhancing beta-cell activity in T2DM patients (Bhansali et al. 2009).

Similarly, the 2014 Liu et al. trial in China demonstrated the safety and efficacy of allogeneic Wharton's jelly-derived mesenchymal stem cells (WJ-MSCs) in 22 T2DM patients. These patients received WJ-MSCs twice via IV injection and directly into the pancreas via the splenic artery. Over 12 months, they exhibited improved immunological control, increased beta-cell function, and reductions in HbA1c levels, blood glucose, insulin doses, and oral anti-diabetic medication requirements. Notably, there were no significant adverse side effects (Liu et al. 2014).

In a 2017 trial conducted in India, 30 T2DM patients were divided into three groups. Two of these groups received either autologous BM-MNCs or bone marrow mesenchymal stromal cells (BM-MSCs) infused into the pancreas, while the third group received a placebo. Following a 12-month follow-up period, both therapy groups showed a significant decrease in the need for exogenous insulin and improvements in various glycemic control parameters, all without experiencing notable side effects (Bhansali et al. 2017).

Furthermore, in a 2013 clinical experiment in China, three T2DM patients received umbilical cord blood stem cells injected into the pancreas via a dorsal pancreatic artery. Over a six-month follow-up period, all patients exhibited elevated C-peptide levels, reduced insulin requirements, and more stable blood glucose readings, with no reported adverse side effects (Tong et al. 2013).

4 Future Directions and Conclusion

HSCT has become known as a potential T2DM therapy option. Before the process can be widely used as a therapeutic option, despite the fact that the findings of multiple studies of a similar nature are encouraging, there are a number of issues that must be resolved. Future studies should concentrate on solving these problems and enhancing the safety and effectiveness of HSCT for T2DM.

References

American Journal of Transplantation. Am J Transplant. 2014;14(12):i.

American Diabetes Association. Diagnosis and classification of diabetes mellitus. Diabetes Care 2014;37(Supplement_1):S81–90.

Bhansali A, Upreti V, Khandelwal N, Marwaha N, Gupta V, Sachdeva N, et al. Efficacy of autologous bone marrow-derived stem cell transplantation in patients with type 2 diabetes mellitus. Stem Cells Dev. 2009;18(10):1407–16.

Bhansali S, Dutta P, Kumar V, Yadav MK, Jain A, Mudaliar S, et al. Efficacy of autologous bone marrow-derived mesenchymal stem cell and mononuclear cell transplantation in type 2 diabetes mellitus: a randomized. Placebo-Controlled Comparative Study Stem Cells Dev. 2017;26(7):471–81.

Cantú-Rodríguez OG, Lavalle-González F, Herrera-Rojas MÁ, Jaime-Pérez JC, Hawing-Zárate JÁ, Gutiérrez-Aguirre CH, et al. Long-term insulin independence in type 1 diabetes mellitus using a simplified autologous stem cell transplant. J Clin Endocrinol Metab. 2016;101(5):2141–8.

Carulli E, Pompilio G, Vinci MC. Human hematopoietic stem/progenitor cells in type one diabetes mellitus treatment: is there an ideal candidate? Cells. 2023;12(7):1054.

Couri CEB. C-peptide levels and insulin independence following autologous nonmyeloablative hematopoietic stem cell transplantation in newly diagnosed type 1 diabetes mellitus. JAMA. 2009;301(15):1573.

Couri CEB, De Oliveira MC, Simões BP. Risks, benefits, and therapeutic potential of hematopoietic stem cell transplantation for autoimmune diabetes. Curr Diab Rep. 2012;12(5):604–11.

Dalla Via V, Halter JP, Gerull S, Arranto C, Tichelli A, Heim D, et al. New-onset post-transplant diabetes and therapy in long-term survivors after allogeneic hematopoietic stem cell transplantation. In Vivo. 2020;34(6):3545–9.

Davidson J, Wilkinson A, Dantal J, Dotta F, Haller H, Hernandez D, et al. New-onset diabetes after transplantation: 2003 international consensus guidelines1: Transplantation 2003;75(Supplement):SS3–24.

Diabetes Prevention Program Research Group. Long-term effects of lifestyle intervention or metformin on diabetes development and microvascular complications over 15-year follow-up: the Diabetes Prevention Program Outcomes Study. Lancet Diabetes Endocrinol. 2015;3(11):866–75.

Engelhardt BG, Jagasia SM, Crowe JE, Griffith ML, Savani BN, Kassim AA, et al. Predicting posttransplantation diabetes mellitus by regulatory T-cell phenotype: implications for metabolic intervention to modulate alloreactivity. Blood. 2012;119(10):2417–21.

Engelhardt BG, Savani U, Jung DK, Powers AC, Jagasia M, Chen H, et al. New-Onset Post-Transplant Diabetes Mellitus after Allogeneic Hematopoietic Cell Transplant Is Initiated by Insulin Resistance, Not Immunosuppressive Medications. Biol Blood Marrow Transplant. 2019;25(6):1225–31.

Fuji S, Rovó A, Ohashi K, Griffith M, Einsele H, Kapp M, et al. How do I manage hyperglycemia/post-transplant diabetes mellitus after allogeneic HSCT. Bone Marrow Transplant. 2016;51(8):1041–9.

Griffith ML, Jagasia M, Jagasia SM. Diabetes mellitus after hematopoietic stem cell transplantation. Endocr Pract. 2010;16(4):699–706.

Griffith ML, Jagasia MH, Misfeldt AA, Chen H, Engelhardt BG, Kassim A, et al. Pretransplantation C-peptide level predicts early posttransplantation diabetes mellitus and has an impact on survival after allogeneic stem cell transplantation. Biol Blood Marrow Transplant. 2011;17(1):86–92.

Gu W, Hu J, Wang W, Li L, Tang W, Sun S, et al. Diabetic ketoacidosis at diagnosis influences complete remission after treatment with hematopoietic stem cell transplantation in adolescents with type 1 diabetes. Diabetes Care. 2012;35(7):1413–9.

Hammer MJ, Casper C, Gooley TA, O'Donnell PV, Boeckh M, Hirsch IB. The contribution of malglycemia to mortality among allogeneic hematopoietic cell transplant recipients. Biol Blood Marrow Transplant. 2009;15(3):344–51.

Kahn SE, Cooper ME, Del Prato S. Pathophysiology and treatment of type 2 diabetes: perspectives on the past, present, and future. The Lancet. 2014;383(9922):1068–83.

Lee SJ, Seaborn T, Mao FJ, Massey SC, Luu NQ, Schubert MA, et al. Frequency of abnormal findings detected by comprehensive clinical evaluation at 1 year after allogeneic hematopoietic cell transplantation. Biol Blood Marrow Transplant. 2009;15(4):416–20.

Li L, Shen S, Ouyang J, Hu Y, Hu L, Cui W, et al. Autologous hematopoietic stem cell transplantation modulates immunocompetent cells and improves β-cell function in Chinese patients with new onset of type 1 diabetes. J Clin Endocrinol Metab. 2012;97(5):1729–36.

Liu X, Zheng P, Wang X, Dai G, Cheng H, Zhang Z, et al. A preliminary evaluation of efficacy and safety of Wharton's jelly mesenchymal stem cell transplantation in patients with type 2 diabetes mellitus. Stem Cell Res Ther. 2014;5(2):57.

Majhail NS, Challa TR, Mulrooney DA, Baker KS, Burns LJ. Hypertension and diabetes mellitus in adult and pediatric survivors of allogeneic hematopoietic cell transplantation. Biol Blood Marrow Transplant. 2009a;15(9):1100–7.

Majhail NS, Flowers ME, Ness KK, Jagasia M, Carpenter PA, Arora M, et al. High prevalence of metabolic syndrome after allogeneic hematopoietic cell transplantation. Bone Marrow Transplant. 2009b;43(1):49–54.

Mathur A, Taurin S, Alshammary S. The safety and efficacy of mesenchymal stem cells in the treatment of Type 2 diabetes—a literature review. Diabetes Metab Syndr Obes. 2023;16:769–77.

Olausson JM, Hammer MJ, Brady V. The impact of hyperglycemia on hematopoietic cell transplantation outcomes: an integrative review. Oncol Nurs Forum. 2014;41(5):E302–12.

Peng BY, Dubey NK, Mishra VK, Tsai FC, Dubey R, Deng WP, et al. Addressing stem cell therapeutic approaches in pathobiology of diabetes and its complications. J Diabetes Res. 2018;2018:1–16.

Snarski E, Milczarczyk A, Torosian T, Paluszewska M, Urbanowska E, Król M, et al. Independence of exogenous insulin following immunoablation and stem cell reconstitution in newly diagnosed diabetes type I. Bone Marrow Transplant. 2011;46(4):562–6.

Subramanian S, Trence DL. Immunosuppressive agents: effects on glucose and lipid metabolism. Endocrinol Metab Clin North Am. 2007;36(4):891–905.

Taskinen M, Saarinen-Pihkala UM, Hovi L, Lipsanen-Nyman M. Impaired glucose tolerance and dyslipidaemia as late effects after bone-marrow transplantation in childhood. The Lancet. 2000;356(9234):993–7.

Tong Q, Duan L, Xu Z, Wang H, Wang X, Li Z, et al. Improved insulin secretion following intrapancreatic UCB transplantation in patients with T2DM. J Clin Endocrinol Metab. 2013;98(9):E1501–4.

Vinci MC, Gambini E, Bassetti B, Genovese S, Pompilio G. When good guys turn bad: bone marrow's and hematopoietic stem cells' role in the pathobiology of diabetic complications. Int J Mol Sci. 2020;21(11):3864.

Voltarelli JC, Couri CEB, Stracieri ABPL, Oliveira MC, Moraes DA, Pieroni F, et al. Autologous nonmyeloablative hematopoietic stem cell transplantation in newly diagnosed type 1 diabetes mellitus. JAMA. 2007;297(14):1568.

Voltarelli JC, Couri CEB, Oliveira MC, Moraes DA, Stracieri ABPL, Pieroni F, et al. Stem cell therapy for diabetes mellitus. Kidney Int Suppl. 2011;1(3):94–8.

Wędrychowicz A, Ciechanowska M, Stelmach M, Starzyk J. Diabetes mellitus after allogeneic hematopoietic stem cell transplantation. Horm Res Paediatr. 2013;79(1):44–50.

Winer S, Chan Y, Paltser G, Truong D, Tsui H, Bahrami J, et al. Normalization of obesity-associated insulin resistance through immunotherapy. Nat Med. 2009;15(8):921–9.

Arrhythmias and Conduction Disorders in HSCT

Majid Haghjoo, Amir Farjam Fazelifar, Farzaneh Ashrafi, Ehsan Zaboli, and Elgar Enamzadeh

Abstract Cardiac complications, particularly arrhythmias and conduction disorders, present significant challenges in the management of patients undergoing hematopoietic stem-cell transplantation (HSCT). Despite the relatively low incidence of these complications, which is reported to be less than 10%, their impact on non-relapse mortality and quality of life is substantial. This study aimed to review the types, risk factors, mechanisms, diagnosis, and management of various arrhythmias seen in HSCT recipients. Supraventricular arrhythmias like atrial fibrillation and atrial flutter are the most common arrhythmias observed, occurring in 7–27% of patients. Chemotherapy drugs used for HSCT conditioning like anthracyclines, alkylating agents, and proteasome inhibitors have been shown to cause arrhythmias through various mechanisms. Factors like older age, pre-existing cardiovascular disease, mediastinal irradiation, graft-versus-host disease increase arrhythmia risk. Detailed evaluation including electrocardiography, ambulatory monitoring aids diagnosis. Management depends on arrhythmia type and involves rate or rhythm control, anticoagulation, drug dose reduction/change based on complications. Close collaboration between cardiologists and oncologists is needed for regular monitoring, prevention, and timely treatment of arrhythmias in HSCT patients given their impact on outcomes.

M. Haghjoo · A. F. Fazelifar (✉)
Cardiac Electrophysiology Center, Rajaie Cardiovascular Medical and Research Center, Iran University of Medical Sciences, Tehran, Iran
e-mail: fazelifar@gmail.com

F. Ashrafi
Department of Hematology Oncology, Isfahan University of Medical Sciences, Isfahan, Iran

E. Zaboli
Department of Hematology and Oncology, Gastrointestinal Cancer Research Center, Mazandaran University of Medical Sciences, Sari, Iran

E. Enamzadeh
Cardio-Oncology Research Center, Rajaie Cardiovascular Medical and Research Center, Iran University of Medical Sciences, Tehran, Iran

© The Author(s), under exclusive license to Springer Nature Switzerland AG 2024   141
A. Alizadehasl et al. (eds.), *Cardiovascular Considerations in Hematopoietic Stem Cell Transplantation*, https://doi.org/10.1007/978-3-031-53659-5_11

1 Introduction

Over the past few years Because of advances and developments in chemotherapy the clinical results of patients receiving hematopoietic stem-cell transplantation (HSCT) have drastically improved (McDonald et al. 2020). In patients receiving hematopoietic stem cell transplantation (HSCT) receiving HSCT the risk of non-relapse mortality (NRM)is evaluated using The Hematopoietic Cell Transplantation Comorbidity Index (HCT-CI),a competent scoring system. The HCT-CI evaluates the presence and seriousness of comorbidities or pre-existing medical conditions based on a patient's past medical history and physical assesses. It has been established that the HCT-CI is an effective tool for forecasting outcomes following HSCT and has gained widespread acceptance in clinical practice.

One of the important comorbidities included in the Hematopoietic Cell Transplantation Comorbidity Index (HCT-CI) is cardiac disease.

In patients undergoing HSCT, existing or improving of Cardiac disease, such as coronary artery disease, heart failure, and arrhythmias, increase the risk of complications and mortality (Sorror et al. 2005).

Studies have shown that cardiac disease is an important predictor of outcomes in patients undergoing HSCT. Patients with cardiac disease are at increased risk of NRM, particularly in the first year after transplantation. In addition, cardiac complications such as arrhythmias, heart failure, and myocardial infarction can occur in the post-transplant period, particularly in patients with pre-existing cardiac disease.

Although cardiac complications are not very usual after HSCT (less than 10%) but because of high rate of mortality and diminished quality of life is an important consideration (Blaes et al. 2016; Murdych and Weisdorf 2001; Tuzovic et al. 2019). The onset is segregated into two different phases: the early and first phase period which begins in first 100 days after HSCT and the late and second phase period which begins several years later. Early cardiovascular problems are dominated by Arrhythmias, particularly SVT and atrial fibrillation/ flutter, preamble cardiometabolic risk factors, atherosclerotic cardiovascular disease (ASCVD), and heart failure generally present afterwards.

- The HSCT comorbidity index is a supplement to the conventional past medical history and physical examination performed by an expert medical practitioner as a part of pretransplant screening.

Early diagnosis and therapeutic intervention are essential since fully recovery of decreased cardiac function is difficult and uncommon.

MAJOR FACTORS AFFECTING CARDIAC COMPLICATIONS Some of the major factors affecting cardiac complications during BMT are:

1. Age: Older patients undergoing BMT have a higher risk of developing cardiac complications than younger patients. This is because aging is associated with a higher risk of cardiac disease.

2. Pre-existing cardiac conditions: presence of prior heart disease in patients receiving HSCT such as coronary artery disease, heart failure, and arrhythmias, increased the risk of further cardiac complications during BMT.

3. Radiation therapy: Radiation therapy, which is often used as part of the BMT conditioning regimen, can damage the heart and blood vessels, leading to cardiac complications.

4. Chemotherapy: Chemotherapy drugs used during the BMT conditioning regimen can also cause cardiac complications, such as heart failure and arrhythmias.

5. Graft-versus-host disease (GVHD): GVHD is a common complication of BMT, in which the transplanted donor cells attack the recipient's tissues. GVHD can affect the heart, leading to cardiac complications.

6. Infection: Infection is a common complication of BMT, and can also lead to cardiac complications, particularly if the infection affects the heart or blood vessels.

7. Fluid overload: Patients undergoing BMT may receive large volumes of fluids and blood products, which can lead to fluid overload and cardiac complications, such as heart failure.

8. Total body irradiation (TBI): TBI is a type of radiation therapy that is often used as part of the BMT conditioning regimen. It can cause damage to the heart and blood vessels, leading to cardiac complications.

9. Nutritional deficiencies: Patients undergoing BMT are often at risk of nutritional deficiencies, which can lead to cardiac complications, such as heart failure and arrhythmias.

It is important to note that the risk and severity of cardiac complications during BMT can vary greatly depending on individual patient factors, such as overall health, medical history, and the specific type of BMT being performed. Close monitoring and early detection of any cardiac complications are crucial in minimizing their impact on patients undergoing BMT.

1.1 Arrhythmia

The most frequent cardiac event associated with HSCT is arrhythmia which usually happens within 10 days after HSCT.

In an analysis of 1177 who received HSCT and, Tonorezos et al. the incidence of arrhythmia during or after the HSCT procedure reported 5% (Tonorezos et al. 2015). The most common and predominant arrhythmias had been atrial flutter, AF, or supraventricular tachycardia, and patients who received allo-HSCT experienced more arrhythmias than those who underwent auto-HSCT. (8% vs. 4%, respectively, P = 0.001). The occurrence of arrhythmia with the median onset of 6–9 days following HSCT was 3.9% and 9.4% according to Two large-scale studies (Singla et al. 2013; Hidalgo et al. 2004). A reduced-intensity conditioning (RIC) regimen was used in a study of 278 HSCT recipients with, and the incidence of arrhythmia was 9% (3%

with atrial flutter and also 6% with AF) (Peres et al. 2010; Chiengthong et al. 2019). Thus, within the specific research, there may also be heterogeneity in the disease type, donor source, conditioning chemotherapy regimen, or pre-transplant cardiac function and performance. Arrhythmias after HSCT cannot be detected or prevented with any certainty at this time (Armenian et al. 2017). electrolyte imbalance and disturbance frequently result from Diarrhea or kidney dysfunction brought on by conditioning chemotherapy regimen toxicity, infection in HSCT regularly induces, GVHD, which can typically results in arrhythmia maybe due to electrolyte disturbances.

1.2 Classes of Chemotherapy Drugs and Arrhythmias

Anthracyclines

Anthracyclines are a class of chemotherapy drugs usually used in the treatment of different cancers, such as those that require bone marrow transplantation (BMT). However, one of the most serious side effects of anthracyclines is their potential to cause heart damage and arrhythmias, which can be particularly problematic for patients undergoing BMT.

The most common cardiac hazard of anthracyclines is Cardiomyopathy; however this class of medications could regularly lead to primary arrhythmias caused by Chemotherapy and secondary arrhythmias caused by cardiomyopathy. The prevalence of anthracycline-related Cardiomyopathy is typically dose related (5%–8% prevalence at 450 mg/m^2 cumulative dose) and it is no longer reversible.

Even at the beginning of the first anthracycline infusion of anthracyclines different and multiple ECG changes and arrhythmias may be observed. Usually non-specific, consist of ST-T phase alterations, T-wave flattening, QRS voltage reduction, and QTc interval prolongation. About 30.3% of patients can experience them ECG changes brought on by anthracyclines appears to be dose independent in contrast to cardiomyopathy (Dindogru et al. 1978).

QTc interval prolongation (QTc > 450 ms) is usual with use of anthracyclines, with 11.5% of individuals experiencing it after receiving their first chemotherapy dose (Horacek 2009). subsequent cycles of chemotherapy treatment with anthracyclines, can increase the prevalence of QT interval prolongation,which can seldom cause ventricular arrhythmias (Iwata et al. 1984). It has been demonstrated that dexrazoxane therapy,reduces the QTc interval correlated with anthracyclines (Galetta et al. 2005).

The use of anthracyclines has been linked to an extend range of arrhythmias, such as sinus tachycardia and bradycardia, premature ventricular complexes (PVCs), premature atrial complexes (PACs), supraventricular tachycardia (SVT), and sustain and non-sustain ventricular tachycardia (VT) according to Holter monitoring. Approximately 65.5% of individuals experienced arrhythmias Even during the primary cycle of chemotherapy (Kilickap et al. 2007). In one study, the rate of arrhythmia repeat was low (3%) during the first hour after administration but multiplied obviously (24%) in first 24 h (Steinberg et al. 1987), it hard to establish that the prevalence of these known arrhythmias are side effect of anthracycline infusion.

The majority of anthracycline-induced arrhythmias are benign, but AF has been a known and also common therapeutic challenge, with occurrence of 10.3% in one report (Kilickap et al. 2007). In a recent research, implantable cardioverter defibrillators were used by survivors with anthracycline-induced cardiomyopathy who experienced episodes of AF, different types of VT, or VF in 73.9%, 56.6%, and 30.4% of cases, respectively. The superiority of arrhythmias was the same as it was in patients who had implantable cardioverter defibrillators placed for cardiomyopathy unrelated to cancer (Mazur, et al. 2017).while the majority of anthracycline related arrhythmias may accompany cardiomyopathy, uncommon malignant ventricular arrhythmias may happen during infusion of anthracycline drugs or during the first period of chemotherapy administration before the beginning of cardiac dysfunction (Wortman et al. 1979). Hypokalemia may be worsen arrhythmias (Kishi et al. 2000), and it is important to preserve electrolyte balances during treatment with anthracyclines. Bradyarrhythmias such as atrioventricular (AV) block have also been reported rarely.

1.3 Alkylating Agent Drugs

A well-known chemotherapeutic drug called cyclophosphamide is frequently included in the conditioning regimen before bone marrow transplantation (BMT). cyclophosphamide can be used to treat graft-versus-host disease (GVHD), a frequent consequence of BMT as well as its role in the conditioning regimen. When treating GVHD, Cyclophosphamide may be administered, either by itself or in combination with other drugs.

One of the most common signs of cyclophosphamide-induced cardiotoxicity is CTIA. ECG changes, such as T-wave or ST-segment alternations, prolongation of QTc, and arrhythmias were detected in 33% of the survivors in a group of patients who underwent cyclophosphamide administration prior to bone marrow transplantation, with 27% growing a prior drop in QRS voltage (Kupari et al. 1990).

Patients who underwent cyclophosphamide therapy have been found to have A wide variety of arrhythmias, such as SVT, AF, PVCs, PACs, and other ventricular arrhythmias with prior prolongation of QTc interval, and AV block, with resulting in sporadic requirement for pacemaker implantation (Ramireddy et al. 1994).

In bone marrow transplantation (BMT) treatments for a number of hematologic malignancies including multiple myeloma, lymphoma, and leukemia, Melphalan is widely used as a conditioning regimen.

AF is the most frequent manifestation of CTIA, which is one of the most common side effects and manifestations of melphalan therapy, happens in 6.6% to 22.5% of patients, especially elderly survivors, experience it (Olivieri et al. 1998; Mileshkin et al. 2005) Supraventricular arrhythmias, such as SVT and AF, were seen in 11%

of individuals In a report of 438 patients who underwent melphalan treatment with history of HSCT, with an incidence of 0–2% with other chemotherapy protocol (Feliz et al. 2011).

A chemotherapeutic medication called BUSULFAN is a drug used in conditioning regimens for bone marrow transplantation (BMT). Busulfan may be included in different and various conditioning regimens, and the one chosen for a patient will depend on the patient's unique circumstances, such as the kind and stage of cancer or blood disorder, age, overall health, and previous treatments.

Busulfan is frequently used in conditioning programs for BMT, including these: Busulfan and cyclophosphamide, Busulfan and fludarabine, Busulfan, fludarabine, and anti thymocyte globulin (ATG).

The precise regimen utilized for a patient having busulfan-assisted BMT is chosen by the patient's medical team and based on the patient's unique factors such as the type and stage of cancer, overall health, and prior treatments.

It has a $\leq$ 6.4% incidence and is also linked to AF,when combined with cyclophosphamide (Ulrickson et al. 2009).

1.4 Proteasome Inhibitors

The proteasome inhibitor chemotherapeutic drug known as Bortezomib is used to treat mantle cell lymphoma and multiple myeloma malignancies. The most usual cardiotoxicity complication correlate with bortezomib is Heart failure, which can result in subsequent arrhythmias. With the usage of bortezomib, isolated occurrences of SVT and AF have been noticed. Seldom, has the need for permanent pacemaker implantation for bradyarrhythmias, such as complete AV block been proven (Diwadkar et al. 2016).

Another proteasome inhibitor drug used to treat patients with refractory or relapsed multiple myeloma is Carfilzomib. Carfilzomib has been seen to excess the heart failure risk similar to how bortezomib does. following cycles of carfilzomib therapy saw a significant decrease in the prevalence of cardiac arrhythmias (Siegel et al. 2013).

Hematopoietic stem cell transplantation (HSCT) is a potential cure for various hematologic malignancies, and rarely non-malignant conditions. Advances in HSCT during the recent years significantly increased cancer survivors in long-term follow-up (Majhail et al. 2013; Solh et al. 2018). However, there is growing recognition that HSCT-related treatment modalities can cause wide spectrum of cardiac arrhythmias, including brady- and tachycardias (Buza et al. 2017). Atrial fibrillation (AF), is the most frequent arrhythmia, however, ventricular arrhythmias, particularly those associated to QT prolongation, and bradycardias may also happen (Fradley et al. 2021). These arrhythmias can increase duration of hospitalization and mortality at one-year (Blaes et al. 2016; Tonorezos et al. 2015). With the awareness of these post-transplant complications and their related morbidity/mortality, the international guidelines also recommended a comprehensive approach for long-term follow-up in survivors after HSCT (Group 2022a).

Despite increased recognition, there is no dedicated prospective data on true incidence of tachyarrhythmias in adult HSCT recipients. Additionally, few studies have adequately investigated necessary evaluation, proper treatment, and follow-up modalities. In this chapter, we present an overview on evaluation, management, and follow-up of cardiac tachyarrhythmias after HSCT.

Tachycardia Classification

A. Supraventricular tachyarrhythmias: Supraventricular arrhythmias (SVTs) are a well-recognized complication after HSCT. Estimates of incidence is between 7 and 27% of adult HSCT recipients (Tonorezos et al. 2015; Singla et al. 2013; Feliz et al. 2011; Sureddi et al. 2012; Arun et al. 2017). AF and atrial flutter (AFL) appear to be most frequent arrhythmia, however, other SVTs occurring in up to 5% of patients has also been reported (Table 1) (Hidalgo et al. 2004; Chiengthong et al. 2019).

1. Atrial fibrillation: AF is the most frequent type of arrhythmia encountered in the human. AF development is linked to the higher risk of stroke, heart failure, dementia, and death (Wolf et al. 1991; Stewart et al. 2002). AF is also reported to happen at a higher incidence in the HSCT setting (Hidalgo et al. 2004; Peres et al. 2010). Nonetheless, the predisposing factors of AF are also not well defined. These factors may include the chemotherapy drugs, prior mediastinal irradiation, transplanted stem cells, or accompanying comorbidities.

Diagnosis: AF is electrocardiographically characterized by an irregular ventricular rate (QRS complexes) and absence of p-waves. There may be fibrillatory waves, or they may be absent. The ventricular response usually varies between 80 and 180 beats/min.

Management: AF treatment in HSCT setting would be identical to the general population with unique issues related to drug-drug interactions (DDI) that may found with definite HSCT-related treatments. The approach to the management of AF requires the usual decisions regarding symptom control with rate or rhythm control and stroke prevention with effective anticoagulation.

Rate versus rhythm control: In asymptomatic AF patients, suggested approach is rate control, with a target of resting heart rate < 110 beats/min (Group 2020). Rate control can be accomplished with beta-adrenergic receptor antagonists, calcium channel blockers such as diltiazem and verapamil, and digoxin or combination of these agents. Beta-blockers are preferred rate-controlling agents in HSCT setting (Group 2022a). Metoprolol is preferred over the carvedilol because of fewer DDIs (Fradley et al. 2021). Calcium antagonists (diltiazem and verapamil) are the second-choice rate controlling drugs; however, they should be used carefully because they inhibit cytochrome P450 CYP3A4 metabolism and will increase plasma concentrations of various chemotherapy agents (Lopez-Fernandez et al. 2019). Digoxin is not

Table 1 Chemotherapy drugs and associated arrhythmias

Drug class	AF	AFL	ST	Other SVT	QT prolongation	VT
Anthracyclines						
Doxorubicin	✓		✓		✓	
Idarubicin	✓		✓		✓	
Mitoxantrone	✓		✓		✓	
Antimetabolites						
Fludarabine						
Alkylating agents						
Cyclophosphamide	✓		✓	✓		✓
Melphalan	✓			✓		✓
Antimicrotubule agents						
Arsenic trioxide					✓	
Immunomodulatory drugs						
Lenalidomide	✓					✓
Thalidomide	✓					✓
Proteasome inhibitors						
Carfilzomib	✓					
Small-molecule tyrosine kinase inhibitors						
Ibrutinib	✓	✓				✓
Chimeric antigen receptor T-cell therapy						
Tisagenlecleucel	✓			✓		
Axicabtagene	✓			✓		
Ciloleucel	✓			✓		
Histone deacetylase inhibitors						
Panobinostat						✓
Immune checkpoint inhibitors						
Nivolumab	✓		✓		✓	✓
Ipilimumab	✓		✓		✓	✓
Pembrolizumab	✓		✓		✓	✓

Abbreviations AF: atrial fibrillation; AFL: atrial flutter; ST: sinus tachycardia; SVT: supraventricular tachycardia; VT: ventricular tachycardia

a preferred rate controlling drug in the HSCT population and should also be used with caution because several chemotherapy agents, including ibrutinib, inhibit P-glycoprotein which can increase plasma levels and likely digoxin toxicity (Ganatra et al. 2018).

In symptomatic patients with AF, including those with tachycardia-induced cardiomyopathy, restoration and maintenance of sinus rhythm by a rhythm control strategy is recommended (Group 2020). Direct current cardioversion (DCV) is the preferred option in hemodynamically unstable AF patients as it has higher efficacy than pharmacological cardioversion (PCV) and results in immediate conversion into sinus rhythm. In stable patients, either DCV or ECV can be attempted; PCV is less effective but does not require sedation. Despite the initial success, AF recurrence is an important long-term issue of the cardioversion mainly due to inflammatory environment of malignancy and continued use of provocative chemotherapy drugs (Farmakis et al. 2014).

There are several antiarrhythmic drug (AAD) options for PCV, however, AADs have not been adequately investigated in the cancer population and must be used with particular attention to DDIs leading to QT prolongation. There are other important considerations regarding AAD use in the HSCT patients. These drugs have potential of interaction with cytochrome P450 CYP3A4 and CYP2D6 systems and gastrointestinal re-secretion over a P-glycoprotein transporter. In addition to these DDIs, certain AADs may not be preferred due to underlying cardiovascular and oncological diseases. Dronedarone should be avoided because of serious DDIs leading to QT prolongation and important interactions with anthracyclines. Although amiodarone is a safer choice compared with dronedarone, DDIs and QT prolongation is still an important issue. Sotalol and dofetilide are good options in patients with preserved left ventricular (LV) function, however, QT prolongation and DDI should be carefully considered (Yamreudeewong et al. 2003). In addition, these drugs should be carefully administered in myeloma patients with concurrent amyloid and those with LV hypertrophy. Class IC AADs (flecainide and propafenone) are alternative options in patients with preserved LV function. Potential issues with these agents are concomitant use with anthracyclines and use in patients with myeloma with concurrent amyloid (Fradley et al. 2021).

Whereas catheter ablation is also a potential and known rhythm control management in symptomatic AF patients, the benefits of this approach in cancer patients (including HSCT population) has not been studied. AF patients with tachycardia-induced cardiomyopathy may benefit from this approach (Group 2020). LV function improvement may help us to proceed with HSCT in a safer environment.

Stroke prevention: Similar to all patients with active cancer, thromboembolic events are commonly observed in HSCT patients. Concomitant AF also boost the risk of stroke and systemic embolism. Anticoagulation is the main pillar of thromboembolism prevention. The CHA2DS2-VASc risk score has been used to determine stroke risk in general population. However, this scoring system not been validated in patients with cancer (D'Souza et al. 2018). Despite this limitation, the CHA2DS2-VASc score is still recommended for determining the stroke risk and lead to anticoagulation decision in cancer patients with AF/AFL0.17 In the absence of contraindications, anticoagulation is necessary for men with a CHA2DS2-VASc score $\geq$ 2 and for women with a score $\geq$ 3 (Group 2020).

Bleeding risk assessment is more complicated in patients with malignancy. Although the HAS-BLED scoring system is commonly used to determine bleeding risk in the general population, it does not consider settings unique to cancer patients such as thrombocytopenia and intracerebral metastases. Therefore, HAS-BLED score cannot be dependable in HSCT patients. It is necessary to dedicate risk prediction models are improved in cancer patients to help anticoagulation decision making in the setting of AF /AFL.

In addition to the CHA2DS2-VASc and HASBLED scores, type of malignancy, stage, prognosis and the other thrombo-embolic or bleeding risk should be considered for long-term oral anticoagulation. There is four options for anticoagulation in AF/AFL patients, including vitamin K antagonists (VKA), low-molecular weight heparins (LMWH), non-VKA oral anticoagulants (NOAC), and left atrial appendage (LAA) occluder devices.

The use of VKAs has many disadvantages in HSCT patients including multiple DDIs, dietary interactions, narrow therapeutic window, need for frequent INR testing, and increased risk of subdural hematomas with ibrutinib (Fradley et al. 2021). Despite all shortcomings, they stay the lonely choice in patients with a mechanical prosthetic valve or more than moderate mitral stenosis.

LMWHs are considered a practical temporarily anticoagulation strategy, exclusively in admitted patients taking myelosuppressive chemotherapy or those with active bleeding. Other favorable situations for LMVHs are severe kidney impairment (CrCl < 15 mL/min), significant DDIs with NOAC, and platelet count < 50,000/μL (Group 2022a). However, there are no studies investigating the LMVH position in preventing stroke in the setting of AF /AFL and their use is extrapolated from their verified effectiveness in patients with venous thromboembolism (Lee et al. 2003). In addition, their use is limited by patient discomfort related to twice daily injections.

Although patients with cancer were either omitted from or inadequately represented in the pivotal NOAC AF trials, secondary analyses of these trials (ROCKET AF, ARISTOTLE, ENGAGE AF-TIMI 48) and observational data suggested similar safety and efficacy when compared with VKAs (Chen, et al. 2019; Melloni, et al. 2017; Fanola et al. 2018; Mariani et al. 2021). All NOACs have interaction with the P-glycoprotein system, but dabigatran is most affected. CYP 3A4 is the major enzyme involved in the metabolization of rivaroxaban and apixaban and therefore, these NOACs should be used cautiously in combination with other drugs such as inducers of this enzyme, including ibrutinib (Fitzgerald and Howes 2016). In addition, NOAC use is limited by severe renal failure, or impaired gastrointestinal absorption.

Considering the difficulties associated with anticoagulation in HSCT patients, early use of LAA occluder devices may be recommended. However, there are no prospective date in these patients, and the likelihood of thrombus induced with devices could hypothetically be augmented in a hypercoagulable setting (Isogai et al. 2021). Similar to general population, antiplatelet therapy is not endorsed for prevention of stroke or systemic thromboembolism in the HSCT patients with AF.

Follow-up and monitoring of AF: According the ESC guidelines, baseline, and serial ECG (3 and 12 months, then yearly) is recommended in HSCT patients (Group 2022a). In case of secondary AF associated with concise factors—include chemotherapy agents, other drugs, or electrolyte abnormalities—a accurate clinical evaluation of the predisposition for AF recurrence is recommended, with need to revisit after a period of 3 months. In this visit, a comprehensive assessment including history taking, physical examination, 12-lead ECG, and 24–48 h ambulatory ECG monitoring. If results of these evaluation showed any evidence of AF recurrence, referral to a cardiac electrophysiologist is recommended.

2. Atrial flutter: AFL is the second most common arrhythmia in the HSCT setting. The pathophysiology of AFL is similar to AF and has been reviewed previously.

Diagnosis: AFL is electrocardiographically characterized by a regular ventricular rate (QRS complexes) and presence of characteristic flutter waves (Fig. 2). Flutter waves may have "sawtooth" appearance, or "notched" upright pattern. Ventricular rate is determined by the atrioventricular (AV) conduction. The most common AV ratio is 2:1, resulting in a ventricular response of 150 bpm. Sometimes, the ventricular response is irregular and may simulate AF.

Management: Management of AFL is comparable to AF with small differences. Rate control is usually ineffective in the AFL setting and rhythm control is needed in most cases. Therefore, referral for catheter ablation is more likely in patients with AFL. Anticoagulation is done with same considerations as in AF patients.

Follow-up and monitoring of AFL: AFL follow-up is similar to AF patients (first time and serial ECG at 3 and 12 months, then yearly). Actually, AF is a common observation in the follow-up of patients with AFL.

3. Other supraventricular tachycardias: Other SVTs are uncommon in the HSCT patients. These arrhythmias may appear as episodes of the sinus tachycardia (ST), focal atrial tachycardia (FAT), or premature atrial complexes. These SVTs mainly related to various factors including pain, anxiety, dehydration, hypoxia, acidosis, inflammation, cardiac tamponade, pulmonary embolism, anemia, sepsis, dysautonomia, and chemotherapy agents. Among the chemotherapy drugs, ST is most commonly observed during treatment with anthracyclines, cyclophosphamide, and arsenic trioxide (Fradley et al. 2021).

Diagnosis: It is important to differentiate between ST and FAT. ECG is the first diagnostic tool to differentiate between these arrhythmias. ST is characterized by ventricular rate of more than 100 beats/min and normal appearing p-waves; however, FAT is and organized arrhythmia having abnormal appearing p-waves with isoelectric baseline (unlike AFL).

Management: The identification and treatment of the underlying etiology of ST is the cornerstone of management. In cases of inappropriate ST, ivabradine and beta-blockers are preferred drugs. Acute management of FAT should follow the same pathway that used in other supraventricular tachycardias (Brugada et al. 2020). Chronic management of recurrent or incessant FAT is catheter ablation. In non-recurrent cases, oral beta-blocker or class IC AADs are recommended.

Follow-up and monitoring of AFL: No monitoring is required for ST and successfully ablated FAT. In medically treated FAT, regular follow-up in arrhythmia or cardiology clinic is recommended (baseline and serial ECG at 3 and 12 months, then yearly).

B. QT prolongation and ventricular arrhythmias: The QT prolongation is reported in up to 22% of different anti-cancer drugs (Porta-Sánchez et al. 2017). Highest incidence of QT prolongation has been associated with arsenic trioxide in 63%

of patients (Soignet et al. 2001). The second highest incidence of QT prolongation (28.8%) has been reported with tyrosine kinase inhibitor drug class (nilotinib: 38.7%, dasatinib: 41.7%) (Abu Rmilah et al. 2020). Direct blockade of the rapid component of the delayed rectifier potassium current (IKr) is the main cause of QTc prolongation. However, QT prolongation may also be related to direct inhibition of the intracellular signaling pathways, including phosphoinositide 3-kinase (Ballou et al. 2015). Although QTc prolongation is common with many chemotherapeutic agents (Table 1), the incidence of life-threatening arrhythmias like torsades de pointes (TdP) is less frequent occurring in less than 1% of patients treated with these drugs (Porta-Sánchez et al. 2017).

Majority of chemotherapy-induced ventricular arrhythmia (VA) are attributed to the QT prolongation and subsequent development of TdP. Nevertheless, sustained monomorphic ventricular tachycardia and fibrillation may rarely occur secondary to administration of chemotherapy drugs (Table 1). The frequency of these VAs is increased in patients with advanced malignancies and associated cardiovascular comorbidities (Alexandre et al. 2018). Mechanism of these arrhythmias comprised of a direct or indirect consequence of the chemotherapeutic drugs on heart muscle, concurrent ischemia, and associated electrolyte abnormalities (Alexandre et al. 2018; Enriquez, et al. 2017).

Diagnosis: To determine the TdP risk, QTc interval is the readily available marker that is well established in clinical trials. The QT interval is best measured on lead II, V5 or V6. To decrease the impression of beat-to-beat variability, it is necessary to measure the average of QT intervals during 3 to 5 beats in sinus arrhythmia is present or during 10 beats in the AF rhythm. In cancer patients, QT correction using Fridericia (QTcF) formula $(QT/3\sqrt{RR})$ is recommended since it is more precise at heart rate extremes.17 For patients with intraventricular conduction delays (paced rhythms or bundle branch block), a practical and user-friendly method is using a modified formula: $mQTcF = (QT-(QRS-120))/3\sqrt{RR}$ (Haghjoo et al. 2021).

QTc intervals of 450 ms in men and 460 ms in women are recommended as the upper cutoff of normal on baseline ECG (Schwartz et al. 1993). The risk of arrhythmia generally growth when the QTc is > 500 ms or the ΔQTc (i.e., change from baseline) is > 60 ms because TdP rarely occur when QTc is < 500 ms (Group 2022b). However, treatment options should not routinely changed by a ΔQTc > 60 ms if the QTc remains < 500 ms (Herrmann et al. 2022).

Management: In general, treatment of HSCT-induced VAs should follow treatment guidelines in the general population (Group 2022b). Consultation with cardiology team is recommended in patients with baseline prolonged QTc interval ($\geq$ 480 ms), patients receiving QT-prolonging drugs, and those who report new presentations such as syncope or pre-syncope, dizziness, or palpitations. monitoring of ECG in therapy period must be considered at baseline, and then 7–15 days after first dose or when changing drug dose, monthly for the first 3 months of treatment and then sporadically after that based on the patient's condition and chemotherapy drug. Patients with diarrhea should be monitored more often, and those receiving arsenic trioxide should be monitored weekly with ECG. Electrolyte abnormalities including hypocalcemia, hypomagnesemia, and hypokalemia should be treated before initiating anti-cancer medications.

QT-prolonging cancer therapy should be temporarily stopped in asymptomatic patients with QTcF $\geq$ 500 ms. Interruption of other QT-prolonging drug and correction of electrolyte disturbance are also recommended. Daily 12-lead ECG should be obtained until resolution of the QT prolongation. In patients with QTcF 480–500 ms, correction of reversible risk factors and weekly ECG monitoring is recommended. If QTcF remains < 500 ms, cancer therapy can be continued at the same or reduced dose (Group 2022a). In patients who experienced TdP, QT prolonging cancer drug should be stopped. IV magnesium is the initial treatment of choice irrespective of serum magnesium concentration. Transvenous pacing (90–110 beats/min) is very effective in preventing recurrence and may be useful in magnesium refractory TdP or when TdP is provoked by bradycardia. IV isoproterenol (titrated to a heart rate 90 beats/min) is additional treatment modality, and it is helpful when temporary pacing is unavailable or while preparing for temporary transvenous pacing. Symptomatic life-threatening sustained monomorphic VT or VF require urgent intervention.

According to the ESC guidelines, a multidisciplinary approach is recommended in indiviuals, who have experienced overt QT prolongation to debate for alternate anti-cancer treatments (Group 2022a). In these patients, resuming the culprit anti-cancer drug may be considered at a reduced dose. After restarting QT prolonging anti-cancer drug, weekly ECG monitoring should be considered for the first 4–6 weeks and then after monthly during treatment.

In asymptomatic patients with self-limited VA, there is no need to discontinue the cancer drug. Symptomatic VA need reduction in anti-cancer drug dose or stopping the culprit drug and patients should be visited by the cardiologist (Group 2022a). starting usage of class IA, IC, and III antiarrhythmic drugs is not permitted because of the risk of DDIs and QTc prolongation. Class IB drugs are drug of choice in this setting because they are less likely to cause DDIs or QT prolongation. Amiodarone is the preferred option in patients with structural heart disease. Beta-blockers are the best choice in patients with symptomatic premature ventricular complexes or non-sustained VT. Prompt defibrillation is indicated in patients with hemodynamically unstable VAs.

There is no dedicated clinical trial investigating the implantable cardioverter defibrillator (ICD) therapy in chemotherapy-induced cardiomyopathy (CIC). However, symptomatic CIC patients and fulfilled the guideline criteria (LV ejection fraction $\leq$ 35%, New York Heart Association class II to III symptoms, and life expectancy > 1 year) may benefit form ICD implantation (Fradley et al. 2017). CIC may also be in the company of intraventricular conduction defect manifesting as a prolonged QRS interval ($\geq$120 ms). Cardiac resynchronization therapy (CRT) has the potential of improving LV function and clinical symptoms in these patients. In a study, the MADIT-CHIC in which prospectively evaluated the role of CRT in CIC patients (Singh et al. 2019). MADIT-CHIC clearly disclosed that CRT significantly improve the clinical condition and echocardiographic parameters in medically refractory CIC patients with reduced LV function (LVEF $\leq$ 35%) and a left bundle-branch block.

1.5 "Bradyarrhythmic Syndrome in HSCT Patients"

Hematopoietic stem cell transplantation (HSCT) is a widely used therapeutic technique to manage various blood cancers and disorders. The process involves the infusion of stem cells from either the patient or a donor after a preparative regimen of chemotherapy, radiation or both depending on the patient's condition. HSCT is usually a safe and tolerable procedure, but it is not without adverse effects, one of which is bradyarrhythmia.

Pacemaker cells in the sinoatrial node in the right atrium persistently generate electrical impulses without resting phase, thereby the normal heart rhythm and rate depends on the correct functioning of this vital element. After pacemaker impulse generation in the atrium and atrial contraction, it should be transferred to the ventricles to initiate the electro-mechanical phase. The conduction system between atriums and ventricles are atrioventricular node (AVN), main His bundle, right and left bundles, and Purkinje system.

Bradyarrhythmia is a condition when the heart produces complexes slower than usual and can be broadly classified into 2 general categories: Sinoatrial node disease (SAD) and atrioventricular node disorder. (AVB). This rhythm disorder can cause loss of conciousness, pre-syncope, dizziness, fainting, or shortness of breath. Any abnormal pulse formation or transfer in the Sino-atrial node (SAN), Atrio-Ventricular Node (AVN), or His- Purkinje system can induce bradyarrhythmia in ECG or heart monitoring (Kusumoto et al. 2019).

To evaluate the risk of bradyarrhythmias, we usually consider the following categories:

A: Sinus Node Disease (SND)

1. Sinus bradycardia (Fig. 1): heart rate of fewer than 50 beats per minute is usually considered bradycardia (Kusumoto et al. 2019). Bradycardia is not always a significant disorder. We should notice if bradycardia is transient or permanent. In well-trained athletes and healthy elderly people, asymptomatic bradycardia can be detected in heart rhythm monitoring. Therefore, multiple factors, especially symptoms, should be recognized and taken into consideration by the individual person. The number of pacemaker cells in the sinoatrial node (SAN) decreases with age. If pacemaker cells in SAN are not able to increase heart rate proportional with incremental activity or demand, in many studies translates to Chronotropic incompetence. The maximum heart rate in these patients is less than 80% of the expected heart rate reserved during exercise. These patients complain of early fatigue, dyspnea, or dizziness.

2. Sinus pause or arrest of more than 3 s can be noticed in patients with sinus node dysfunction but is not always a distinct abnormality (Fig. 2). Rhythm monitoring can reveal many episodes of sinus pauses during sleepiness or vagotonia. Symptomatic sinus pauses are defined as Sick Sinus Syndrome (SSS)

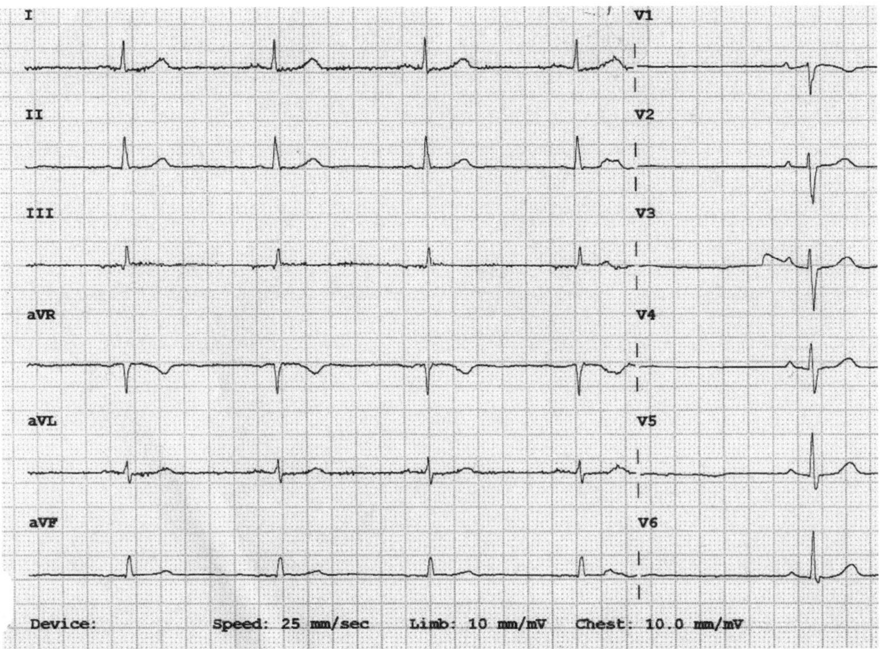

Fig. 1 Sinus node bradycardia

3. Tachycardia-bradycardia syndrome (Fig. 3): Sinus bradycardia or sinus pause appears after periods of abnormal atrial tachyarrhythmias such as atrial fibrillation or flutter. This rhythm disorder is a subgroup of SSS and is usually named the "tachy-brady" syndrome. Sometimes bradycardia disappears after tachyarrhythmia management, but usually, bradyarrhythmia needs therapeutic intervention, independent of tachyarrhythmia abolishment.

Any SAN function is not only dependent on the pacemaker cells' function but also the autonomic system. Sympathetic stimulation increases heart rate and parasympathetic activity vice versa. Any interpretation of sinus bradycardia or SAN disorders is incomplete without consideration of the autonomic system's function. Transient disturbances in atrial impulse generation are, in most cases, related to hyperactivity of the parasympathetic system (Fig. 4).

B: Atrio-Ventricular Block (AVB)

After pacemaker impulse generation in the atrium and atrial contraction, it should be transferred to the ventricles to initiate the electro-mechanical phase. The conduction system between atriums and ventricles are atrioventricular node (AVN), main His bundle, right and left bundles, and Purkinje system. AVN is a sensitive subendocardial structure and it is easily affected by aging, ischemic changes or drugs and loses its normal function. For this reason, it is not uncommon to see disorders related to AVN conduction in elderly people. Therefore any significant reduction in heart rate could

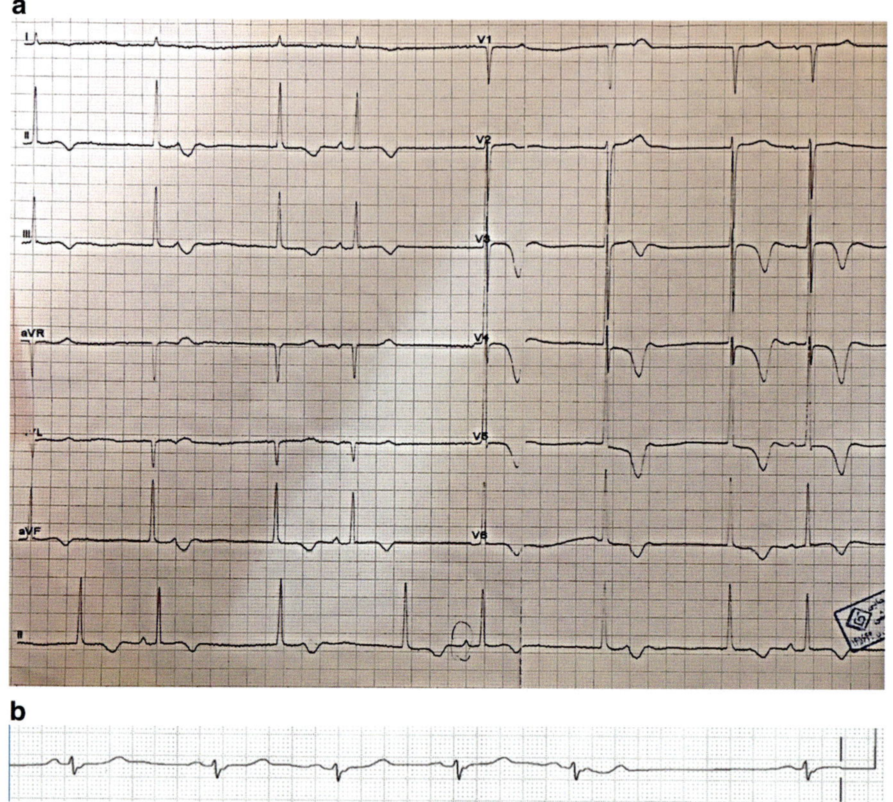

Fig. 2 **a** Sinus pause with junctional escape rhythm in a patient with history of hypertrophic cardiomyopathy. **b** Sinus node wenckebach pattern

be related to the conduction impairment between the atriums and ventricles. Healthy AVN can generate an escape rate between 40 to 60 bpm instead of SAN and usually save a patient's life. Still, in patients with crucial underlying heart diseases, other pacemaker cells below AVN are untrustworthy and threaten the patient. Although, invasive electrophysiology studies can show the diseased part of the conduction system with a significant percentage of confidence, evaluation of ECG and rhythm monitoring helps to define the abnormal part of the conduction system. Different types of AVBs are summarized as follows:

1. First-degree AVB (Fig. 5): P waves conduct to the ventricles in a 1:1 pattern with a PR interval (P wave to the initial part of QRS) of more than 200 ms (ms). The problem is almost always in AVN. In a few cases, PR Interval of more than 300 ms may cause symptoms such as shortness of breath, heart palpitations, and even dizziness and/or syncope, but it is often considered a benign disorder (Crisel et al. 2011).

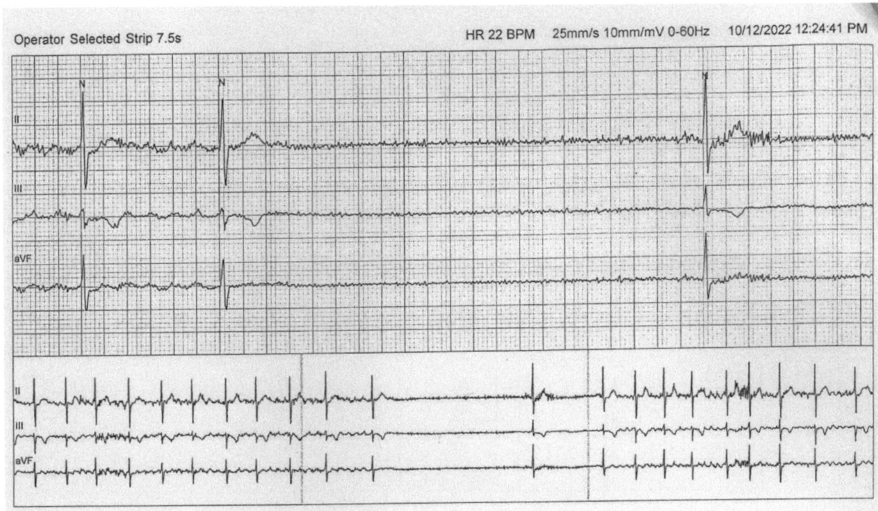

Fig. 3 Tachy- brady syndrome, one of features of sick sinus syndrome

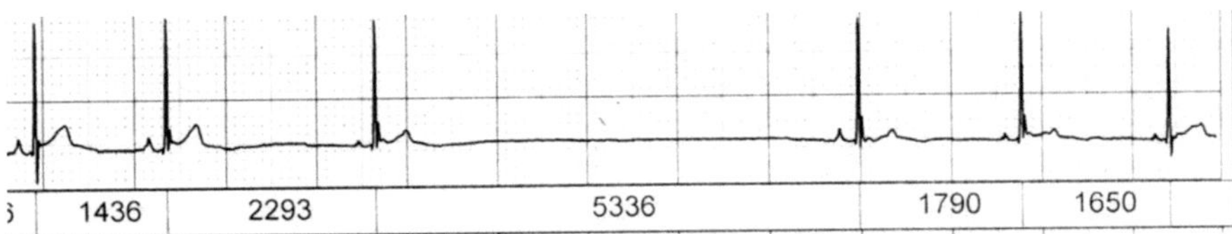

Fig. 4 Sinus bradycardia and pause is seen during sleepiness due to the parasympathic over activation

Second-degree atrioventricular block: P waves conduct to the ventricles, but not in 1 to 1 manner. ECG shows P waves that are not conducted to the ventricles in well-defined patterns, and according to these patterns and autonomic system manipulation, the physician can predict the abnormal part of the conduction system and stratify the risk of asystole, syncope, or cardiac death. The following patterns are categorized in second-degree AVB:

1.1 Mobitz type I (Fig. 6): PR interval increases gradually and there is a significant PR interval difference between the PR interval before and after the nonconducted P wave. This pattern is usually related to AVN disorder and AVB improves with exercise and injection of Atropine, Adrenaline, Isoproterenol, and other sympathomimetic drugs, and deteriorates with carotid sinus massage or AVN drug blockers. This type of block usually has a benign course. Correction of the reversible causes can improve the AVB and the patient's outcome. Hypervagotonia can induce and maintain this type of AVB and should be considered in the patient's evaluation.

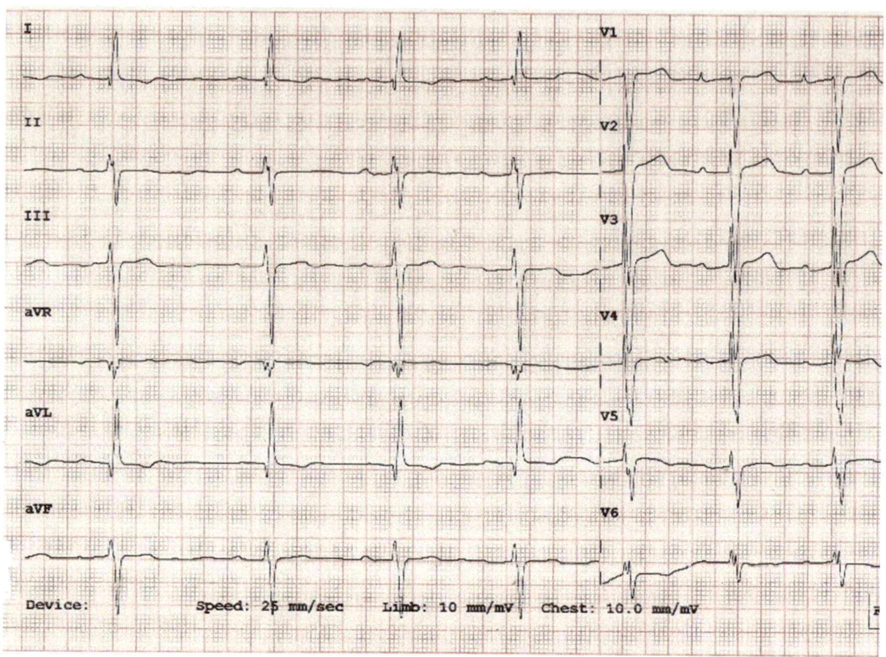

Fig. 5 Prolonged PR interval more than 200 ms, is named as first degree AV block or delay

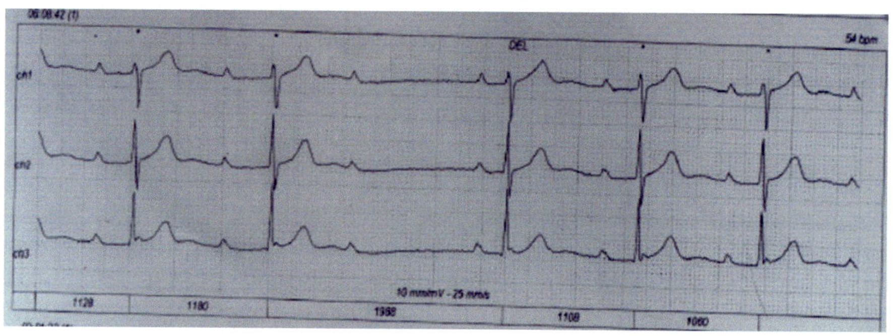

Fig. 6 Second degree AV block, Mobitz type 1. PR interval are significantly different before and after blocked P wave

1.2. Mobitz type II (Fig. 7): In the setting of a constant PR interval, a P wave does not conduct to the ventricle abruptly. In this block pattern, if the QRS is normal, the site of disturbance is in the His bundle, and in the presence of a wide QRS complex, in or below of the His region. Unlike AVN, the His bundle and the Purkinje system are not affected by the autonomic system, but are influenced by the amount of atrial rhythm input to these parts. Since the escape rhythm is not reliable, patients usually experience worse clinical symptoms such as

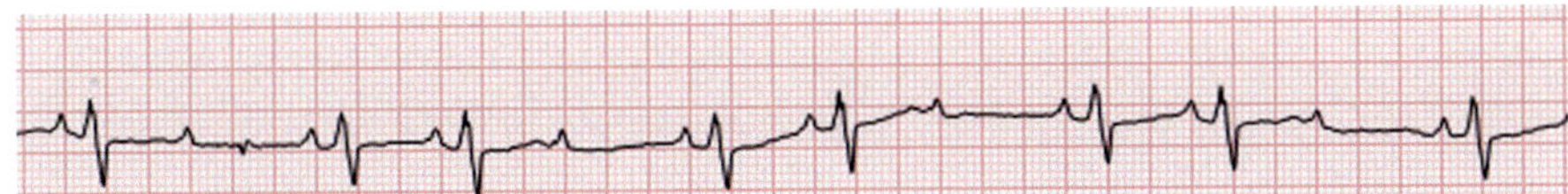

Fig. 7 Second degree AVB, Mobits type 2

syncope, and if not diagnosed in time, are more at risk of sudden cardiac death. This disorder is usually transient, and common treatments for bradycardia, such as atropine or epinephrine injection, can make clinical conditions worse. Therefore, the main treatment is to pay attention to the underlying factors and use temporary pacemakers.

1.3. 2 to 1 atrioventricular block (Fig. 8a, b): This type of heart block is commonly seen. According to the previous explanations, if the PR interval in the conducted P wave is less than 160 milliseconds and the QRS complex is narrow, the location of the disorder is probably in the His bundle, and if the PR interval is more than 200 milliseconds, the site of injury is in most cases AVN. In the presence of the wide QRS complex, the probability of His-Purkinje disease increases. Autonomic system maneuvers help to recognize the abnormal segment of the conduction system. Carotid sinus massage improves 2 to 1 AVB if the problem is in the His-Purkinje system and AVB deteriorates in AVN conduction disorder. Atropine injection and exercise have opposite effects. It is important to pay attention to the fact that atropine should not be injected in a patient who has syncope and whose ECG pattern is suspected to be Hiss bundle disorder. Atropine injection may lead to asystole and a more severe reduction of ventricular rate (Fig. 9).

2. Advanced, high-degree or high-grade AVB (Fig. 10) means two or more consecutive P waves are blocked at a stable physiologic rate and some of the normal atrial activities conduct to the ventricle. This type of AVB is usually related to the His-Purkinje disease and patients are usually admitted with syncope or high-risk symptoms.

3. Complete or third-degree AVB (Fig. 11): according to the escape rate, QRS widening, and underlying disease the level of AVB is usually predictable. Wider QRS complexes and slower escape rates are more compatible with His-Purkinje disease. It should be kept in mind that patients fibrillation in atrial chambers also present with heart block. In such a situation, not only the ventricular response is greatly reduced, but also the irregular ventricular response is replaced with the regular pattern (Fig. 12).

Apart from childhood and adolescence when bradyarrhythmia occurs due to congenital AVB or hereditary cardiac channelopathies (Baruteau et al. 2016), in most cases bradyarrhythmia occurs due to known factors such as old age and fibrosis in the conduction system, ischemia, and vascular causes, myocarditis, drugs It is used in the treatment of high blood pressure and heart failure, heart surgery and less common cases such as heart trauma.

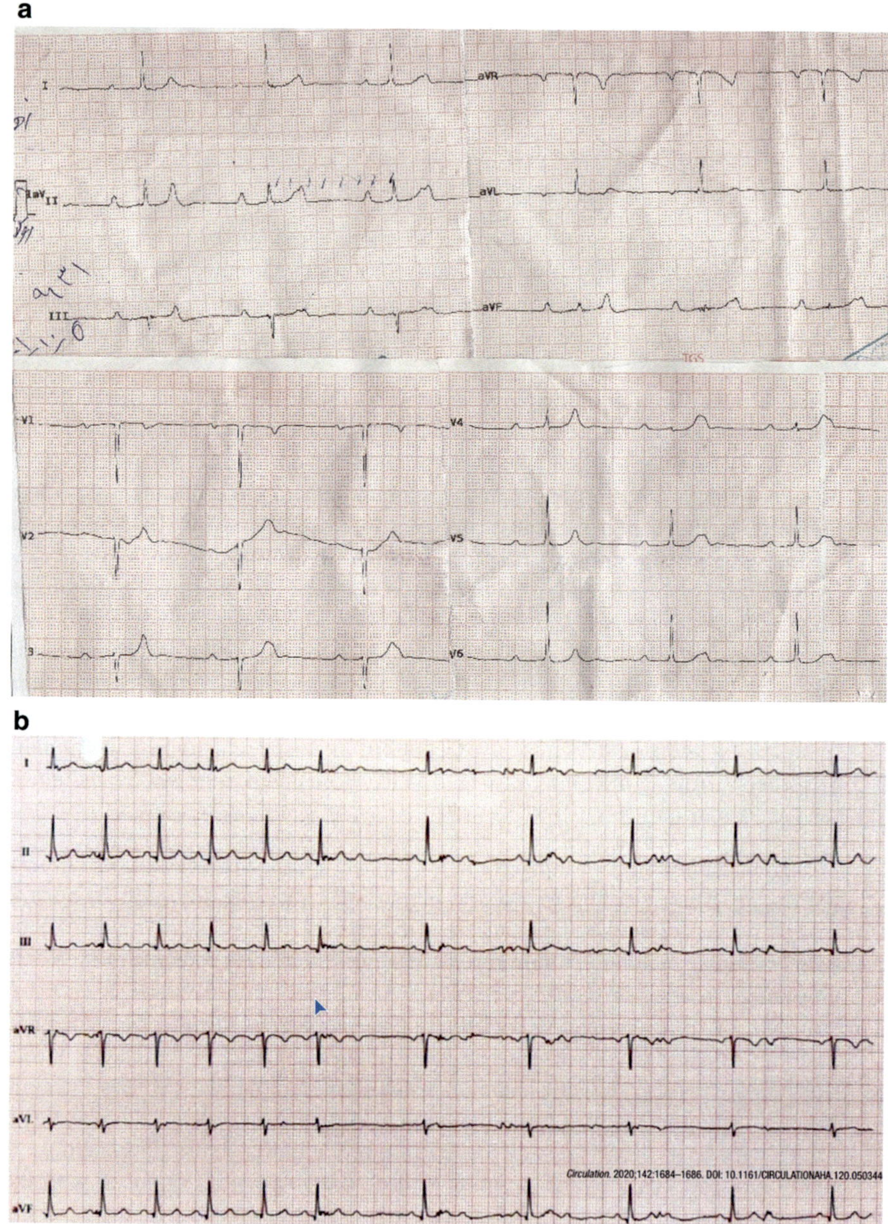

Fig. 8 **a** 2:1 AVB with prolonged PR interval in conducted P wave. AV node is most probably the level of block. **b** 2:1 AVB with PR interval less than 160 ms. Intra His (narrow QRS) or infra His bundle (wide QRS) injury is most probably the level of conduction system damage

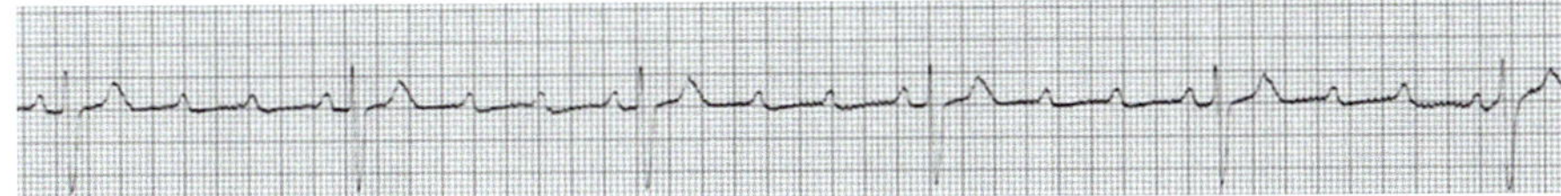

Fig. 9 Bradyarrhythmia aggravation after atropine injection in a patient with intra His 2 to 1 AVB

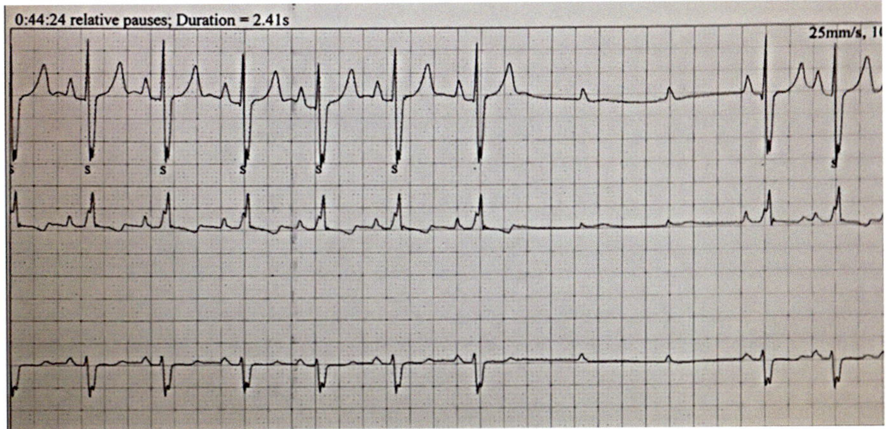

Fig. 10 Advance AV block. 2 or more consecutive P waves don't conduct to the ventricles

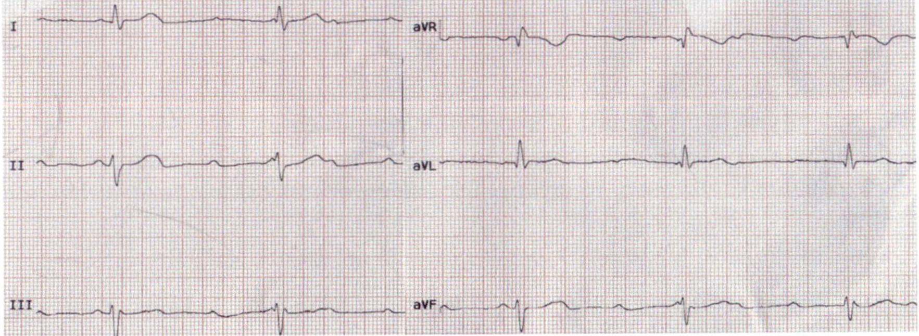

Fig. 11 Complete AVB with regular ventricular escape rhythm and dissociated from sinus atrial activation

Many HSCT patients are at an age where, in addition to the aforementioned factors due to the nature of the blood cancer and involvement of the conduction system by cancer cells, chemotherapy drugs, autonomic system disorders secondary to cancer involvement, or chemotherapy side effects, electrolyte disturbances caused by chemotherapy and mediastinal radiotherapy, are more at risk of progressive bradyarrhythmia syndrome.

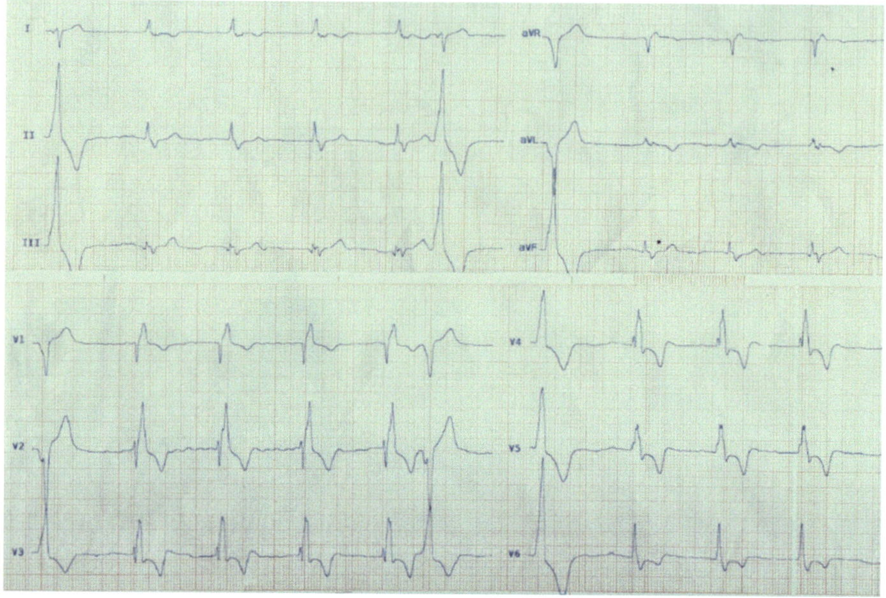

Fig. 12 Atrial fibrillation with regular wide escape rhythm, compatible with complete AV block in a patient with history of congenital heart disease. There are infrequent PVCs in the ECG

According to the mentioned cases, it is necessary to pay attention to the role of blood cancers, chemotherapy for the HSCT process, and the continuation of drug treatments after that and mediastinal irradiation in the occurrence of cardiac bradyarrhythmic syndrome. These factors can cause or aggravate cardiac arrhythmias independently or in the context of underlying heart diseases.

Consultation between cardiologists and oncologists during the therapy process of HSCT patients is necessary to prevent the occurrence of cardiac bradyarrhythmia syndrome.

Some of the most important things to consider are mentioned below:

1. Direct effect of blood cancer cells: Infiltration of the AV nodes by lymphoma can start heart rate reduction syndrome (Lal et al. 2016). Microscopic leukemic cardiac involvement is detected in 30–37% of necropsy series and is usually associated with leukemic appearance in the blood (Bekkers et al. 2004). Bundle branch block and QRS widening, AVB and bradyarrhythmia, and ventricular tachyarrhythmias are one of the side effects of blood cells attacking the heart.
2. HSCT chemotherapy drugs: chemotherapy drugs in different classes can induce bradyarrhythmia syndrome.

    2.1. This complication can be due to the direct drug effects on the heart or the aggravation of the underlying heart rhythm disorder.

2.2. Until recently, long-term cardiac rhythm monitoring was not common for these patients. With the development of long-term cardiac monitoring facilities, it is possible to better comment on the side effects of chemotherapy drugs. For this reason, personal experiences in the occurrence of cardiac bradyarrhythmic events should be considered.

2.3. Paying attention to the chemotherapy drug's dose-time category is very important in investigating the possibility of rhythm disorder. Bradycardia may be observed in higher doses of a chemotherapy drug, or maybe the rhythm disorder is only limited to the time of drug injection, and the patient's problem is solved by terminating the drug administration.

2.4. It is very important to pay attention to other drugs used by the patient, such as cardiac drugs and others, to check for drug interactions. For this purpose, we usually use drug databases for dangerous drug interactions with more than 80% sensitivity and specificity reports (Marcath et al. 2018). This warning is more vital for patients who take beta blockers, calcium channel blockers, digoxin or new chemotherapy drugs (Fonseca et al. 2021).

In the following, we mention some of the most important drugs used in HSCT patients who have the risk of cardiac bradyarrhythmia syndrome (Fonseca et al. 2021; Tamargo et al. 2015):

Alkylating Agents Alkylating agents are a type of chemotherapy medication that functions by attaching chemical groups to cancer cells' DNA molecules, ultimately resulting in the death of the cancer cells. Leukemia, lymphoma, and solid tumors are just a few of the various cancers these medications treat. Myocardial ischemia, direct cardiotoxicity to myocytes, endothelial damage, and elevated vagal tone are the mechanisms implicated in developing bradyarrhythmic cardiac syndromes with these medicines. Some of this group's most effective medications are listed below.

Cyclophosphamide: This is one of those drugs whose side effects can be dose-dependent, and its side effects often decrease significantly in a short time after stopping the drug While this medication can cause both sinus bradycardia and AVB, it can also cause other rhythm problems, such as atrial fibrillation, atrial flutter, VT, and QT prolongation (Kupari et al. 1990; Ramireddy et al. 1994; Morandi et al. 2005; Gottdiener et al. 1981; Cazin, et al. 1986; Agarwal and Burkart 2013).

Busulfan: This drug has been reported in case reports to cause Atrioventricular block (AVB) in some patients who have received high doses of this medication (Ulrickson et al. 2009; Buza et al. 2017).

Ifosfamide: There are case reports for sinus bradycardia and AV block after drug administration, but this drug is more known for inducing left ventricle dysfunction and atrial fibrillation (Buza et al. 2017; Fonseca et al. 2021; Tamargo et al. 2015; Quezado et al. 1993).

There is limited information about other alkylating agents, such as Melphalan, Thiotepa, Carmustine, and Bendamustine on inducing sinus bradycardia or AV block. Although there have been reports of Melphalan-induced bradycardia and atrioventricular (AV) block (Ma et al. 2020).

Anthracyclines These drugs work by interfering with the DNA of cancer cells, therefore they are not able to divide and multiply. Anthracyclines are usually administered intravenously. Cancer treatment-induced arrhythmia was first described in patients receiving anthracyclines (Buza et al. 2017; Ma et al. 2020).

Cardiomyopathy and QT prolongation is the most common cardiac side effects. Sinus tachycardia, atrial fibrillation, PVC, and ventricular tachycardia are more frequent in patients receiving anthracyclines but may be secondary to the ventricular dysfunction (Horacek 2009; Buza et al. 2017).

Sinus bradycardia and AV block case reports have been published, but electrolyte evaluation, especially hypokalemia should be noticed to prevent arrhythmic complications (Kishi et al. 2000; Buza et al. 2017).

The mechanisms for cardiac complications are abnormal Ca^{2+} homeostasis, cardiac apoptosis, mitochondrial injury, and oxidative stress (Fonseca et al. 2021).

Doxorubicin: The incidence of bradyarrhythmic adverse events is up to 3.4% and is usually transient.

Daunorubicin: Sinus bradycardia, AV block, and bifascicular block have been induced by this anthracycline drug (Tamargo et al. 2015; Mandal et al. 2016).

Epirubicin: This anticancer agent is not frequently used, but there are a few reports about sinus bradycardia, sinus exit block, and AV block induction (Tamargo et al. 2015; Usnarska-Zubkiewicz et al. 1992).

Idarubicin: This drug is less cardiotoxic than anthracyclines. Limited information is available regarding this drug.

Mitoxantrone: This agent is a derivation of dihydroxyanthracenedione, that seems to be an effective and better-tolerated alternative to the anthracycline component of standard regimens for most hematologic cancer. However, there are reports regarding the occurrence of bradycardia due to the use of these drugs (Benekli et al. 1997).

Antimetabolites Antimetabolites are chemotherapeutic drugs that disrupt cancer cells' ability to synthesize DNA and RNA. They achieve this by mimicking the structure of naturally occurring molecules that are required for DNA and RNA synthesis, such as nucleotides. When cancer cells take up these antimetabolites, they incorporate them into DNA and RNA during replication, causing errors and ultimately preventing cancer cells from dividing and multiplying. The mechanism for adverse effects is through Coronary ischemia (vasospasm, thrombosis due to endothelial damage (Fonseca et al. 2021).

Fludarabine: This antimetabolite agent has been reported to induce bradycardia and atrioventricular (AV) block in some patients. However, this is a rare side effect and may occur only in a small number of patients (Chung-Lo et al. 2010).

Clofarabine: Data for cardiac bradyarrhythmia and AV block after Clofarabine prescription is very limited, but there are few cases reports for sinus bradycardia induction (Buza et al. 2017).

Cytarabine: Pacemaker implantation after a bradyarrhythmia event due to a Cytarabine prescription has been reported (Tamargo et al. 2015; Cil et al. 2007).

Gemcitabine: Sinus bradycardia and AV block after Gemcitabine usage are rare and arrhythmic side effects of this drug are usually atrial and ventricular tachyarrhythmias.

The risk of sinus bradycardia after the 5-FU prescription is higher than other antimetabolites, but it is not used routinely in HSCT patients (Fonseca et al. 2021).

Plant Alcaloid These are antitumor agents derived from plants. The specific act of this drug is blocking the potency of a cancer cell to divide and become two cells.

Etoposide: This drug, which is also called VP-16, is monitored for side effects such as hypotension and myocardial ischemia during infusion (Tamargo et al. 2015). Etoposide induces bradyarrhythmia rarely. There is only one reported case in the literature (Gill et al. 2017). The complete mechanism of bradycardia is not defined but may be secondary to Etoposide deposition in the SA node as the main structure.

Platinum Agent These drugs are important agents or drugs used for cancer chemotherapy. Cisplatin, carboplatin, and oxaliplatin are neutral platinum structures with two amine ligands and two additional ligands that can be acquainted for further binding with DNA (Chen et al. 2013).

Carboplatin: Scientific research for the development of bradycardia or heart block when taking carboplatin is Insufficient and limited to a case report (Hartmann and Lipp 2003). and an important complication of the drug is kidney or liver toxicity, peripheral neuropathy or electrolyte disturbances such as hyponatremia. Meanwhile, another drug of the same family, namely cisplatin, can cause sinus bradycardia and heart block with comorbidities such as coronary artery spasms, electrolyte disorders, and autonomic system disorders. The incidence of bradyarrhythmia complications with this drug reaches 1% (Darling 2015).

Anti T Lymphocyte Antibody Therapies These drugs help to treat HSCT patients by manipulating the immune system.

Alemtuzumab: There are several case reports about severe bradycardia during drug infusion and sinus bradyarrhythmia incidence is about 1–10% (Buza et al. 2017; Fonseca et al. 2021; Nerrant et al. 2017).

Alemtuzumab infusion has an association with hypertension, atrial fibrillation, and ventricular arrhythmias (Buza et al. 2017; Tamargo et al. 2015).

Lymphocyte Immune Globulin: There aren't adequate reports about bradyarrhythmic effects during drug administration.

Antithymocyte globulin: A few case reports are available about transient severe bradycardia during ATG treatment (Elazhary and Alawyat 2022).

Anti-CD20 Monoclonal Antibodies These are used to attain B cell depletion, and at the start were developed to manage B cell proliferative disorders, including non-Hodgkin's lymphoma (NHL) and chronic lymphocytic leukemia (CLL).

Rituximab: There are published data about reversible and irreversible sinus bradycardia and AV block with rituximab infusion. The side effect mechanism could be the Inhibition of calcium channel attributes of the CD20 antigen on cardiac myocytes (Buza et al. 2017; Mariani et al. 2021; Fonseca et al. 2021; Torosoff et al. 2018; Ko Ko et al. 2020).

3. Thoracic radiation

Many patients with Hodgkin's lymphoma receive mediastinal radiation. All types of bradyarrhythmic syndromes, including, sick sinus syndrome, nodal and infranodal AV block, and right or left bundle branch block might be seen acutely or with more incidence, years or decades after thoracic radiation (Desai, et al. 2018). The underlying mechanism for this complication is inflammation resulting from reactive oxygen species after radiation exposure (Donnellan et al. 2019). Chronic and progressive fibrosis, endothelial dysfunction, stenosis of the sinus node and AV node arterioles, and compression from adjacent valvular calcification damage to the conduction system (Desai, et al. 2018; Donnellan et al. 2019). HSCT patients should be monitored for the occurrence of arrhythmic heart disorders caused by radiotherapy in the past. conduction system injuries in these patients, sometimes have transient and temporary clinical manifestations, and this is less noticed than other side effects caused by radiotherapies such as calcification of valves or progressive narrowing of heart vessels. In case of symptoms such as dizziness, syncope, or pre-syncope, and especially in the presence of bundle branch block, it is recommended to use long-term Holter monitoring, a loop recorder, or an electrophysiology study.

4. Autonomic System Dysfunction: One of the most important causes of bradyarrhythmia at the SAN or AVN level is the overactivity of the parasympathetic system. One of the most important characteristics of this disorder is the gradual decrease in electrical activity in the SAN and sometimes the AVN at the same time. Sometimes, before or after the attack of bradyarrhythmia, hyperactivity of the sympathetic system can be seen to some extent. Disturbances in the autonomic system may occur in the following ways:

5. Tumor invasion or compression on the carotid sinus body. In case of damage to the autonomic system in the neck or parasympathetic ganglia in other parts of the body, including areas close to the heart, the patient can show bradyarrhythmic syndrome (Toscano, et al. 2020).

6. The side effect of chemotherapy drugs: Chemotherapy can either directly cause more vagus stimulation, such as Alkylating agents or thalidomide drugs, or lead to hyperactivity of the parasympathetic system due to the occurrence of digestive side-effects and nausea and vomiting caused by the use of chemotherapy drugs. In such a context, transient sinus bradycardia or heart block can be seen (Fonseca et al. 2021).

7. The side effect of thoracic radiotherapy: Weakness caused by mediastinal radiotherapy acutely can cause hypervagotonia and sinus bradycardia, and damage to the cardiac parasympathetic ganglia causes sinus tachycardia in the long term (Desai, et al. 2018).

Interaction between cardiovascular drugs and anticancer agents: Many cancer patients with a history of heart disease or doctors' advice to use these drugs to protect the heart during chemotherapy use beta-blockers or other drugs that affect SAN or AVN. For this reason, the risk of drug interactions increases in these patients, and in different ways, drugs may show more side effects when taken together. These drug interactions may be seen at different stages such as absorption, distribution, metabolism, or elimination of another drug. It is very important to pay attention to the function of the liver, and kidney, and the absorption and excretion of drugs. Referring to drug database banks can provide good information to the doctor in these cases.

Diagnosis and Management:

When bradyarrhythmic syndrome occurs in patients, the following points should be considered:

1. Apart from the heart rate, the patient's symptoms are very important. Any high-risk clinical signs such as syncope or pre-syncope should be investigated with high importance. Due to the transient nature of some conduction system disorders, the patient may have a normal heart rate when visiting the doctor, but cardiac syncope can occur as a result of a severe and temporary decrease in heart rate due to trauma to the conduction system caused by drugs or radiotherapy. For example, a young HSCT patient with a history of radiotherapy in childhood suffers syncope during activity, which is probably due to damage in the bundle of His system. After resting, this person may show no evidence of a conduction system disorder on the ECG. Because in these patients, the intensification of sympathetic activity leads to heart block.

2. ECG recording or heart rhythm monitoring is very important at the time of brad-yarrhythmic syndrome. ECG and heart rhythm changes can reliably determine the center of the rhythm disorder. For example, long-term sinus node asystole in the context of hypervagotonia has a lower risk and a better prognosis compared to transient severe bradycardia caused by sick sinus syndrome (Fig. 13). The first case can be controlled when clinical symptoms appear with measures such as hydration, changing the dose or duration of chemotherapy drug administration, or atropine injection, but in the second case, depending on the conditions and type of underlying heart disease, pacemaker implantation should be considered. A cardiologist can use a heart rhythm Holter, loop recorder, or electrophysi-ology study for risk stratification. The use of chemotherapy drugs can turn an asymptomatic bradyarrhythmic syndrome into its symptomatic type, and for this reason, it is necessary to intervene by the medical team. During the consultation with the cardiologist for asymptomatic bradyarrhythmia, the oncologist should mention this point and consider the risk of aggravating the rhythm disorder due to the use of chemotherapy drugs.

3. In case of symptomatic bradyarrhythmia, it is recommended to use atropine or sympathomimetic drugs such as isoproterenol, dopamine, epinephrine, and norepinephrine (Ref. (McDonald et al. 2020)). But it should be kept in mind that in cases where the heart block is due to a disorder in the Purkinje system, the use of these drugs not only does not help to improve the patient's condition but there is a possibility of intensifying the heart block and reducing the heart rate to smaller amounts. For this reason, it is necessary for the medical personnel who are involved in the treatment of these patients, to be fully aware of the operation

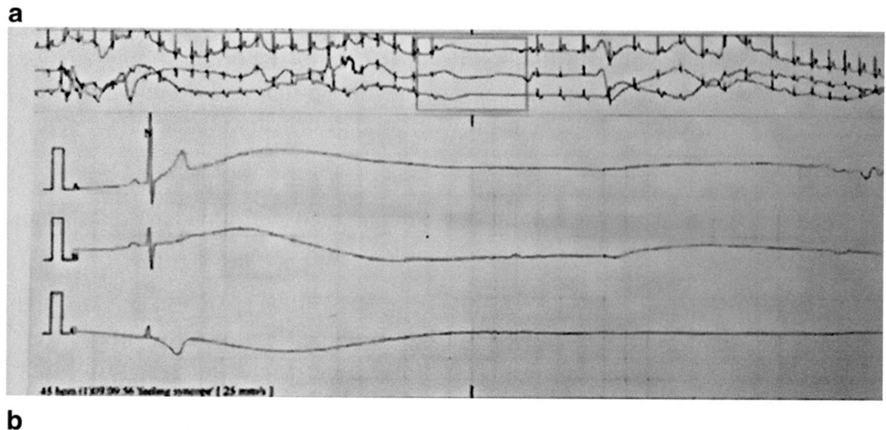

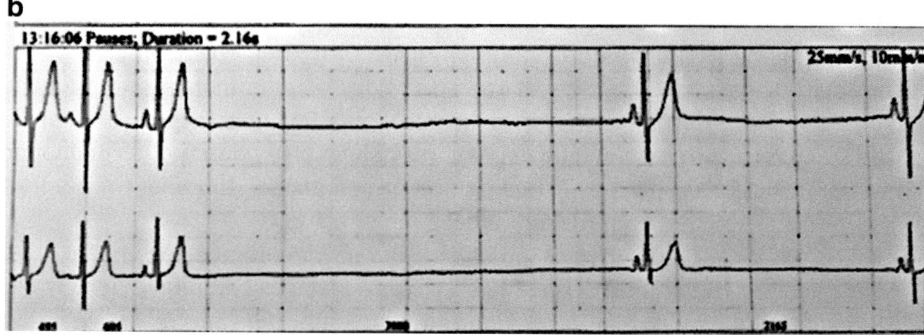

Fig. 13 Rhythm monitoring of 2 patients with syncopal attack. **a** Sinus tachycardia and gradual bradycardia and asystole, terminates after a long time and again converts to sinus tachycardia. These findings are compatible with vasovagal syncope. **B** Abrupt asystole after tachycardia termination, compatible with sick sinus syndrome

and use of the percutaneous pacemaker. Based on the clinical conditions, the cardiologist will implant a temporary or permanent pacemaker if necessary.

2 Conclusion

Today, the success rate and survival rate of HSCT patients have increased greatly, but some side effects, although small, can seriously endanger the lives of these patients. Bradyarrhythmic syndrome is one of these serious complications that can diagnose at any time from the onset of the cancer to chemotherapy, radiotherapy, bone marrow transplantation or years after the completion of the treatment process. Production and development of new drugs and paying attention to the electrophysiological characteristics of these drugs can increase the awareness of cardiologists and oncologists to the risk of this dysrhythmia. Unfortunately, despite all these advances, drug interactions, the emergence of new cardiac drugs, the administration of chemotherapy drugs in different doses and duration of injection, or the simultaneous use of several different chemotherapy drugs, can cause the occurrence or exacerbation of cardiac brady or tachyarrhythmias or abnormal increase in QT interval. It seems that the closer cooperation of oncologist and cardiologist and the use of new cardiac monitoring systems that allow for longer and more accurate heart monitoring can help to improve the conditions of HSCT patients.

Se.

References

Abu Rmilah AA, et al. Risk of QTc prolongation among cancer patients treated with tyrosine kinase inhibitors. Int J Cancer. 2020;147(11):3160–7.

Agarwal N, Burkart TA. Transient, high-grade atrioventricular block from high-dose cyclophosphamide. Tex Heart Inst J. 2013;40(5):626.

Alexandre J, et al. Anticancer drug-induced cardiac rhythm disorders: current knowledge and basic underlying mechanisms. Pharmacol Ther. 2018;189:89–103.

Armenian SH, et al. National institutes of health hematopoietic cell transplantation late effects initiative: the cardiovascular disease and associated risk factors working group report. Biol Blood Marrow Transplant. 2017;23(2):201–10.

Arun M, et al. The incidence of atrial fibrillation among patients with AL amyloidosis undergoing high-dose melphalan and stem cell transplantation: experience at a single institution. Bone Marrow Transplant. 2017;52(9):1349–51.

Ballou LM, Lin RZ, Cohen IS. Control of cardiac repolarization by phosphoinositide 3-kinase signaling to ion channels. Circ Res. 2015;116(1):127–37.

Baruteau A-E, et al. Evaluation and management of bradycardia in neonates and children. Eur J Pediatr. 2016;175:151–61.

Bekkers B, Denarié B, Bos G. Massive cardiac involvement in acute lymphatic leukemia. Heart. 2004;90(3):354.

Benekli M, Kars A, Guler N. Mitoxantrone-induced bradycardia. Ann Intern Med. 1997;126(5):409.

Blaes A, Konety S, Hurley P. Cardiovascular complications of hematopoietic stem cell transplantation. Curr Treat Options Cardiovasc Med. 2016;18:1–10.

Brugada J, et al. 2019 ESC guidelines for the management of patients with supraventricular tachycardia the task force for the management of patients with supraventricular tachycardia of the European society of Cardiology (ESC) developed in collaboration with the association for European paediatric and congenital Cardiology (AEPC). Eur Heart J. 2020;41(5):655–720.

Buza V, Rajagopalan B, Curtis AB. Cancer treatment–induced arrhythmias: focus on chemotherapy and targeted therapies. Circ: Arrhythmia Electrophysiol. 2017;10(8):e005443.

Cazin B et al. Cardiac complications after bone marrow transplantation. A report on a series of 63 consecutive transplantations. Cancer. 1986;57(10): 2061–69.

Chen X, et al. Platinum-based agents for individualized Cancer Treatmen. Curr Mol Med. 2013;13(10):1603–12.

Chen ST et al. Efficacy and safety of rivaroxaban versus warfarin in patients with non-valvular atrial fibrillation and a history of cancer: observations from ROCKET AF. Eur Heart J Qual Care Clin Outcomes. 2019;5(2):145–52.

Chiengthong K, et al. Arrhythmias in hematopoietic stem cell transplantation: a systematic review and meta-analysis. Eur J Haematol. 2019;103(6):564–72.

Chung-Lo W, et al. Fludarabine-induced bradycardia in a patient with refractory leukemia. Ann Saudi Med. 2010;30(3):246–7.

Cil T, et al. Cytosine-arabinoside induced bradycardia in patient with non-Hodgkin lymphoma: a case report. Leuk Lymphoma. 2007;48(6):1247–9.

Crisel RK, et al. First-degree atrioventricular block is associated with heart failure and death in persons with stable coronary artery disease: data from the Heart and Soul Study. Eur Heart J. 2011;32(15):1875–80.

D'Souza M, et al. CHA2DS2-VASc score and risk of thromboembolism and bleeding in patients with atrial fibrillation and recent cancer. Eur J Prev Cardiol. 2018;25(6):651–8.

Darling H. Cisplatin induced bradycardia. Int J Cardiol. 2015;182:304–6.

Desai MY et al. Radiation-associated cardiac disease: a practical approach to diagnosis and management. JACC: Cardiovasc imaging. 2018;11(8): 1132–149.

Dindogru A, et al. Electrocardiographic changes following adriamycin treatment. Med Pediatr Oncol. 1978;5(1):65–71.

Diwadkar S, Patel AA, Fradley MG. Bortezomib-induced complete heart block and myocardial scar: the potential role of cardiac biomarkers in monitoring cardiotoxicity. Case Rep Cardiol. 2016. 2016.

Donnellan E, Jellis CL, Griffin BP. Radiation-associated cardiac disease: from molecular mechanisms to clinical management. Curr Treat Options Cardiovasc Med. 2019;21:1–13.

Elazhary S, Alawyat HA. Bradycardia associated with antithymocyte globulin treatment of a pediatric patient with sickle cell disease: a case report and literature review. Hematol Transfus Cell Therapy. 2022;44:284–7.

Enriquez A et al. Increased incidence of ventricular arrhythmias in patients with advanced cancer and implantable cardioverter-defibrillators. JACC: Clin Electrophysiol 2017;3(1): 50–6.

Fanola CL, et al. Efficacy and safety of edoxaban in patients with active malignancy and atrial fibrillation: analysis of the ENGAGE AF-TIMI 48 trial. J Am Heart Assoc. 2018;7(16): e008987.

Farmakis D, Parissis J, Filippatos G. Insights into onco-cardiology: atrial fibrillation in cancer. J Am Coll Cardiol. 2014;63(10):945–53.

Feliz V, et al. Melphalan-induced supraventricular tachycardia: Incidence and risk factors. Clin Cardiol. 2011;34(6):356–9.

Fitzgerald JL, Howes LG. Drug interactions of direct-acting oral anticoagulants. Drug Saf. 2016;39:841–5.

Fonseca M, et al. Bradyarrhythmias in Cardio-Oncology. South Asian J Cancer. 2021;10(03):195–210.

Fradley MG, et al. Rates and risk of arrhythmias in cancer survivors with chemotherapy-induced cardiomyopathy compared with patients with other cardiomyopathies. Open Heart. 2017;4(2): e000701.

Fradley MG, et al. Recognition, prevention, and management of arrhythmias and autonomic disorders in cardio-oncology: a scientific statement from the American Heart Association. Circulation. 2021;144(3):e41–55.

Galetta F, et al. Effect of epirubicin-based chemotherapy and dexrazoxane supplementation on QT dispersion in non-Hodgkin lymphoma patients. Biomed Pharmacother. 2005;59(10):541–4.

Ganatra S et al. Ibrutinib-associated atrial fibrillation. JACC: Clin Electrophysiol. 2018;4(12): 1491–500

Gill D, et al. Rare cause of cardiotoxicity. Arch Med. 2017;9:1.

Gottdiener JS, et al. Cardiotoxicity associated with high-dose cyclophosphamide therapy. Arch Intern Med. 1981;141(6):758–63.

Group ESD. 2020 ESC guidelines for the diagnosis and management of atrial fibrillation developed in collaboration with the European Association of Cardio-Thoracic Surgery (EACTS). Eur Heart J. 2020.

Group ESD. 2022 ESC guidelines on cardio-oncology developed in collaboration with the European Hematology Association (EHA), the European Society for Therapeutic Radiology and Oncology (ESTRO) and the International Cardio-Oncology Society (IC-OS). European Heart J Cardiovasc Imaging. 2022a;23(10): E333–E465.

Group ESD. 2022 ESC guidelines for the management of patients with ventricular arrhythmias and the prevention of sudden cardiac death. Eur Heart J. 2022b. 43(40): 3997–4126.

Haghjoo M, et al. Effect of COVID-19 medications on corrected QT interval and induction of torsade de pointes: Results of a multicenter national survey. Int J Clin Pract. 2021;75(7): e14182.

Hartmann JT, Lipp H-P. Toxicity of platinum compounds. Expert Opin Pharmacother. 2003;4(6):889–901.

Herrmann J, et al. Defining cardiovascular toxicities of cancer therapies: an International Cardio-Oncology Society (IC-OS) consensus statement. Eur Heart J. 2022;43(4):280–99.

Hidalgo J, et al. Supraventricular tachyarrhythmias after hematopoietic stem cell transplantation: incidence, risk factors and outcomes. Bone Marrow Transplant. 2004;34(7):615–9.

Horacek J et al. Assessment of anthracycline-induced cardiotoxicity with electrocardiography. Exper Oncol. 2009.

Isogai T, et al. Procedural and short-term outcomes of percutaneous left atrial appendage closure in patients with cancer. Am J Cardiol. 2021;141:154–7.

Iwata N, et al. Aclarubicin-associated QTc prolongation and ventricular fibrillation 1, 2. Cancer Treat Rep. 1984;68(3):527.

Kilickap S, et al. Early and late arrhythmogenic effects of doxorubicin. South Med J. 2007;100(3):262–6.

Kishi S, et al. Torsade de pointes associated with hypokalemia after anthracycline treatment in a patient with acute lymphocytic leukemia. Int J Hematol. 2000;71(2):172–9.

Ko Ko NL, Minaskeian N, El Masry HZ. A case of irreversible bradycardia after rituximab therapy for diffuse large B-cell lymphoma. Cardio-Oncology. 2020;6:1–4.

Kupari M, et al. Cardiac involvement in bone marrow transplantation: electrocardiographic changes, arrhythmias, heart failure and autopsy findings. Bone Marrow Transplant. 1990;5(2):91–8.

Kusumoto FM, et al. 2018 ACC/AHA/HRS guideline on the evaluation and management of patients with bradycardia and cardiac conduction delay: a report of the American College of Cardiology/American Heart Association Task Force on Clinical Practice Guidelines and the Heart Rhythm Society. J Am Coll Cardiol. 2019;74(7):e51–156.

Lal KS, Tariq RZ, Okwuosa T. Haemodynamic instability secondary to cardiac involvement by lymphoma. Case Rep 2016;2016:bcr2016215775.

Lee AY, et al. Low-molecular-weight heparin versus a coumarin for the prevention of recurrent venous thromboembolism in patients with cancer. N Engl J Med. 2003;349(2):146–53.

Lopez-Fernandez T, et al. Atrial fibrillation in active cancer patients: expert position paper and recommendations. Revista Española De Cardiología (english Edn). 2019;72(9):749–59.

Ma L-L, et al. A case report: high dose melphalan as a conditioning regimen for multiple myeloma induces sinus arrest. Cardio-Oncology. 2020;6(1):1–4.

Majhail NS, et al. Prevalence of hematopoietic cell transplant survivors in the United States. Biol Blood Marrow Transplant. 2013;19(10):1498–501.

Mandal PK, Kumar M, Bhattyacharyya M. A rare cause of bifascicular block: Daunorubicin induced cardiotoxicity. Archives of Medicine and Health Sciences. 2016;4(1):64–6.

Marcath LA, et al. Comparison of nine tools for screening drug-drug interactions of oral oncolytics. J Oncol Practice. 2018;14(6):e368–74.

Mariani MV, et al. Direct oral anticoagulants versus vitamin K antagonists in patients with atrial fibrillation and cancer a meta-analysis. J Thromb Thrombolysis. 2021;51:419–29.

Mazur M et al. Burden of cardiac arrhythmias in patients with anthracycline-related cardiomyopathy. JACC: Clinic Electrophysiol. 2017;3(2):139–50.

McDonald GB, et al. Survival, nonrelapse mortality, and relapse-related mortality after allogeneic hematopoietic cell transplantation: comparing 2003–2007 versus 2013–2017 cohorts. Ann Intern Med. 2020;172(4):229–39.

Melloni C et al. Efficacy and safety of apixaban versus warfarin in patients with atrial fibrillation and a history of cancer: insights from the ARISTOTLE trial. Am J Med. 2017;130(12) 1440–448. e1.

Mileshkin L, et al. Cardiovascular toxicity is increased, but manageable, during high-dose chemotherapy and autologous peripheral blood stem cell transplantation for patients aged 60 years and older. Leuk Lymphoma. 2005;46(11):1575–9.

Morandi P, et al. Cardiac toxicity of high-dose chemotherapy. Bone Marrow Transplant. 2005;35(4):323–34.

Murdych T, Weisdorf D. Serious cardiac complications during bone marrow transplantation at the University of Minnesota, 1977–1997. Bone Marrow Transplant. 2001;28(3):283–7.

Nerrant E, Thouvenot E, Castelnovo G. Severe bradycardia: An unreported adverse infusion-associated reaction (IAR) with alemtuzumab. Revue Neurologique. 2017;173(3):175–6.

Olivieri A, et al. Paroxysmal atrial fibrillation after high-dose melphalan in five patients autotransplanted with blood progenitor cells. Bone Marrow Transplant. 1998;21(10):1049–53.

Peres E, et al. Cardiac complications in patients undergoing a reduced-intensity conditioning hematopoietic stem cell transplantation. Bone Marrow Transplant. 2010;45(1):149–52.

Porta-Sánchez A, et al. Incidence, diagnosis, and management of QT prolongation induced by cancer therapies: a systematic review. J Am Heart Assoc. 2017;6(12): e007724.

Quezado ZM, et al. High-dose ifosfamide is associated with severe, reversible cardiac dysfunction. Ann Intern Med. 1993;118(1):31–6.

Ramireddy K, Kane KM, Adhar GC. Acquired episodic complete heart block after high-dose chemotherapy with cyclophosphamide and thiotepa. Am Heart J. 1994;127(3):701–4.

Schwartz PJ, et al. Diagnostic criteria for the long QT syndrome. An update. Circulation. 1993;88(2):782–4.

Siegel D, et al. Integrated safety profile of single-agent carfilzomib: experience from 526 patients enrolled in 4 phase II clinical studies. Haematologica. 2013;98(11):1753.

Singh JP, et al. Association of cardiac resynchronization therapy with change in left ventricular ejection fraction in patients with chemotherapy-induced cardiomyopathy. JAMA. 2019;322(18):1799–805.

Singla A, et al. Incidence of supraventricular arrhythmias during autologous peripheral blood stem cell transplantation. Biol Blood Marrow Transplant. 2013;19(8):1233–7.

Soignet SL, et al. United States multicenter study of arsenic trioxide in relapsed acute promyelocytic leukemia. J Clin Oncol. 2001;19(18):3852–60.

Solh MM, et al. Long term survival among patients who are disease free at 1-year post allogeneic hematopoietic cell transplantation: a single center analysis of 389 consecutive patients. Bone Marrow Transplant. 2018;53(5):576–83.

Sorror ML, et al. Hematopoietic cell transplantation (HCT)-specific comorbidity index: a new tool for risk assessment before allogeneic HCT. Blood. 2005;106(8):2912–9.

Steinberg JS, et al. Acute arrhythmogenicity of doxorubicin administration. Cancer. 1987;60(6):1213–8.

Stewart S, et al. A population-based study of the long-term risks associated with atrial fibrillation: 20-year follow-up of the Renfrew/Paisley study. Am J Med. 2002;113(5):359–64.

Sureddi RK, et al. Atrial fibrillation following autologous stem cell transplantation in patients with multiple myeloma: incidence and risk factors. Ther Adv Cardiovasc Dis. 2012;6(6):229–36.

Tamargo J, Caballero R, Delpón E. Cancer chemotherapy and cardiac arrhythmias: a review. Drug Saf. 2015;38:129–52.

Tonorezos ES, et al. Arrhythmias in the setting of hematopoietic cell transplants. Bone Marrow Transplant. 2015;50(9):1212–6.

Torosoff M, et al. Resolution of sinus bradycardia, high-grade heart block, and left ventricular systolic dysfunction with rituximab therapy in Henoch-Schonlein purpura. Intern Med J. 2018;48(7):868–71.

Toscano M et al. Carotid sinus syndrome in a patient with head and neck cancer: a case report. Cureus. 2020;12(2)

Tuzovic M, et al. Cardiac complications in the adult bone marrow transplant patient. Curr Oncol Rep. 2019;21:1–16.

Ulrickson M, et al. Busulfan and cyclophosphamide (Bu/Cy) as a preparative regimen for autologous stem cell transplantation in patients with non-Hodgkin lymphoma: a single-institution experience. Biol Blood Marrow Transplant. 2009;15(11):1447–54.

Usnarska-Zubkiewicz L, et al. Effect of epirubicin on the heart conduction system in patients with Hodgkin's disease. Pol Arch Med Wewn. 1992;87(3):173–82.

Wolf PA, Abbott RD, Kannel WB. Atrial fibrillation as an independent risk factor for stroke: the Framingham Study. Stroke. 1991;22(8): 983–88.

Wortman JE, et al. Sudden death during doxorubicin administration. Cancer. 1979;44(5):1588–91.

Yamreudeewong W, et al. Potentially significant drug interactions of class III antiarrhythmic drugs. Drug Saf. 2003;26:421–38.

Coronary Artery Disease in HSCT

Seyed Mohsen Razavi, Mohammad Sarraf, Ata Firouzi, Sayeh Parkhideh, Masoud Sayad, and Azam Yalameh

Abstract A curative possibility for certain people with hematologic malignancies and sporadically for others with benign hematologic diseases is hematopoietic stem cell transplantation (HSCT). The advancement of HSCT techniques and supportive care has enabled a growing number of people to survive long-term. Nevertheless, these survivors are more likely to develop chronic, debilitating illnesses over the long term, such as early cardiovascular disease. Patients with cancer are more likely to experience side effects than the general population, and there is a strong correlation between traditional cardiovascular risk factors, and the risk of developing cardiovascular diseases. The HSCT can be divided into two categories: autologous and allogeneic. After allogenic HSCT, late adverse effects are more common. The occurrence of late-occurring vascular events, including atherosclerosis, myocardial infarction, cerebrovascular accidents, peripheral arterial disease, vasculitis, or cardiac events, including left ventricular dysfunction, myocarditis, valvular heart disease, conduction disorders, and pericardial diseases. These issues frequently arise earlier than would be anticipated. All patients should be screened with clinical history and diagnostic tools by a cardiologist and preferably a cardio-oncologist specialist. Conventional cardiovascular (CV) risk factors should be managed thoroughly pre-HSCT.

S. M. Razavi
Hematology and Oncology Department, Firoozgar Hospital, Iran University of Medical Sciences, Tehran, Iran

M. Sarraf
Department of Cardiovascular Medicine, Mayo Clinic, Rochester, MI, USA

A. Firouzi
Cardiovascular Intervention Research Center, Rajaie Cardiovascular Medical and Research Center, Iran University of Medical Sciences, Tehran, Iran, Tehran, Iran

S. Parkhideh
Hematopoietic Stem Cell Research Center, Shahid Beheshti University of Medical Sciences, Tehran, Iran

M. Sayad (✉) · A. Yalameh
Cardio-Oncology Research Center, Rajaie Cardiovascular Medical and Research Center, Iran University of Medical Sciences, Tehran, Iran
e-mail: msdsayad@gmail.com

© The Author(s), under exclusive license to Springer Nature Switzerland AG 2024   173

A. Alizadehasl et al. (eds.), *Cardiovascular Considerations in Hematopoietic Stem Cell Transplantation*, https://doi.org/10.1007/978-3-031-53659-5_12

Keywords Allogeneic HSCT · Atherosclerosis · Autologous HSCT · Coronary artery disease · Hematopoietic stem cell transplantation

Abbreviations

CAC score	Coronary artery calcium score
CAD	Coronary artery disease
CT	Computed tumography
CTA	CT angiography
CV	Cardiovascular
CVA	Cerebrovascular accidents
DM	Diabetes mellitus
ECG	Electrocardigram
FRS	Framingham Risk Score
GvHD	Graft versus host disease
HSCT	Hematopoietic stem cell transplantation
HTN	Hypertension
IHD	Ischemic heart disease
LV	Left ventricular
MI	Myocardial infarction
MRI	Magnetic resonance imaging
PAD	Peripheral arterial disease
RT	Radiation therapy
TBI	Total body irradiation

1 Coronary Artery Disease

1.1 Definition

Hematopoietic stem cell transplantation (HSCT) can be used for treatment of hematologic malignancies, bone marrow failure syndromes, hemoglobinopathies, immunodeficiencies, autoimmune diseases, and some solid cancers. After high-dose chemotherapy or radiation therapy, autologous or allogeneic hematopoietic cells that have been previously gathered are injected. Because of advancements in HSCT techniques and supportive care, the survival of HSCT recipients has dramatically increased (Bhatia et al. 2007; Weisdorf et al. 2005). The chance of late complications, such as CV diseases, increases as individuals live longer (Armenian et al. 2017). HSCT is classified according to the source of infused hematopoietic cells. The mobilization, collection, and cryopreservation of patient's own hematopoietic cells are the components of autologous HSCT. Following the administration of high-dose chemotherapy or radiation therapy, these cells are then reinfused as "rescue" hematopoiesis cells. After administering high doses of chemotherapy or radiation, allogeneic HSCT involves collecting hematopoietic cells from a healthy individual (the "donor") and then injecting them into the patient (the "recipient").

Numerous patients being considered for HSCT have recently received cardiotoxic disease treatment (ie, anthracyclines, cyclophosphamide, chest radiation) and are at expanded risk for cardiac dysfunction. Patients over the age of 60 are more likely to undergo HSCT because of improvements in HCT techniques over the past decade (Majhail et al. 2013; Mitchell et al. 2022). These older patients are also more likely to have cardiovascular comorbidities like HTN, hyperlipidemia, arteriosclerosis, and diabetes. Some HSCT indications, such as light chain amyloidosis, thalassemia, and systemic sclerosis, are associated with cardiovascular disease.

Each patient should undergo a 12-lead ECG, a chest x-ray, a transthoracic echocardiogram, and a clinical history. Patients with high-risk features (signs or symptoms of angina) should be referred to a cardio-oncologist for further evaluation and risk factor modification (Mitchell et al. 2022).

Atherosclerosis is an inflammatory process and endothelial lesions happen a long time before clinical signs like stroke, coronary artery disease (CAD) or peripheral arterial disease (PAD). CAD is one of the major CV diseases affecting the worldwide populace and is a major cause of mortality in both developed and developing countries (Tichelli et al. 2007; Geng et al. 2016; Hansson 2005; Stoll and Bendszus 2006; Armenian and Chow 2014). This chapter summarizes the principles of HSCT and CAD as a complication.

1.2   Diagnosis

- The Framingham Risk Score (FRS), can be used to determine an individual's risk of developing CAD over the course of ten-year follow up, classifying them as low(10% risk), intermediate(10–20% risk), or high(>20%), is frequently used to guide primary prevention of CAD in the general population. It is important to note that, while some subgroups are at a much higher risk, the average risk of CAD among allogeneic HSCT patients is similar to that of persons in the general population with low-intermediate risk (Tichelli et al. 2007, 2008; Armenian and Chow 2014; Chow et al. 2011; Armenian et al. 2012; Greenland et al. 2010). HSCT survivors appear to develop CAD on average earlier than the normal population (Armenian et al. 2012).

  The gold standard for determining CAD is conventional coronary angiography. Although coronary angiography shows the exact size of the vessels and the degree of stenosis with acceptable accuracy, it is still an invasive method and therefore non-invasive methods are used especially in asymptomatic patients (Voros, et al. 2011; Woods et al. 2012).

Computed tomography (CT) scan: CT-imaging, such as CAC score and CT angiography, is a precise, non-invasive modality for diagnosis of CAD for intermediate risk patients in general population. Each plaque is visually evaluated and classified according to the degree of stenosis, and is either calcified, thrombotic or mixed in those who are at lower risk (Voros, et al. 2011; Woods et al. 2012). This non-calcified plaque results in ulceration and plaque disruption, as opposed to calcified plaque, which forces the vessel wall to outward remodeling (Stoll and Bendszus 2006).

CTA enables evaluation of plaque morphology and observation of ulcerated lesions. Significant coronary lesions may still exist in the absence of calcification, despite the fact that the degree of coronary artery calcification offers useful predictive information on the risk of CAD (Hansson 2005; Stoll and Bendszus 2006). Coronary artery calcium or CTA has not been thoroughly researched for its effectiveness in detecting coronary atherosclerosis in cancer patients who are prone highly to CAD. Data limitations has hindered the regular use of screening methods for those patients who are most vulnerable to CAD. So, we need more studies for better screening in this population.

- Because they are aware of the constraints in resources and the accessibility of diagnostic tools among treatment facilities, the ACC/AHA have devised alternate screening measures (Greenland et al. 2010) for asymptomatic patients who are at risk for CAD. These recommendations take into account the individual's FRS category also any comorbidities which can change their total risk. The predictive value of many of these demonstrative tools in HCT recipients is unknown, and the role of these tests in CAD in these patients still needs to be fully established, similar to CT-based imaging. If HSCT may be safely postponed, pre-HSCT stress testing and/or coronary catheterization are advised for patients with unstable angina to enable dual antiplatelet therapy after coronary revascularization (Mitchell et al. 2022).

Blood biomarkers: In a wide age range and across a variety of ethnic groups, elevated high-sensitivity C-reactive protein (hs-CRP) levels are linked to a higher risk of coronary artery disease (CAD). While traditional risk variables like HTN, DM, obesity, or inflammation (such as infection or GvHD) can partially impact the predictive value of hs-CRP, it has an important role in detection of patients at the lower risk for CAD (Madjid and Willerson 2011). People have been classified as "very low risk" when their Hs-CRP levels are less than 1 ng/ml and their CAC score is less than 100 (Yeboah et al. 2012; Erbel et al. 2010). The etiopathogenesis of vascular injury in HCT survivors may be better understood by using different endothelial injury indicators, such as vWF.

In individuals with GvHD, vascular endothelial cells have been thought to be a key target for alloreactive cytotoxic T-lymphocytes. Concentrates in small amounts of HSCT patients have demonstrated that constant GvHD can cause microvessel injury and reduced microvessel thickness, alterations that are related to increased vWF from endothelial cells (Biedermann et al. 2002; Armenian and Bhatia 2008).

1.3 Symptoms

Patients with IHD present with a wide range of symptoms. In most cases of chronic (stable) angina, unstable angina, microvascular angina, and acute MI, chest pain is the most common symptom. However, signs of IHD additionally happen in which chest pain is silent or not significant, like heart failure, asymptomatic (silent) myocardial ischemia, cardiac arrhythmias, and sudden death. Notably, IHD may also present with anginal equivalents like epigastric pain, dyspnea, effort intolerance, excessive fatigue, and other symptoms that are more common in women, older people, and diabetics (Armenian and Bhatia 2008).

CV side effects ought to be completely investigate with cardiovascular biomarkers, transthoracic echocardiography, Holter monitoring, or potentially stress testing as clinically showed (Mitchell et al. 2022).

1.4 Epidemiology

- When compared to the general population, these patients have a four times higher risk of developing CV disease, that typically begins 14 years earlier. The incidence of vascular diseases such as significant CAD or CVA in allogeneic HSCT patients is 10% after 15 years, and this risk increased 20% after 20 years (Tichelli et al. 2007, 2008a, b; Armenian and Chow 2014; Majhail et al. 2012). In this population, the median age at first MI is as low as 53 years (range 35–66 years), much earlier than would be expected for the general population (67 years) (Greenland et al. 2010) in comparison to autologous HSCT survivors (61 years) (Armenian et al. 2010, 2012).
- Tichelli, A. et al., revealed in a single-center study, based on 265 long-term HSCT survivors, that cumulative incidence of arterial events after allogeneic HSCT, such as CVA, CAD, and PAD is 22% at 25 years (Tichelli et al. 2007).
- An arterial event was linked to allogeneic HSCT and the incidence of more than one conventional CV risk factors in multivariate analysis (Tichelli et al. 2007).

1.5 Etiology and Risk Factors

Conventional CV risk factors, allogenic HSCT, radiotherapy and cardiotoxic chemotherapy agents are all risk factors for vascular events in HSCT survivors (Mitchell et al. 2022; Leger et al. 2018). HSCT recipients have higher rates of all CV risk factors than the general population. As mentioned, HSCT survivors have a high risk of CV events, the process of atherosclerosis accelerates by exposure to ionizing radiation (Armenian and Chow 2014). So, there are additional non-conventional CV risk factors (chemotherapy and/or radiation-therapy side effects) in these patients which caused late CV complications.

- Total body radiation used in HSCT conditioning regimens damages endothelial cells, promotes their proliferation, thickens the intima, creates medial scarring, lipid deposits, and induces adventitial fibrosis (Lee and Mallik 2005), all of which increase the risk of arterial vascular disease (CAD, peripheral vascular disease, and cerebrovascular disease). Armenian., et al., revealed patients who received chest irradiation prior to transplantation had a 9.5-fold increased incidence of CAD (Zhao et al. 2022). Coronary CTA revealed coronary artery damage in 12 of 31 (39%) adult survivors of childhood Hodgkin lymphoma who had received RT (30 Gy in 48% of the recipients) prior to transplantation (Mulrooney et al. 2014). Conditioning with TBI is related with a higher incidence of hyperlipidemia and DM in both juvenile and adult HSCT recipients. How TBI increases the likelihood of these CV risk factors is unclear. The combined effects of abdominal radiation and post-HSCT gonadal dysfunction may contribute to the elevated risk of DM and hyperlipidemia in HSCT patients who underwent TBI (Armenian and Chow 2014).

- It is important to keep in mind that HSCT performed after radiation conditioning could result in excessive iron accumulation from pack cell transfusions, which could lead to cardiomyopathy through the production of free radicals and reactive oxygen species (ROS) (Zhao et al. 2022). Patients who recieved abdominal radiation for pancreatic or hepatic cancers developed insulin resistance and metabolic syndrome, that could be radiation-induced (Armenian and Chow 2014).

- Long-term immunosuppression with calcineurin inhibitors, corticosteroids, and/or mTOR inhibitors is required for GVHD patients. Hyperlipidemia, HTN, and diabetes mellitus, all of which are risk factors for cardiovascular disease, are side effects of these medications. Drugs for management of GvHD also increase the risk of CV risk factors. Dyslipidemia reported in up to 80% of solid organ transplantation recipients who is on immunosuppressive drugs, and insulin resistance and hypertension are much of the time experienced results of GvHD drugs like cyclosporine, tacrolimus and steroids. History of GvHD has related with higher risk for every one of the three CV risk factors (hypertension: $RR = 9.1, p < 0.01$; diabetes: $RR = 5.8$; dyslipidemia: $p < 0.01$).

  Von-Willebrand Factor (vWF), a biomarker of endothelial damage, exhibits a strong correlation with chronic GvHD, indicating that an immunologic pathway may be associated with progression of atherosclerosis in HSCT recipients (Armenian and Chow 2014). When combined, these data serve as the foundation for screening methods for periodic monitoring as well as more restricted CV risk factor management. Screening strategies for targeted surveillance are based on these findings.

- The risk of arterial disease is most noteworthy in whom that have different CV risk factors (total occurrence [CI]: 10–12% at 10 years); prior to the HSCT, exposure to cardiotoxic therapies, such as chest radiation, raises this risk to 15% 10 years after the HSCT.

  Although chronic "cardiac" GvHD is uncommon and difficult to define, it has significant indirect CV effects (Armenian and Chow 2014).

- The cardiovascular adverse effects of all immunosuppressant medications used to treat chronic GvHD predispose to adverse cardiovascular events. Patients with chronic GvHD have been found to have an increased risk of arterial vascular disease, including CAD, peripheral vascular disease, and cerebrovascular disease, as well as impaired diastolic function and increased LV mass (Mitchell et al. 2022).
- Patients with chronic GVHD require aggressive CV risk factor modification and prompt investigation of symptoms indicative of arterial vascular disease.
- When compared to autologous HCT recipients and those with similar age and sex in the general population, allogeneic HSCT recipients was prone to developing CV risk factors like HTN, DM, and hyperlipidemia. According to a recent review companion study (Armenian and Chow 2014), the cumulative incidence of HTN, DM, and hyperlipidemia over ten years was 37.7%, 18.1%, and 46.7%, respectively, in allogeneic HSCT patients. The risk of additional (2) CV risk factors estimated 40% (compared to 26% in autologous HSCT patients).
- Clonal hematopoiesis of indeterminate potential is a specific property that has recently been proved as a cardiovascular risk factor after HSCT. Clonal hematopoiesis, which begins with a single, undeveloped hematopoietic cell, is the extension of platelets and is related with the improvement of myeloid neoplasms and cardiovascular disease in the general population. In this condition, if it coincides with systemic inflammation, which is likely, the risk of cardiovascular disease increases (Goldberg et al. 1998; Versluys et al. 2018).
- Several important risk factors that contribute to HSCT-related cardiovascular problems include: (1) Chest radiation or anthracycline exposure prior to HSCT; (2) a large dosage of cyclophosphamide as part of the conditioning regimen; and (3) cyclophosphamide treatment of GvHD following transplantation (4) co-occurring conditions including HTN, diabetes, dyslipidemia, or GvHD; (5) Additional drugs that were taken before transplantation for the management of the underlying disease, during transplant or conditioning, or after transplantation for maintenance, such as Tyrosine-kinase inhibitors and corticosteroids. Based on 1828 HSCT patients who lived for more than a year, Armenian et al. developed a risk prediction model for cardiovascular disease (Zhao et al. 2022). Age, anthracycline cumulative dose, chest irradiation, HTN, DM,smoking and other factors were taken into consideration when calculating risk scores.

The cumulative incidences of heart failure and CAD over a 10-year period were significantly correlated with the risk scores. Patients who are at low risk (three risk factors or less), at intermediate risk (four to six risk factors), and at high risk (six or more risk factors) have 10-year cumulative rates of CV disease of 3.7, 9.9, and 26.2% (Zhao et al. 2022).

Despite of etiologic factors implicated are still controversial, immunological mechanism appear to be involved in the progression of atherosclerosis (Tichelli et al. 2007).

2 Treatment

At present, there is no distinct strategy for observing long-term survivors for CV complications. Efforts should focus on prevention and early treatment of CV risk factors (Armenian et al. 2012; Madjid and Willerson 2011).

Routine clinical evaluation for all HSCT survivors is suggested annually. It is necessary to implement aggressive CV risk factor modification (smoking cessation, controlling blood pressure, managing lipids, weight control, glycemic control, and therapeutic lifestyle modification) (Mitchell et al. 2022).

Counseling and education about healthy diet and regular exercise should be provided to patients (Mitchell et al. 2022).

- During treatment, ASCO advises routine cardiac surveillance and frequent patient monitoring. The American Society of Echocardiography (ASE) and the European Association of Cardiovascular Imaging (EACVI) also recommend that patients who received anthracycline have their cardiac function evaluated prior to, at the end of, and six months following the transplant. Five to 10 years following transplantation, an cchocardiography and coronary artery function test should be done based on the risk factor prediction. If the patient developed cardiac symptoms, such as shortness of breath, prompt evaluation should be done. Additionally, patients with a history of chest radiation prior to HSCT should be visited by a cardiologist annually, and also patients with symptomatic heart failure or those at high risk for cardiovascular complications should consider consulting a cardiologist (Zhao et al. 2022).
- This risk prediction model will therefore be used as a guide to choose the most appropriate conditioning regimens, preventative therapies, and LV function monitoring for HSCT patients throughout the transplant procedure. For long-term follow-up in HSCT survivors, it is recommended to screen by biomarkers and imaging modalities such as CAC score by cardiac MRI, cardiac CT scan, echocardiography, lipid profiles, HbA1c, and BNP/NT-proBNP levels every two years (for high risk patients) and every five years (for intermediate risk patients), and every ten years (for low risk patients) (Zhao et al. 2022).
- Seven international bone marrow transplant organizations produced a useful guideline and for managing long-term complications in HSCT patients according to retrospective studies and expert opinion. A thorough clinical evaluation for cardiovascular risk factors, including a fasting lipid profile and fasting blood sugar, was recommended for all HSCT survivors one year after transplantation and every year after that. Prior to transplant, the risk of anthracycline cardiotoxicity or radiation-induced cardiotoxicity should be carefully tracked and assessed, and if necessary, an echocardiography screening should be carried out. Traditional cardiovascular risk factors include DM, HTN and hyperlipidemia, all of which need to be promptly managed (Zhao et al. 2022).
- The best primary and secondary prevention measures should be taken for patients who have accelerated atherosclerosis or at risk of vasculotoxicity or who are at high

risk of developing it (for example, as a result of chest radiation exposure, BCR-ABL inhibitors, and cisplatin treatment). Even if hyperlipidemia is not present, these include high-dose statins (Herrmann and Cardio-Oncology, 2022). During HSCT, statin therapy should be administered to all patients with known coronary artery disease (Mitchell et al. 2022).

References

Armenian SH, Bhatia S. Cardiovascular disease after hematopoietic cell transplantation–lessons learned. Haematologica. 2008;93(8):1132–6.

Armenian SH, Chow EJ. Cardiovascular disease in survivors of hematopoietic cell transplantation. Cancer. 2014;120(4):469–79.

Armenian SH, et al. Predictors of late cardiovascular complications in survivors of hematopoietic cell transplantation. Biol Blood Marrow Transplant. 2010;16(8):1138–44.

Armenian SH, et al. Cardiovascular risk factors in hematopoietic cell transplantation survivors: role in development of subsequent cardiovascular disease. Blood J Am Soc Hematol. 2012;120(23):4505–12.

Armenian SH, et al. National institutes of health hematopoietic cell transplantation late effects initiative: the cardiovascular disease and associated risk factors working group report. Biol Blood Marrow Transplant. 2017;23(2):201–10.

Bhatia S, et al. Late mortality after allogeneic hematopoietic cell transplantation and functional status of long-term survivors: report from the Bone Marrow Transplant Survivor Study. Blood J Am Soc Hematol. 2007;110(10):3784–92.

Biedermann BC, et al. Endothelial injury mediated by cytotoxic T lymphocytes and loss of microvessels in chronic graft versus host disease. The Lancet. 2002;359(9323):2078–83.

Chow EJ, et al. Cardiovascular hospitalizations and mortality among recipients of hematopoietic stem cell transplantation. Ann Intern Med. 2011;155(1):21–32.

Erbel R, et al. Coronary risk stratification, discrimination, and reclassification improvement based on quantification of subclinical coronary atherosclerosis: the Heinz Nixdorf Recall study. J Am Coll Cardiol. 2010;56(17):1397–406.

Geng H-H, et al. The relationship between C-reactive protein level and discharge outcome in patients with acute ischemic stroke. Int J Environ Res Public Health. 2016;13(7):636.

Goldberg SL, et al. Value of the pretransplant evaluation in predicting toxic day-100 mortality among blood stem-cell and bone marrow transplant recipients. J Clin Oncol. 1998;16(12):3796–802.

Greenland P, et al. 2010 ACCF/AHA guideline for assessment of cardiovascular risk in asymptomatic adults: a report of the American College of Cardiology Foundation/American Heart Association task force on practice guidelines developed in collaboration with the American Society of Echocardiography, American Society of Nuclear Cardiology, Society of Atherosclerosis Imaging and Prevention, Society for Cardiovascular Angiography and Interventions, Society of Cardiovascular Computed Tomography, and Society for Cardiovascular Magnetic Resonance. J Am Coll Cardiol. 2010;56(25):e50–103.

Hansson GK. Inflammation, atherosclerosis, and coronary artery disease. N Engl J Med. 2005;352(16):1685–95.

Herrmann J. Cardio-oncology practice manual: a companion to Braunwald's heart disease e-book. Elsevier Health Sciences; 2022.

Lee PJ, Mallik R. Cardiovascular effects of radiation therapy: practical approach to radiation therapy-induced heart disease. Cardiol Rev. 2005;13(2):80–6.

Leger KJ, et al. Lifestyle factors and subsequent ischemic heart disease risk after hematopoietic cell transplantation. Cancer. 2018;124(7):1507–15.

Madjid, M, Willerson JY. Inflammatory markers in coronary heart disease. Br Med Bull. 2011;100(1).

Majhail NS, et al. Prevalence of hematopoietic cell transplant survivors in the United States. Biol Blood Marrow Transplant. 2013;19(10):1498–501.

Majhail N et al., Sociedade Brasileira de Transplante de Medula Ossea. Recommended screening and preventive practices for long-term survivors after hematopoietic cell transplantation. Biol Blood Marrow Transplant. 2012;47(3):348–71.

Mitchell JD, Lenihan DJ. Cardio-oncology: essentials for effective consultation. In: The Washington Manual Cardiology Subspecialty Consult, 2022.

Mulrooney DA, et al. Coronary artery disease detected by coronary computed tomography angiography in adult survivors of childhood Hodgkin lymphoma. Cancer. 2014;120(22):3536–44.

Stoll G, Bendszus M. Inflammation and atherosclerosis: novel insights into plaque formation and destabilization. Stroke. 2006;37(7):1923–32.

Tichelli A, et al. Premature cardiovascular disease after allogeneic hematopoietic stem-cell transplantation. Blood J Am Soc Hematol. 2007;110(9):3463–71.

Tichelli A, Bhatia S, Socié G. Cardiac and cardiovascular consequences after haematopoietic stem cell transplantation. Br J Haematol. 2008;142(1):11–26.

Tichelli A et al. Late cardiovascular events after allogeneic hematopoietic stem cell transplantation: a retrospective multicenter study of the Late Effects Working Party of the European Group for Blood and Marrow Transplantation. Hhaematologica 2008;93(8):1203–210.

Versluys A, et al. Predictors and outcome of pericardial effusion after hematopoietic stem cell transplantation in children. Pediatr Cardiol. 2018;39:236–44.

Voros S et al. Coronary atherosclerosis imaging by coronary CT angiography: current status, correlation with intravascular interrogation and meta-analysis. JACC: Cardiovasc Imag 2011;4(5): 537–48.

Weisdorf DJ et al. Late mortality in survivors of autologous hematopoietic cell. 2005.

Woods KM, et al. The prognostic significance of coronary CT angiography. Curr Cardiol Rep. 2012;14:7–16.

Yeboah J, et al. Comparison of novel risk markers for improvement in cardiovascular risk assessment in intermediate-risk individuals. JAMA. 2012;308(8):788–95.

Zhao Y, et al. Cardiovascular complications in hematopoietic stem cell transplanted patients. J Personal Med. 2022;12(11):1797.

Peripheral Arterial Disease in HSCT

Behrooz Najafi, Bahram Mohebbi, Jamal Moosavi, Parisa Firoozbakhsh, Negar Dokhani, and Mohammad Dabiri

Abstract Arterial diseases, mainly presenting as coronary, cerebrovascular, and peripheral artery diseases, are a group of late-onset complications of stem cell transplantation which are mostly attributed to atherosclerosis. Besides conventional risk factors, there are some transplantation-related factors that predispose recipients of allogeneic HST to atherosclerosis, including endothelial damage caused by conditioning regimen ± total body irradiation or graft-versus-host disease. Allogeneic hematopoietic stem cell transplant is accompanied with an increased incidence and intensity of conventional cardiovascular risk factors, including hypertension, diabetes, and dyslipidemia, eventually leading to increased risk of arterial disorder. Detailed pathogenesis of arterial disease following HST and proper screening and preventive strategies are presented in the current chapter.

Keywords Cancer · Chemotherapy · Radiation · Vascular side effect

Abbreviations

CV	Cardiovascular
CVRF	Cardiovascular risk factor
DLP	Dyslipidemia

B. Najafi
Department of Oncology, Guilan University of Medical Sciences, Rasht, Iran

J. Moosavi
Cardiovascular Intervention Research Center, Rajaie Cardiovascular Medical and Research Center, Iran University of Medical Sciences, Tehran, Iran

B. Mohebbi (✉) · P. Firoozbakhsh · N. Dokhani
Cardio-Oncology Research Center, Rajaie Cardiovascular Medical and Research Center, Iran University of Medical Sciences, Tehran, Iran
e-mail: roodbar@yahoo.com

M. Dabiri
Hematology-Oncology Department, Cancer Institute, Imam Khomeini Hospital, School of medicine, Tehran University of Medical Sciences, Tehran, Iran

DM   Diabetes mellitus
GH   Growth hormone
GVHD   Graft-versus-host disease
HDL   High-density lipoprotein
HST   Hematopoietic stem cell transplantation
HTN   Hypertension
TBI   Total body irradiation
TG   Triglyceride

Allogenic hematopoietic stem cell transplantation (HST) is the main cure of various diagnosed neoplastic and non-neoplastic hematologic disorders (Tichelli et al. 2007). Due to significant advances in the outcomes and long-term prognosis of HST and increased number of recipients during the past decades, immediate survival is no longer the main concern and general health status and late-onset side effects of HST are in the center of attention) (Ades et al. 2002; Socié et al. 2003; Tichelli and Socié 2005). HST is still accompanied with long-term morbidity and mortality, and theoretically, any organ can be affected by late-onset complications. Occurrence of secondary malignancies, cataract, pulmonary side effects, infertility, and endocrine dysfunction following transplantation has been thoroughly investigated (Kolb et al. 1999; Legault and Bonny 1999; Salooja et al. 2001; Soubani et al. 1996; Tichelli et al. 1993). Cardiovascular (CV) events are another group of late-onset complications that are hard to be directly related to HST. They could be underestimated due to their high prevalence in general population, their low incidence rate following transplantation, and the long time interval needed for the occurrence of their clinical manifestations.

Arterial diseases, mainly presenting as coronary, cerebrovascular, and peripheral artery disease, are a subgroup of cardiovascular complications involving arterial blood vessels and are mostly attributed to atherosclerosis. Atherosclerosis is a chronic inflammatory mechanism with an indolent course in which the endothelial lesions exist decades prior to occurrence of clinical manifestations (Elkind 2006; Hansson 2005; Stoll and Bendszus 2006). Common risk factors for atherosclerosis that could also be present in general population include hypertension (HTN), dyslipidemia (DLP), diabetes mellitus (DM), smoking, obesity, sedentary lifestyle, and positive family history (Lloyd-Jones et al. 2006). Beside these well-described risk factors, there are some transplantation-related predisposing factors that make recipients of allogeneic HST more prone to atherosclerosis, including endothelial damage which can occur after conditioning regimen ± total body irradiation (TBI) (Basavaraju and Easterly 2002; Schultz-Hector 1992) or as a result of graft-versus-host disease (GVHD) (Biedermann et al. 2002). Elevated incidence and intensity of conventional risk factors, like HTN, DLP, and glucose intolerance in recipients can be a consequence of endocrine dysfunction following transplantation, prolonged treatment with immunosuppressive drugs, including cyclosporine, tacrolimus, sirolimus, mycophenolate, and corticosteroids (Couriel et al. 2005), or sedentary lifestyle after HST (Taskinen et al. 2000).

There are some pre- and post-transplantation factors that can precipitate the process of atherosclerosis and eventually lead to premature arterial disease after HST. According to the findings of a large collaborative prospective study conducted in United States, allogeneic HST, in contrast to autologous HST, is associated with an increase in the prevalence of DM and HTN, especially when patients are conditioned with TBI, and this can potentially increase the risk of CV adverse effects (Scott Baker et al. 2007). Another study also demonstrated a 2.2 fold increase in the risk of metabolic syndrome in patients receiving allogeneic HST, compared with age- and gender-matched controls (Majhail et al. 2009). Cardiovascular risk factors (CVRFs) that appear after HST, including HTN, DM, DLP, smoking, and physical inactivity appear to be more responsible for early occurrence of arterial diseases, compared to those existing when transplantation is done. A high rate of subclinical CV injury was detected in the HST recipients transplanted in pediatric age, which was proved by increased thickness of carotid intima media and elevated velocity of aortic pulse wave. Abdominal obesity and physical inactivity could independently accelerate the process of arterial stiffening which was worsened by time after HST (Borchert-Mörlins et al. 2018).

- **Pathogenesis of arterial disorders following HST**

Arterial disorders following HST have a multi-factorial mechanism, with CVRFs playing the main role in premature development of atherosclerosis and additional direct and indirect modifying factors accelerating the process (Armenian et al. 2017). Chemo-radiotherapy and GVHD can lead to a direct endothelial injury (Carreras and Diaz-Ricart 2011). Conditioning, in addition to GVHD and its treatment are indirectly implicated in the pathogenesis of arterial disorders by causing endocrine impairment, including growth hormone deficiency, hypothyroidism, and gonadal dysfunction (Armenian et al. 2017). Patients' individual characteristics, including age, gender, past medical history, comorbid factors, and genetic predisposition can accelerate the risk of arterial disorders after HST (Musunuru et al. 2015).

A. **Endocrine dysfunction**

Growth hormone (GH) deficiency can impair insulin sensitivity and lead to elevated triglyceride (TG), decreased high-density lipoprotein (HDL) cholesterol, and HTN (Murray and Shalet 2005). GH deficiency can eventually lead to increased risk of metabolic syndrome in pediatric patients treated with allogeneic HST (Taskinen et al. 2007). Hypothyroidism, which is a frequent late-onset side effect of HST, can cause metabolic syndrome-like changes in those received HST, especially when conditioned with TBI, and increase the risk of arterial disease by disrupting the lipid profile (Nuver et al. 2002). Gonadal dysfunction is a common complication of HST among female recipients, especially when conditioned with TBI, and is strongly associated with metabolic syndrome, severe insulin resistance, and dyslipidemia (Steffens et al. 2008). Elevations in serum levels of leptin, which can be due to obesity-induced leptin resistance can also lead to development of metabolic syndrome (Airaghi et al. 2011).

B. Endothelial injury

Endothelial injury, which can be the first event causing early onset atherosclerosis and subsequent arterial disorder in long term survivors of HST, can be caused by either radiation or GVHD. Radiation exposure can directly trigger microvasculature endothelial injury and cause vascular stiffening, and lead to arterial disease (Carreras and Diaz-Ricart 2011). Animal studies have proven a dose-dependent relationship between radiation exposure and reductions in endothelial cell outgrowth, inhibiting re-endothelialization or angiogenesis after vascular injury, ultimately leading to radiation-induced endothelial dysfunction, leaving arteries prone to upcoming vascular injuries (Soucy et al. 2010).

GVHD is another potential cause of endothelial injury. The host's vascular endothelial cells are the first group of cells that come into contact with the donor's T-lymphocyte and can be damaged by being the direct target of graft-versus-host reaction. This transplant-induced endothelial injury can initiate an inflammatory process and accelerate the development of atherosclerosis by an immune-mediated mechanism (Biedermann et al. 2002).

C. Additional factors

Magnesium deficiency which can occur as a result of reduced dietary intake, intestinal loss during chronic gastrointestinal GVHD, or impaired renal reabsorption caused by calcineurin inhibitors, can be potentially participated in occurrence of HTN and metabolic syndrome following HST (Aisa et al. 2005; Champagne 2008).

• Screening and preemptive recommendations for at-risk HST survivors

The main goal of screening and preemptive strategies is to prevent the occurrence of arterial disorders by managing modifiable risk factors and suggesting healthful lifestyle routines. The first step in optimizing screening and preemptive strategies is the proper diagnosis of patients at risk of arterial disorder. This includes those having prior history of mediastinal, head or neck irradiation (Dorresteijn et al. 2002; Smith et al. 2008), as well as those conditioned with TBI (Tichelli et al. 2007, 2008). Patients developing endocrine dysfunction (Steffens et al. 2008; Taskinen et al. 2000, 2007) or chronic GVHD after HST, and those receiving long-lasting immunosuppressive therapy are also at increased risk of arterial diseases.

The best approach for preventing or postponing arterial complications is the appropriate management of CVRFs. The main focus should be concerned on modifiable and treatable risk factors, including obesity, diabetes, dyslipidemia, hypertension, smoking, sedentary lifestyle, and unhealthy diet. Screening for these risk factors should be initiated as soon as possible and shouldn't be delayed until withdrawal of immunosuppressive drugs.

There are two main approaches for treating these modifiable predisposing factors, which can be either primarily treating the underlying CVRF, which resembles the treatment in general population but with some specific precautions taken on drug interactions or organ toxicity, or treating the underlying endocrine dysfunction that maybe the indirect cause of some risk factors. Regular counseling on pursuing a healthful lifestyle and education on the significance of adherence to the treatment are essential in conducting preemptive practice.

References

Ades L, Guardiola P, Socie G. Second malignancies after allogeneic hematopoietic stem cell transplantation: new insight and current problems. Blood Rev. 2002;16(2):135–46.

Airaghi L, Usardi P, Forti S, Orsatti A, Baldini M, Annaloro C, Lambertenghi DG. A comparison between metabolic syndrome post-hematopoietic stem cell transplantation and spontaneously occurring metabolic syndrome. J Endocrinol Invest. 2011;34:e6–11.

Aisa Y, Mori T, Nakazato T, Shimizu T, Yamazaki R, Ikeda Y, Okamoto S. Effects of immunosuppressive agents on magnesium metabolism early after allogeneic hematopoietic stem cell transplantation. Transplantation. 2005;80(8):1046–50.

Armenian SH, Chemaitilly W, Chen M, Chow EJ, Duncan CN, Jones LW, Pulsipher MA, Remaley AT, Rovo A, Salooja N. National institutes of health hematopoietic cell transplantation late effects initiative: the cardiovascular disease and associated risk factors working group report. Biol Blood Marrow Transplant. 2017;23(2):201–10.

Basavaraju SR, Easterly CE. Pathophysiological effects of radiation on atherosclerosis development and progression, and the incidence of cardiovascular complications. Med Phys. 2002;29(10):2391–403.

Biedermann BC, Sahner S, Gregor M, Tsakiris DA, Jeanneret C, Pober JS, Gratwohl A. Endothelial injury mediated by cytotoxic T lymphocytes and loss of microvessels in chronic graft versus host disease. The Lancet. 2002;359(9323):2078–83.

Borchert-Mörlins B, Memaran N, Sauer M, Maecker-Kolhoff B, Sykora K-W, Blöte R, Bauer E, Schmidt BM, Melk A, Beier R. Cardiovascular risk factors and subclinical organ damage after hematopoietic stem cell transplantation in pediatric age. Bone Marrow Transplant. 2018;53(8):983–92.

Carreras E, Diaz-Ricart M. The role of the endothelium in the short-term complications of hematopoietic SCT. Bone Marrow Transplant. 2011;46(12):1495–502.

Champagne CM. Magnesium in hypertension, cardiovascular disease, metabolic syndrome, and other conditions: a review. Nutr Clin Pract. 2008;23(2):142–51.

Couriel D, Saliba R, Escalon M, Hsu Y, Ghosh S, Ippoliti C, Hicks K, Donato M, Giralt S, Khouri I. Sirolimus in combination with tacrolimus and corticosteroids for the treatment of resistant chronic graft-versus-host disease. Br J Haematol. 2005;130(3):409–17.

Dorresteijn LD, Kappelle AC, Boogerd W, Klokman WJ, Balm AJ, Keus RB, van Leeuwen FE, Bartelink H. Increased risk of ischemic stroke after radiotherapy on the neck in patients younger than 60 years. J Clin Oncol. 2002;20(1):282–8.

Elkind MS. Inflammation, atherosclerosis, and stroke. Neurologist. 2006;12(3):140–8.

Hansson GK. Inflammation, atherosclerosis, and coronary artery disease. N Engl J Med. 2005;352(16):1685–95.

Kolb H, Socié G, Duell T, Van Lint MT, Tichelli A, Apperley JF, Nekolla E, Ljungman P, Jacobsen N, Van Weel M. Malignant neoplasms in long-term survivors of bone marrow transplantation. Ann Intern Med. 1999;131(10):738–44.

Legault L, Bonny Y. Endocrine complications of bone marrow transplantation in children. Pediatr Transplant. 1999;3(1):60–6.

Lloyd-Jones DM, Leip EP, Larson MG, d'Agostino RB, Beiser A, Wilson PW, Wolf PA, Levy D. Prediction of lifetime risk for cardiovascular disease by risk factor burden at 50 years of age. Circulation. 2006;113(6):791–8.

Majhail NS, Flowers ME, Ness KK, Jagasia M, Carpenter PA, Arora M, Arai S, Johnston L, Martin PJ, Baker KS. High prevalence of metabolic syndrome after allogeneic hematopoietic cell transplantation. Bone Marrow Transplant. 2009;43(1):49–54.

Murray RD, Shalet SM. Insulin sensitivity is impaired in adults with varying degrees of GH deficiency. Clin Endocrinol. 2005;62(2):182–8.

Musunuru K, Hickey KT, Al-Khatib SM, Delles C, Fornage M, Fox CS, Frazier L, Gelb BD, Herrington DM, Lanfear DE. Basic concepts and potential applications of genetics and genomics for cardiovascular and stroke clinicians: a scientific statement from the American heart association. Circ: Cardiovasc Genet. 2015;8(1):216–42.

Nuver J, Smit AJ, Postma A, Sleijfer DT, Gietema JA. The metabolic syndrome in long-term cancer survivors, and important target for secondary preventive measures. Cancer Treat Rev. 2002;28(4):195–214.

Salooja N, Szydlo R, Socie G, Rio B, Chatterjee R, Ljungman P, Van Lint M, Powles R, Jackson G, Hinterberger-Fischer M. Pregnancy outcomes after peripheral blood or bone marrow transplantation: a retrospective survey. The Lancet. 2001;358(9278):271–6.

Schultz-Hector S. Radiation-induced heart disease: review of experimental data on dose reponse and pathogenesis. Int J Radiat Biol. 1992;61(2):149–60.

Scott Baker K, Ness KK, Steinberger J, Carter A, Francisco L, Burns LJ, Sklar C, Forman S, Weisdorf D, Gurney JG. Diabetes, hypertension, and cardiovascular events in survivors of hematopoietic cell transplantation: a report from the bone marrow transplantation survivor study. Blood. 2007;109(4):1765–72.

Smith GL, Smith BD, Buchholz TA, Giordano SH, Garden AS, Woodward WA, Krumholz HM, Weber RS, Ang K-K, Rosenthal DI. Cerebrovascular disease risk in older head and neck cancer patients after radiotherapy. J Clin Oncol. 2008;26(31):5119.

Socié G, Salooja N, Cohen A, Rovelli A, Carreras E, Locasciulli A, Korthof E, Weis J, Levy V, Tichelli A. Nonmalignant late effects after allogeneic stem cell transplantation. Blood. 2003;101(9):3373–85.

Soubani AO, Miller KB, Hassoun PM. Pulmonary complications of bone marrow transplantation. Chest. 1996;109(4):1066–77.

Soucy KG, Attarzadeh DO, Ramachandran R, Soucy PA, Romer LH, Shoukas AA, Berkowitz DE. Single exposure to radiation produces early anti-angiogenic effects in mouse aorta. Radiat Environ Biophys. 2010;49:397–404.

Steffens M, Beauloye V, Brichard B, Robert A, Alexopoulou O, Vermylen C, Maiter D. Endocrine and metabolic disorders in young adult survivors of childhood acute lymphoblastic leukaemia (ALL) or non-Hodgkin lymphoma (NHL). Clin Endocrinol. 2008;69(5):819–27.

Stoll G, Bendszus M. Inflammation and atherosclerosis: novel insights into plaque formation and destabilization. Stroke. 2006;37(7):1923–32.

Taskinen M, Saarinen-Pihkala UM, Hovi L, Lipsanen-Nyman M. Impaired glucose tolerance and dyslipidaemia as late effects after bone-marrow transplantation in childhood. The Lancet. 2000;356(9234):993–7.

Taskinen M, Lipsanen-Nyman M, Tiitinen A, Hovi L, Saarinen-Pihkala UM. Insufficient growth hormone secretion is associated with metabolic syndrome after allogeneic stem cell transplantation in childhood. J Pediatr Hematol Oncol. 2007;29(8):529–34.

Tichelli A, Socié G. Considerations for adult cancer survivors. Hematol Am Soc Hematol Educ Prog. 2005:516–22.

Tichelli A, Gratwohl A, Egger T, Roth J, Prunte A, Nissen C, Speck B. Cataract formation after bone marrow transplantation. Ann Intern Med. 1993;119(12):1175–80.

Tichelli A, Bucher C, Rovó A, Stussi G, Stern M, Paulussen M, Halter J, Meyer-Monard S, Heim D, Tsakiris DA. Premature cardiovascular disease after allogeneic hematopoietic stem-cell transplantation. Blood, J Am Soc Hematol. 2007;110(9):3463–71.

Tichelli A, Passweg J, Wójcik D, Rovó A, Harousseau J-L, Masszi T, Zander A, Békássy A, Crawley C, Arat M. Late cardiovascular events after allogeneic hematopoietic stem cell transplantation: a retrospective multicenter study of the late effects working party of the European group for blood and marrow transplantation. Haematologica. 2008;93(8):1203–10.

Thrombotic Disease in Thrombosis in Hematopoietic Stem Cell Transplantation (HSCT) Recipients

Parham Sadeghipour, Abbas Hajfathali, Farid Rashidi, and Abolghsem Allahyari

Abstract In patients with hematopoietic stem cell transplantation (HSCT), different risk factors can be numbered based on the type of thrombotic event. The risk factors that may increase incidence of venous thromboembolism (VTE) includes' cancers, prolonged immobility caused by hospitalization, cytotoxic chemotherapy or radiotherapy, infectious diseases, traumatic brain events, and graft-versus-host disease (GVHD). In the case of Sinusoidal obstruction syndrome (SOS), risk factors are divided into pre- and post-transplantation factors, and some pharmaceutical factors are also involved. Young age of the recipient, liver damage or a previous liver transplantation, underlying diseases, history of abdominal or liver radiation, and exposure to hepatotoxic drugs. Regarding the incidence of various thrombotic events, it can be different depending on the type of risk factors and the type of transplant as well as the diagnostic criteria. In general, the incidence of thrombosis in allogeneic transplants is higher. In the studies conducted, the incidence of thrombotic events in HSCT patients may vary between 8 and 20%. No specific recommendations exist for diagnosing and treating HSCT recipients complicated by VTE, with the consensus extrapolated from cancer-associated VTE guidelines. Several prediction models to enhance diagnostic accuracy have been presented; nonetheless, none of them has externally been validated. The higher bleeding tendency and incidence of thrombocytopenia necessitate a balance between bleeding and thrombotic risks. Although low-molecular-weight heparin is still considered the classic anticoagulation agent

P. Sadeghipour
Cardiovascular Intervention Research Center, Rajaie Cardiovascular Medical and Research Center, Iran University of Medical Sciences, Tehran, Iran

A. Hajfathali
Shahid Beheshti University of Medical Sciences, Tehran, Iran

F. Rashidi (✉)
Tuberculosis and Lung Disease Research Center, Tabriz University of Medical Sciences, Tabriz, Iran
e-mail: fr2652@yahoo.com

A. Allahyari
Hematology Department, Faculty of Medicine, Imam Reza Hospital, Mashhad University of Medical Sciences, Mashhad, Iran

for primary and secondary prophylaxis, recent studies have shown promising efficacy and safety for direct-acting oral anticoagulants. In symptomatic catheter-related thrombosis, anticoagulation is usually enough, and catheter removal is reserved for concomitant infection or catheter dysfunction.

Keywords Hematopoietic stem cell transplant · Thrombosis · Venous thromboembolism · Catheter-related thrombosis

Abbreviations

CRT	Catheter-related thrombosis
DOAC	Direct-acting oral anticoagulants
GVHD	Graft-versus-host disease
HSCT	Hematopoietic stem cell transplantation
LMWH	Low-molecular-weight heparin
SOS	Sinusoidal obstruction syndrome
TA-TMA	Transplant-associated thrombotic microangiopathy
VKA	Vitamin K antagonist
VOD	Veno-occlusive disease
VTE	Venous thrombembolism

1 Highlights

1. There are four type of thrombosis events in HSCT.
2. The risk factors based on thrombosis type may be different.
3. The incidence of thrombosis events is different based on type of transplant.
4. The incidence of thrombotic events varies between 8 and 20%.
5. No specific recommendations are available for diagnosing and treating hematopoietic stem cell transplantation recipients.
6. Recommendations for treating and managing VTE are extrapolated from cancer-associated VTE guidelines.
7. Bleeding tendency and accompanied thrombocytopenia are formidable challenges in treating VTE in this population.
8. Low-molecular-weight heparin and direct-acting oral anticoagulants show acceptable efficacy and safety in VTE treatment.
9. Catheter-related thrombosis is predominantly asymptomatic, and between 1 and 4% of patients become complicated.
10. Anticoagulation is usually enough for catheter-related thrombosis, with catheter removal reserved for concomitant infection and catheter dysfunction.

2 Epidemiology

Thrombotic events have always been among the most important challenges in various medical fields. Since lack of understanding and inattention to thrombotic events can lead to irreparable complications and increased risk of mortality, such events should be seriously considered at all medical levels. The most important strategy in dealing with thrombotic events is to first identify their causes and then find ways to prevention.

One of the first steps in preventing thrombotic events is to consider the previously identified risk factors for such events. To this end, adequate and accurate knowledge of the epidemiology of thromboembolic events in various diseases can be very helpful. Since it is of particular importance to determine the type of thrombotic and thromboembolic events a patient may face, adequate knowledge of the type and epidemiology of thromboembolic events can be very helpful in prevention and treating patients.

Bone marrow stem cell transplantation is a standard procedure that has been increasingly used for treating various benign and malignant blood disorders during the last two decades (Labrador et al. 2013). Considering the high cost of bone marrow transplantation and the importance of promoting the recovery of these patients and reducing their mortality rate, some studies have shown thromboembolic events among the serious causes of mortality and morbidity in these patients (Gerber et al. 2008; Tsakiris and Tichelli 2009).

Hemostatic changes as well as thrombotic events, frequently observed in patients undergoing stem cell transplantation may be either venous or arterial in origin. Based on different pathophysiological findings, thrombotic events in these patients can generally be divided into 4 groups: catheter-related thrombosis, sinusoidal obstructive syndrome (SOS), which is also referred to as veno-occlusive disease, and transplant-associated thrombotic microangiopathy (TA-TMA) (Carreras et al. 2019; Copelan 2006; Kansu 2012).

Each of these thrombotic events may be caused by different risk factors. The review of previous studies shows that there are various risk factors for thrombotic events in these patients, such as cancers, prolonged immobility caused by hospitalization, cytotoxic chemotherapy or radiotherapy, infectious diseases, traumatic brain events, and graft-versus-host disease (GVHD) (Gerber et al. 2008; Gonsalves et al. 2008). However, the most important risk factors for thrombotic events among these patients are probably venous thromboembolism (VTE) and graft-versus-host disease (GVHD) (Kansu 2012). Meanwhile, some treatment regimens can also cause thrombosis in these patients in certain cases. For example, high-dose cytotoxic chemotherapy can cause SOS (Helmy 2006), or TMA may be secondarily caused by calcineurin inhibitors, sirolimus, and even rarely caused secondarily by GVHD (Cutler et al. 2005; Kojouri and George 2007).

Since diagnosing the risk factors for SOS is critical for initiation of the treatment or early prophylaxis, they require special attention. The risk factors for SOS can generally be divided into two categories: Pre-transplant patient characteristics and transplant-related factors (Dalle and Giralt 2016).

Major pre-transplant factors associated with increased risk of VOD/SOS include young age of the recipient, liver damage or a previous liver transplantation, underlying diseases (especially advanced cancers), a record of abdominal or liver radiation, and exposure to hepatotoxic drugs, especially gemtuzumab ozogamicin: calicheamicin-conjugated humanized anti-CD33 monoclonal antibody (Carreras et al. 2011; Chao 2014; Cheuk et al. 2007; Cutler et al. 2010; Dignan et al. 2013; McDonald et al. 1993; Pihusch et al. 2005). In addition, increased baseline levels of liver enzymes and bilirubin before HSCT are another important risk factor for VOD/SOS (Bearman 1995; Hägglund et al. 1998). In general, any disorder in the functions of other vital organs can increase the risk of these events; e.g., lung or kidney disorders after a kidney transplant (Barker et al. 2003; Maximova et al. 2014). Using some hormonal agents such as progestin and norethisterone can also increase the risk of such events in young women (Helmy 2006). Some infectious agents, e.g., active viral hepatitis, or taking some antibiotics, e.g., third-generation cephalosporins, can be considered minor risk factors because they have been associated with development of biliary sludge (Matute-Bello et al. 1998).

The second category of risk factors in VOD/SOS patients consists of transplant-related factors, the most important of which is transplant type. For example, the risk of VOD/SOS is higher in unrelated donors and non-T cell-depleted transplants and in patients with allogeneic HSCT (Barker et al. 2003; Cutler et al. 2010; Fisher et al. 1998; Richardson et al. 2002). Some medications such as cyclophosphamide and busulfan can increase the post-transplant risk of VOD/SOS (Nagler et al. 2014; Tsirigotis et al. 2014).

As mentioned earlier, risk factors are classified according to the type of thrombosis after HSCT based on different pathophysiological findings on the development of thrombosis. Thrombotic microangiopathy (TMA) is another type of thrombosis caused by HSCT. The main risk factors for TA-TMA are acute GVHD, old age, female gender, advanced primary hematologic malignancy, non-myeloablative transplant, and high doses of busulfan (Chapin et al. 2014).

Since different risk factors and pathogens are involved in the development of thrombotic events, studies may report conflicting results regarding the incidence of these events. According to one of the most important meta-analyses, the incidence of VTE in HSCT patients is around 5% (Copelan 2006), whereas the incidence rate in GVHD patients was slightly higher. However, heterogeneity makes it difficult to achieve a definitive judgment in this case. The occurrence of VTE can also vary depending on the type of transplant. A study on 589 patients undergoing either autologous or allogeneic HSCT showed higher incidence rate in patients undergoing allogeneic HSCT (Gonsalves et al. 2008).

Different incidence of VOD/SOS has been reported in different studies. One of the most important reasons for this could be that these studies used different diagnostic criteria. This variance may have also resulted from differences in the type of transplantation, study population and prior treatments of patients. Depending on the diagnostic criteria, the mean incidence rate reported in the studies ranged from 8 to 14% (Coppell et al. 2010).

The incidence of TA-TMA in these patients also varies depending on risk factors, such as conditioning regimens, immunosuppressive agents, GVHD, and HLA mismatch (Cho et al. 2010; Martinez et al. 2005). TA-TMA occurs in 10–20% of allogeneic HSCT patients, but the incidence is higher in patients undergoing autologous HSCT (George et al. 2004; Martinez et al. 2005).

3   Diagnosis

Data are scarce on the diagnosis and treatment of venous thromboembolism (VTE) in hematopoietic stem cell transplantation (HSCT) recipients. Consequently, we focus merely on VTE management in HSCT recipients in the present chapter.

No discrete recommendations exist for the diagnosis of VTE in patients with a history of HSCT, and diagnostic decisions should be based on available guidelines (Konstantinides et al. 2020; Mazzolai et al. 2022; Ortel et al. 2020). Nevertheless, especially during the acute phase, various conditions could masquerade VTE in HSCT recipients, underscoring the role of condition-specific prediction tools. Several investigators have suggested various predictors, such as a history of catheter-related thrombosis (CRT), a minimum hospitalization period of 30 days, grade III–IV GvHD, a history of pulmonary embolism or lower extremity deep vein thrombosis, lymphoma diagnosis, a body mass index exceeding 35, a white cell count of higher than 11 (HIGH-2-LOW score), (Martens et al. 2021) a history of stroke, chronic GvHD, hypertension, male sex, and peripheral blood stem cells (the HiGHS2 risk model) (Gangaraju et al. 2021). Although the mentioned scoring systems have shown acceptable internal validity, their external validity needs verification in more comprehensive studies.

4   Treatment

No guidelines or expert consensus have suggested specific therapeutic recommendations for HSCT recipients complicated by VTE. Starting full-dose therapeutic anticoagulation should be the primary goal unless a contraindication exists. Still, several features of this population render the antithrombotic treatment delicate. Firstly, HSCT recipients have a higher bleeding risk, which might compromise antithrombotic therapy. Secondly, the optimal duration of therapeutic anticoagulation needs definition. The risk of VTE recurrence remains high in some patients, and there is no clear strategy on how anticoagulation therapy should be continued 3 to 6 months after standard therapy (Chiu and Lazo-Langner 2023). Thirdly, the implication of direct-acting oral anticoagulants (DOACs) in HSCT recipients should be elucidated.

DOACs need no routine monitoring, have fewer drug interactions, and due to their shorter half-life, can be more easily interrupted. Accordingly, they are considered a better choice than vitamin K antagonists. Despite the promising results of the use of edoxaban (Raskob et al. 2018), rivaroxaban (Young et al. 2018), and apixaban (Agnelli et al. 2020) for secondary VTE prophylaxis, patients with a history of hematological malignancies are generally excluded in pivotal studies, and no data-driven recommendations could, therefore, be suggested.

The higher incidence of thrombocytopenia, myeloablative regimens, the risk of invasive bacterial and fungal infection, the higher incidence of drug interactions, and many other clinical scenarios witnessed in HSCT recipients increase the bleeding tendency of this group of patients. In the absence of a validated bleeding score, certain characteristics increase the bleeding risk. For instance, different bleeding risks have been reported for autologous versus allogeneic transplantation recipients (Schimmer et al. 1998). Anticoagulation is generally well-tolerated in patients with autologous HCST, and major hemorrhagic events are rare (Hegerova et al. 2018). In contrast, intense myeloablative conditioning regimens result in thrombocytopenia and GvHD and increase the risk of major bleeding events in allogeneic transplantation recipients (Labrador et al. 2013). Notably, a specific risk model has been defined to balance the net survival benefit of VTE treatment in the allogeneic population (Deng et al. 2023).

Aside from the type of transplantation, thrombocytopenia is a major concern in HSCT recipients. Although mild thrombocytopenia is largely well-tolerated, there is a direct correlation between platelet counts and bleeding events (Chiu and Lazo-Langner 2023). One should be mindful that most recommendations regarding anti-coagulation treatment in thrombocytopenic patients are extracted from studies on patients with cancer (Samuelson Bannow et al. 2018). Traditionally, low-molecular-weight heparin (LMWH) is the first option as an anticoagulation agent. However, despite the paucity of data, DOACs look promising in the treatment of VTE patients complicated by thrombocytopenia. A practical algorithm for anticoagulation treatment in patients with thrombocytopenia is presented in Fig. 1.

Notwithstanding the mentioned higher bleeding tendency, no specific indication exists for inferior vena cava placement in HSCT recipients, and contraindications to anticoagulation still constitute the most agreed-upon indication.

5   CRT

Although CRT is mostly asymptomatic, it increases the rate of bacteremia and can cause long-term sequelae due to central vein obstruction (Geerts 2014). In addition, in 1–4% of patients, CRT becomes symptomatic and might cause catheter dysfunction. Three to 6 months of anticoagulation therapy with no need for immediate catheter removal is commonly advised for CRT patients, with urgent catheter removal mostly limited to infected thrombosis and dysfunctional catheters not amenable/responsive to pharmacomechanical treatment. Even in these conditions, a short course of anti-coagulation therapy is the usual recommendation. Nevertheless, no recommendation exists for the duration of anticoagulation therapy after catheter removal in patients with a history of CRT.

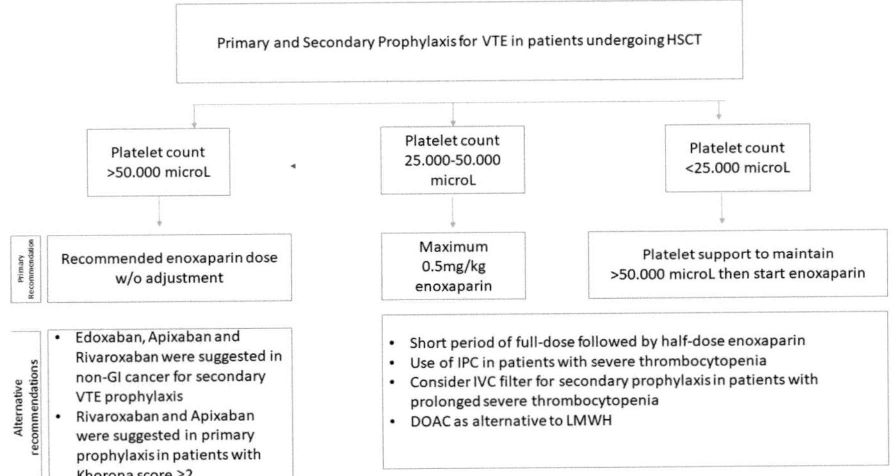

Fig. 1 Primary and Secondary Prophylaxis for VTE in patients undergoing HSCT (Chiu and Lazo-Langner 2023; Streiff et al. 2021). GI, gastrointestinal, HSCT, haematopoietic stem cell transplantation, VTE, venous thromboembolism

6　VTE Prophylaxis

No specific recommendations are available regarding VTE prophylaxis in HSCT recipients, and the course of action vis-à-vis patients with cancer is usually extrapolated from the available evidence. The consensus recommends against routine VTE prophylaxis in ambulatory cancer patients not receiving chemotherapy (Lyman et al. 2015). Patients at risk of VTE who are under chemotherapy might benefit from prophylaxis. Likewise, several risk estimators, such as the Khorana risk score, have been validated in the risk stratification of high-risk patients (Lyman et al. 2015). For hospitalized HSCT patients, VTE prophylaxis is generally recommended. Similar to VTE treatment, a balance between bleeding and thrombotic risks is crucial. As is shown in Fig. 1, a platelet count exceeding 50×10^3 is acceptable for VTE chemoprophylaxis. LMWH is the most studied prophylactic agent; nonetheless, recent studies have shown the non-inferior efficacy and safety of DOACs (Li et al. 2019). For patients at high risk of bleeding, intermittent pneumatic compression might be drawn upon as a bridging solution.

Finally, no studies have shown the benefit of prophylactic anticoagulation in preventing CRT in patients with indwelling catheters.

References

Agnelli G, Becattini C, Meyer G, et al. Apixaban for the treatment of venous thromboembolism associated with cancer. N Engl J Med. 2020.

Barker CC, Butzner J, Anderson R, Brant R, Sauve R. Incidence, survival and risk factors for the development of veno-occlusive disease in pediatric hematopoietic stem cell transplant recipients. Bone Marrow Transplant. 2003;32:79–87.

Bearman S. The syndrome of hepatic veno-occlusive disease after marrow transplantation. Blood. 1995;85:3005–20.

Carreras E, Díaz-Beyá M, Rosiñol L, Martínez C, Fernández-Avilés F, Rovira M. The incidence of veno-occlusive disease following allogeneic hematopoietic stem cell transplantation has diminished and the outcome improved over the last decade. Biol Blood Marrow Transplant. 2011;17:1713–20.

Carreras E, Dufour C, Mohty M, Kröger N. The EBMT handbook: hematopoietic stem cell transplantation and cellular therapies. 2019.

Chao N. How I treat sinusoidal obstruction syndrome. Blood. 2014;123:4023–6.

Chapin J, Shore T, Forsberg P, Desman G, Van Besien K, Laurence J. Hematopoietic transplant-associated thrombotic microangiopathy: case report and review of diagnosis and treatments. Clin Adv Hematol Oncol. 2014;12:565–73.

Cheuk D, Wang P, Lee T, et al. Risk factors and mortality predictors of hepatic veno-occlusive disease after pediatric hematopoietic stem cell transplantation. Bone Marrow Transplant. 2007;40:935–44.

Chiu J, Lazo-Langner A. Venous thromboembolism in hematopoietic stem cell transplantation: a narrative review. Thrombosis Res. 2023.

Cho B-S, Yahng S-A, Lee S-E, et al. Validation of recently proposed consensus criteria for thrombotic microangiopathy after allogeneic hematopoietic stem-cell transplantation. Transplantation. 2010;90:918–26.

Copelan EA. Hematopoietic stem-cell transplantation. N Engl J Med. 2006;354:1813–26.

Coppell JA, Richardson PG, Soiffer R, et al. Hepatic veno-occlusive disease following stem cell transplantation: incidence, clinical course, and outcome. Biol Blood Marrow Transplant. 2010;16:157–68.

Cutler C, Henry NL, Magee C, et al. Sirolimus and thrombotic microangiopathy after allogeneic hematopoietic stem cell transplantation. Biol Blood Marrow Transplant. 2005;11:551–7.

Cutler C, Kim HT, Ayanian S, et al. Prediction of veno-occlusive disease using biomarkers of endothelial injury. Biol Blood Marrow Transplant. 2010;16:1180–5.

Dalle J-H, Giralt SA. Hepatic veno-occlusive disease after hematopoietic stem cell transplantation: risk factors and stratification, prophylaxis, and treatment. Biol Blood Marrow Transplant. 2016;22:400–9.

Deng RX, Zhu XL, Zhang AB, et al. Machine learning algorithm as a prognostic tool for venous thromboembolism in allogeneic transplant patients. Transplant Cell Ther. 2023;29:57.e1-57.e10.

Dignan FL, Wynn RF, Hadzic N, et al. BCSH/BSBMT guideline: diagnosis and management of veno-occlusive disease (sinusoidal obstruction syndrome) following haematopoietic stem cell transplantation. Br J Haematol. 2013;163:444–57.

Fisher D, Vredenburgh J, Petros W, et al. Reduced mortality following bone marrow transplantation for breast cancer with the addition of peripheral blood progenitor cells is due to a marked reduction in veno-occlusive disease of the liver. Bone Marrow Transplant. 1998;21:117–22.

Gangaraju R, Chen Y, Hageman L, et al. Late-occurring venous thromboembolism in allogeneic blood or marrow transplant survivors: a BMTSS-HiGHS2 risk model. Blood Adv. 2021;5:4102–11.

Geerts W. Central venous catheter–related thrombosis. Hematology 2014, the American society of hematology education program book. 2014;2014:306–11.

George JN, Li X, McMinn JR, Terrell DR, Vesely SK, Selby GB. Thrombotic thrombocytopenic purpura-hemolytic uremic syndrome following allogeneic HPC transplantation: a diagnostic dilemma. Transfusion. 2004;44:294–304.

Gerber DE, Segal JB, Levy MY, Kane J, Jones RJ, Streiff MB. The incidence of and risk factors for venous thromboembolism (VTE) and bleeding among 1514 patients undergoing hematopoietic stem cell transplantation: implications for VTE prevention. Blood. 2008;112:504–10.

Gonsalves A, Carrier M, Wells PS, McDiarmid SA, Huebsch LB, Allan DS. Incidence of symptomatic venous thromboembolism following hematopoietic stem cell transplantation. J Thromb Haemost. 2008a;6:1468–73.

Hägglund H, Remberger M, Klaesson S, Lönnqvist B, Ljungman P, Ringdén O. Norethisterone treatment, a major risk-factor for veno-occlusive disease in the liver after allogeneic bone marrow transplantation. Blood J Am Soc Hematol. 1998;92:4568–72.

Hegerova L, Bachan A, Cao Q, et al. Catheter-related thrombosis in patients with lymphoma or myeloma undergoing autologous stem cell transplantation. Biol Blood Marrow Transplant. 2018;24:e20–5.

Helmy A. Updates in the pathogenesis and therapy of hepatic sinusoidal obstruction syndrome. Aliment Pharmacol Ther. 2006;23:11–25.

Kansu E. Thrombosis in stem cell transplantation. Hematology. 2012;17:s159–62.

Kojouri K, George JN. Thrombotic microangiopathy following allogeneic hematopoietic stem cell transplantation. Curr Opin Oncol. 2007;19:148–54.

Konstantinides SV, Meyer G, Becattini C, et al. 2019 ESC guidelines for the diagnosis and management of acute pulmonary embolism developed in collaboration with the European respiratory society (ERS). Eur Heart J. 2020;41:543–603.

Labrador J, Lopez-Anglada L, Perez-Lopez E, et al. Analysis of incidence, risk factors and clinical outcome of thromboembolic and bleeding events in 431 allogeneic hematopoietic stem cell transplantation recipients. Haematologica. 2013;98:437–43.

Li A, Kuderer NM, Garcia DA, et al. Direct oral anticoagulant for the prevention of thrombosis in ambulatory patients with cancer: a systematic review and meta-analysis. J Thromb Haemost. 2019;17:2141–51.

Lyman GH, Bohlke K, Khorana AA, et al. Venous thromboembolism prophylaxis and treatment in patients with cancer: American society of clinical oncology clinical practice guideline update 2014. J Clin Oncol. 2015;33:654.

Martens KL, da Costa WL, Amos CI, et al. HIGH-2-LOW risk model to predict venous thromboembolism in allogeneic transplant patients after platelet engraftment. Blood Adv. 2021;5:167–75.

Martinez MT, Bucher C, Stussi G, et al. Transplant-associated microangiopathy (TAM) in recipients of allogeneic hematopoietic stem cell transplants. Bone Marrow Transplant. 2005;36:993–1000.

Matute-Bello G, McDonald G, Hinds M, Schoch H, Crawford S. Association of pulmonary function testing abnormalities and severe veno-occlusive disease of the liver after marrow transplantation. Bone Marrow Transplant. 1998;21:1125–30.

Maximova N, Ferrara G, Minute M, et al. Experience from a single paediatric transplant centre with identification of some protective and risk factors concerning the development of hepatic veno-occlusive disease in children after allogeneic hematopoietic stem cell transplant. Int J Hematol. 2014;99:766–72.

Mazzolai L, Ageno W, Alatri A, et al. Second consensus document on diagnosis and management of acute deep vein thrombosis: updated document elaborated by the ESC working group on aorta and peripheral vascular diseases and the ESC working group on pulmonary circulation and right ventricular function. Eur J Prev Cardiol. 2022;29:1248–63.

McDonald GB, Hinds MS, Fisher LD, et al. Veno-occlusive disease of the liver and multiorgan failure after bone marrow transplantation: a cohort study of 355 patients. Ann Intern Med. 1993;118:255–67.

Nagler A, Labopin M, Gorin N-C, et al. Intravenous busulfan for autologous stem cell transplantation in adult patients with acute myeloid leukemia: a survey of 952 patients on behalf of the acute leukemia working party of the European Group for blood and marrow transplantation. Haematologica. 2014;99:1380.

Ortel TL, Neumann I, Ageno W, et al. American society of hematology 2020 guidelines for management of venous thromboembolism: treatment of deep vein thrombosis and pulmonary embolism. Blood Adv. 2020;4:4693–738.

Pihusch M, Wegner H, Goehring P, et al. Diagnosis of hepatic veno-occlusive disease by plasminogen activator inhibitor-1 plasma antigen levels: a prospective analysis in 350 allogeneic hematopoietic stem cell recipients. Transplantation. 2005;80:1376–82.

Raskob GE, van Es N, Verhamme P, et al. Edoxaban for the treatment of cancer-associated venous thromboembolism. N Engl J Med. 2018;378:615–24.

Richardson PG, Murakami C, Jin Z, et al. Multi-institutional use of defibrotide in 88 patients after stem cell transplantation with severe veno-occlusive disease and multisystem organ failure: response without significant toxicity in a high-risk population and factors predictive of outcome. Blood, J Am Soc Hematol. 2002;100:4337–43.

Samuelson Bannow B, Lee A, Khorana A, et al. Management of cancer-associated thrombosis in patients with thrombocytopenia: guidance from the SSC of the ISTH. J Thromb Haemost. 2018;16:1246–9.

Schimmer AD, Stewart AK, Keating A, et al. Safety of therapeutic anticoagulation in patients with multiple myeloma receiving autologous stem cell transplantation. Bone Marrow Transplant. 1998;22:491–4.

Streiff MB, Holmstrom B, Angelini D, et al. Cancer-associated venous thromboembolic disease, version 2.2021, NCCN clinical practice guidelines in oncology. J Natl Compr Canc Netw 2021;19:1181–201.

Tsakiris DA, Tichelli A. Thrombotic complications after haematopoietic stem cell transplantation: early and late effects. Best Pract Res Clin Haematol. 2009;22:137–45.

Tsirigotis P, Resnick I, Avni B, et al. Incidence and risk factors for moderate-to-severe veno-occlusive disease of the liver after allogeneic stem cell transplantation using a reduced intensity conditioning regimen. Bone Marrow Transplant. 2014;49:1389–92.

Young AM, Marshall A, Thirlwall J, et al. Comparison of an oral factor Xa inhibitor with low molecular weight heparin in cancer patients with venous thromboembolism: results of a randomized trial (SELECT-D). J Clin Oncol. 2018;36:2017–23.

Pulmonary Hypertension in HSCT

Marzieh Mirtajaddini, Mohammad Sahebjam, Mohsen Esfandbod, Farhad Shahi, and Mina Mohseni

Abstract Pulmonary hypertension (PH) is one of the important complications after hematopoietic stem cell transplantation (HSCT). PH after HSCT can occur with different causes and mechanisms. Increased pulmonary vascular resistance can cause PH, which can lead to right ventricular failure. In this chapter, we discuss the causes of PH, its classification and how to diagnose and treat it in patients after HSCT.

Keywords Pulmonary hypertension · Hematopoietic stem cell transplantation · Right ventricular failure · Pulmonary veno-occlusive disease · Right heart catheterization · Pulmonary vascular resistance · Vasoreactivity test · Echocardiography

M. Mirtajaddini
Rajaie Cardiovascular Medical and Research Center, Iran University of Medical Sciences, Tehran, Iran
e-mail: m.mirtajadini@gmail.com

M. Sahebjam
Department of Clinical Pharmacy, School of Pharmacy, Tehran University of Medical Sciences, Tehran, Iran
e-mail: msahebjam@yahoo.com

M. Esfandbod
Department of Hematology and Oncology, Imam Khomeini Hospital Complex, Tehran University of Medical Sciences, Tehran, Iran
e-mail: Sfandbod@sina.tums.ac.ir

F. Shahi
Department of Hematology and Medical Oncology, Cancer Research Center, Cancer Institute, Imam Khomeini Hospital Complex, Tehran University of Medical Sciences, Tehran, Iran
e-mail: fshahi@yahoo.com

M. Mohseni (✉)
Cardio-Oncology Research Center, Rajaie Cardiovascular Medical and Research Center, Iran University of Medical Sciences, Tehran, Iran
e-mail: mina.mohseni8694@gmail.com

© The Author(s), under exclusive license to Springer Nature Switzerland AG 2024   201
A. Alizadehasl et al. (eds.), *Cardiovascular Considerations in Hematopoietic Stem Cell Transplantation*, https://doi.org/10.1007/978-3-031-53659-5_15

Abbreviations

BNP	Brain natriuretic peptide
CCBs	Calcium channel blockers
CMR	Cardiac magnetic resonance imaging
CO	Cardiac output
CTPA	Computed tomography of pulmonary arteries
CXR	Chest X-ray
ECG	Electrocardiogram
ECMO	Extracorporeal membrane oxygenation
HSCT	Hematopoietic stem cell transplantation
IVC	Inferior vena cava
JVP	Jugular venous pressure
LV	Left ventricle
LVEF	Left ventricular ejection fraction
MIOP	Malignant osteopetrosis
mPAP	Mean pulmonary artery pressure
PA	Pulmonary artery
PAH	Pulmonary arterial hypertension
PAP	Pulmonary arterial pressure
PAWP	Pulmonary artery wedge pressure
PCWP	Pulmonary capillary wedge pressure
PH	Pulmonary hypertension
PVOD	Pulmonary veno-occlusive disease
PVR	Pulmonary vascular resistance
RA	Right atrium
RHC	Right heart catheterization
RV	Right ventricular
RVOT AT	Right ventricular outflow tract acceleration time
sPAP	Systolic PAP
TAPSE	Tricuspid annular plane systolic excursion
TA-TMA	Transplant associated- thrombotic microangiopathy
TRV	Tricuspid regurgitation velocity
WU	Wood Units

1 Definition

Pulmonary hypertension (PH) is an important complication after hematopoietic stem cell transplantation (HSCT) that is underestimated. The hemodynamic definition of PH is a mean pulmonary artery pressure (mPAP) threshold of > 20 mmHg at rest, confirmed by right heart catheterization (RHC). A lower value of > 2 Wood Units

(WU) of the pulmonary vascular resistance (PVR) defines a pre-capillary pattern of PH (Cullivan et al. 2023; Jodele et al. 2013; Levy et al. 2019; Mandras et al. 2020).

Various pulmonary complications can occur after HSCT that can cause an increase in pulmonary arterial pressure (PAP). This issue can cause pulmonary arterial hypertension (PAH) to be overlooked. Patients who develop respiratory symptoms after HSCT should be evaluated for PH. Increased pulmonary artery pressure can lead to permanent pulmonary vascular changes, right ventricular failure, and death (Dandoy et al. 2013).

2 Epidemiology

The prevalence of pulmonary hypertension (PH) is approximately 1% of the world's population. PH can develop after HSCT in both adults and children. PH is not a common complication after HSCT. The prevalence of PH after HSCT varies from 2.4–28% in several studies, and considering the small sample size of the studies and the retrospective nature of most of them, it seems that this percentage of prevalence needs larger, prospective, and multicenter studies. PH following HSCT can be potentially fatal. Considering that the prevalence of pulmonary hypertension in patients after HSCT is not low, routine echocardiography is important for the evaluation of right heart function after HSCT (Cullivan et al. 2023; Dandoy et al. 2013; Kawashima et al. 2021; Levy et al. 2019; Tanaka et al. 2021).

3 Hemodynamics and Classification of PH

Delivering venous blood to the pulmonary arteries for oxygen exchange is the right ventricle's primary function. The right ventricle(RV) is sensitive to an increase in afterload, which is primarily controlled by pulmonary vascular resistance. Right ventricular filling and contraction are disturbed by the increase in afterload brought on by PH. Patients may experience repeated episodes of acute right ventricular decompensation and eventually cardiac dysfunction and death (Dandoy et al. 2013).

Five clinical subgroups are used to categorize PH:

Pulmonary arterial hypertension (including idiopathic PAH), left-sided heart disease-related PH (including heart failure with reduced or preserved LVEF and heart valve disease), PH caused by lung conditions and/or hypoxia, PH as a result of pulmonary artery obstructions (including chronic thromboembolic PH), and PH characterized by unknown or multiple mechanisms (including hematological disorders). Clinical classification and hemodynamic definitions of PH are summarized in Fig. 1 and Table 1 (Dandoy et al. 2013; Mandras et al. 2020; Simonneau et al. 2019).

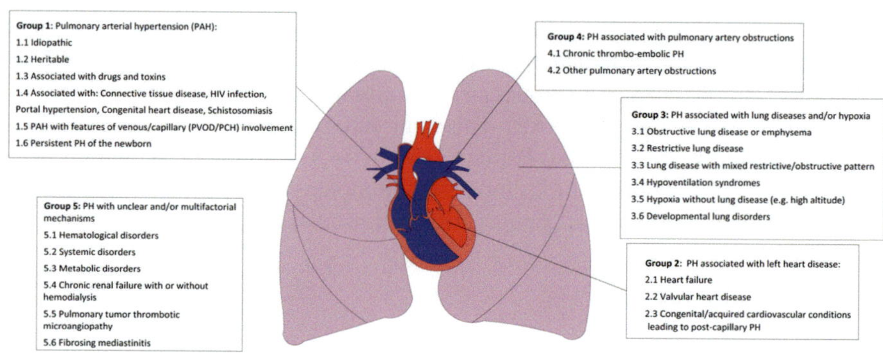

Fig. 1 Clinical classification of PH

Table 1 Hemodynamic definitions of PH

1. Pre-capillary PH	Mean pulmonary arterial pressure > 20 mmhg Pulmonary artery wedge pressure $\leq$ 15 mmhg Pulmonary vascular resistance > 2 Wood units
2. Isolated post-capillary PH	Mean pulmonary arterial pressure > 20 mmhg Pulmonary artery wedge pressure > 15 mmhg Pulmonary vascular resistance $\leq$ 2 Wood units
3. Combined post- and pre-capillary PH	Mean pulmonary arterial pressure >20 mmhg Pulmonary artery wedge pressure >15 mmhg Pulmonary vascular resistance >2 Wood units

Humbert et al. (2022), Peter Libby et al. (2021)

3.1 Pathophysiology

Patients after HSCT had variable pulmonary vasculature involvement. Among the group 1 PH classifications, PAH and pulmonary veno-occlusive disease (PVOD) are most commonly observed in post-HSCT patients. In the next order, mixed arteriolar-venous pathology is more common. Large and medium-sized pulmonary arteries and arterioles are a hallmark of PAH, which is a condition that is pre-capillary in nature. There is intimal hyperplasia with proliferation of pulmonary artery smooth muscle cells, adventitial hypertrophy and hyperplasia. As a result, blood flow is impeded, and the resistance of the pulmonary vessels increases. One of the PAH's early pathological hallmarks is pulmonary vasoconstriction. PVOD belongs to the PAH group and predominantly involves the pulmonary vessels at the post-capillary level. Smooth muscle cell hyperplasia and intimal fibrosis are pathological features of PVOD, leading to luminal narrowing or obstruction of venules and septal veins. Veins and arteries can be involved in both PVOD and PAH, as a result, PAH and PVOD are believed to be two facets of the same illness spectrum rather than two separate diseases. However, PVOD and PAH must be properly differentiated from each other because their treatments are different from each other. Vasodilators, the primary PAH treatment option, can exacerbate pulmonary capillary bed congestion and lead to progressive pulmonary edema. But invasive histological examinations in PVOD patients can be life-threatening and cannot be performed in many cases. The occurrence of PVOD after HSCT is fatal. PVOD is less common than PAH after HSCT. Increased pulmonary artery pressure can occur as a result of endothelial dysfunction of small lung vessels and changes in inflammatory and thrombotic conditions after HSCT. Endothelial dysfunction can occur in both pre- and post-pulmonary capillaries (Dandoy et al. 2013; Ghigna and Dorfmüller 2019; Kawashima et al. 2021; Tuder et al. 2009).

3.2 Etiology and Risk Factors

Transplant-associated thrombotic microangiopathy (TA-TMA) after HSCT may lead to severe pulmonary vascular injury, which manifests itself as respiratory failure and PAH. The transplant complications (TA-TMA and PAH) should be thoroughly examined in pediatric patients with post-HSCT hypoxemia of unclear etiology. A severe, complicated HSCT consequence is TA-TMA. Endothelial injury that results in platelet consumption and microangiopathic hemolytic anemia, as well as fibrin deposition and thrombosis in the microcirculation and end-organ destruction, is what causes TA-TMA. TA-TMA is a well-known complication following HSCT associated with increased death and morbidity. However, TA-TMA may be associated with significant pulmonary vascular damage and may lead to respiratory failure caused by hypoxia following HSCT. In TA-TMA, pulmonary arteriolar narrowing can occur as a result of the damage caused by the fragments of red blood cells and platelets and the formation of microthrombi and fibrin deposition. Persistent inflammation and vasoconstriction cause proliferation of the vascular wall, which causes the elevated vascular resistance observed in PAH. PAH has been reported in most forms of hemolytic anemia. Active hemolysis, for example, is an acute stressor that might exacerbate vasospasm. This causes hemodynamic collapse, fulminant right ventricular failure, and an acute rise in PA resistance. Endothelial damage to pulmonary vessels may disrupt the normal production of nitric oxide and alter the functional response of the endothelin receptor, thereby causing pulmonary vasoconstriction and the development of PH. Underlying disease may be a risk factor for developing PH after HSCT. In various studies, causes such as the use of busulfan as a preconditioning and underlying diseases before HSCT, such as malignant osteopetrosis (MIOP), have also been proposed as possible risk factors for PAH. Different types of PH can be brought on by myeloproliferative disorders as well as HSCT-related issues such as heart failure and thrombosis (Dandoy et al. 2013; Jodele et al. 2013; Kawashima et al. 2019; Levy et al. 2019; Steward et al. 2004).

4 Clinical Presentation

Patients with PH are sometimes asymptomatic or manifest vague symptoms such as fatigue and dyspnea with exertion, especially in the early stages. This factor delays the diagnosis of PH and worsens the prognosis of patients. PH symptoms are typically non-specific and this makes the diagnosis challenging. Dyspnea is the most characteristic symptom of PH. Other signs and symptoms include hypoxia, tachypnea, tachycardia/bradycardia, fatigue, weakness, chest pain, syncope, exertional dyspnea, fatigue, weakness, light-headedness/ syncope and, less frequently, cough and respiratory distress. In advanced stages, a sudden cardiac death could happen to a patient, especially during exercise. Attention should be paid to signs and symptoms related to infectious diseases and pneumonia, such as fever and pulmonary crackles. Evidence of pulmonary edema due to left heart diseases and thromboembolic complications should also be considered as causes of dyspnea and hypoxia. Symptoms of progressive right heart failure that occur in later stages of PAH or advanced disease include edema, ascites, and abdominal distension. Clinical findings include increased second heart sound, prominent JVP, hepatojugular reflux, ascites, hepatomegaly, edema, tricuspid or pulmonary regurgitant murmurs, and S3 gallop (Dandoy et al. 2013; Frost et al. 2019; Jodele et al. 2013; Kawashima et al. 2021; Steward et al. 2004; Topyła-Putowska et al. 2021).

5 Diagnosis

The ECG may show right atrial (RA) enlargement or RV hypertrophy. Chest radiographs are abnormal in 90% of patients with PAH. Evidence on CXR in favor of increased pulmonary artery pressure includes central pulmonary artery dilatation, RA and RV enlargement, and peripheral dearborization. In CXR, other findings such as lung hyperinflation, pneumonia and primary lung disease, as well as features suggesting pulmonary emboli, may help to diagnose the cause of dyspnea or hypoxemia. When a patient exhibits symptoms or signs that suggest PH, echocardiography is the best primary test for determining the probability of PH. Transthoracic echocardiography is a non-invasive diagnostic tool for evaluating RV pressure, and PAP, and cardiac function in patients undergoing HSCT. Before HSCT, echocardiography is advised for all patients. Serial echocardiography after HSCT is recommended at least 3 and 12 months later, and in some cases more than 12 months later. Echocardiography alone is not enough to diagnose the cause and classify the type of PH. Tricuspid regurgitation with a peak velocity greater than 2.8 m/s may indicate PH. However, TR velocity alone cannot accurately determine whether PH is present or absent. The likelihood of PH is low when the peak tricuspid regurgitation velocity (TRV) is less than 2.8 m/s, which corresponds to an estimated systolic PAP (sPAP) of less than 35 mmHg, and there are no other symptoms. In the advanced stages of PAH in echocardiography, the RV is larger than the Left ventricle (LV), and flattening of the interventricular septum occurs. Other findings in favor of PH in echocardiography are IVC diameter > 21 mm with decreased inspiratory collapse, RA area (end-systole) > 18 cm², RVOT AT < 105 ms and/or mid-systolic notching, PA diameter > 25 mm, and TAPSE/sPAP ratio < 0.55 mm/mmHg (Fig. 2). Other diagnostic procedures that help to diagnose the cause of pulmonary hypertension include pulmonary function test, computed tomography of pulmonary arteries (CTPA) and ventilation-perfusion (V/Q) scan, cardiac magnetic resonance (CMR) imaging, positron emission tomography, and exercise testing. RHC is the gold standard for the diagnosis of PH. If mPAP is higher than 20 mmHg and PCWP is lower than 15 mmHg, pre-capillary hypertension is considered, and in these cases, it is recommended to perform Vasoreactivity test in RHC. Vasoreactivity test is performed to evaluate the response to treatment with calcium channel blockers (CCBs). Thrombocytopenia and coagulation disorders in HSCT recipients can prevent invasive diagnostic procedures like heart catheterization and lung biopsy in pulmonary hypertension patients. Patients who have RV failure and/or PH symptoms should be promptly evaluated by a cardiology/pulmonary hypertension team. Increases in the brain natriuretic peptide (BNP) and N-terminal pro-BNP indicate a worse prognosis and have a relationship with right ventricular overload (Dandoy et al. 2013; Frost et al. 2019; Humbert et al. 2022; Jodele et al. 2013; Kawashima et al. 2021; Lyon et al. 2022; Peter Libby et al. 2021; Topyła-Putowska et al. 2021).

6 Clinical Management Strategies and Treatment

Two common types of PH after HSCT are PAH and PVOD, which have a high mortality rate. However, on-time pulmonary hypertension therapy can improve patients' survival. Pulmonary vasodilators are the cornerstone of PAH medical therapy that include:

- Calcium channel blockers: Calcium channel blockers (e.g. nifedipine, diltiazem, and amlodipine) are used in patients who respond to vasoreactivity test. However, the response to calcium channel blockers is limited in most patients with PAH after HSCT.
- Endothelin receptor antagonists: Endothelin receptor antagonists (e.g. ambrisentan, Bosentan and Macitentan) inhibit pulmonary artery vasoconstriction and proliferation of smooth muscle cells. The use of these drugs might be limited in HSCT because of adverse hepatic mechanisms and hepatic complications, such as sinusoidal obstruction syndrome.
- Phosphodiesterase 5 inhibitors and guanylate cyclase stimulators: phosphodiesterase 5 inhibitors (e.g. sildenafil and tadalafil) and guanylate cyclase stimulators (e.g. riociguat) regulate the CGMP pathway and lead to pulmonary artery relaxation.

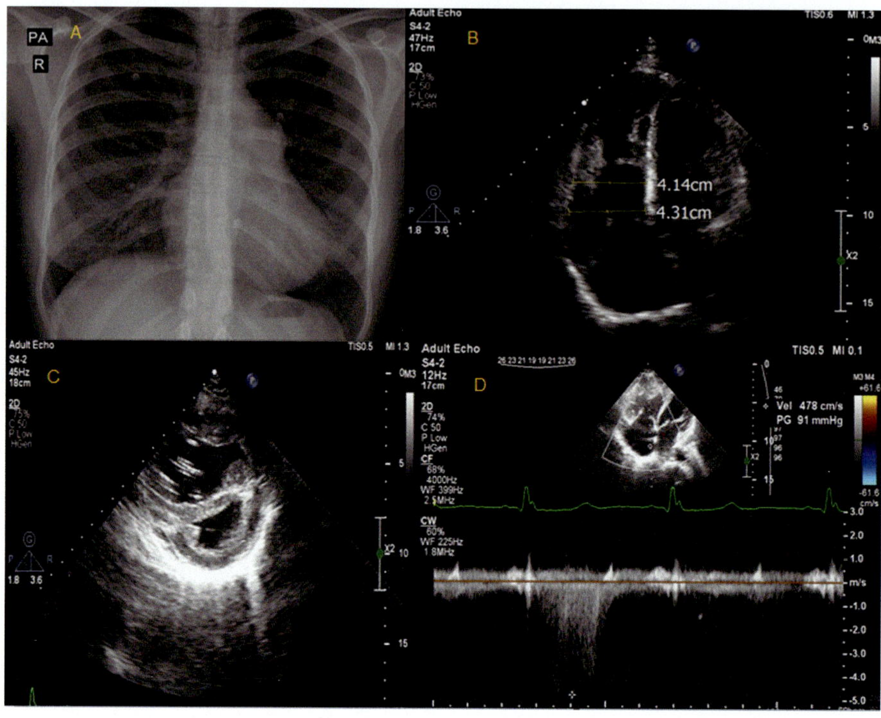

Fig. 2 A, chest X-ray in pulmonary arterial hypertension; B, Right ventricular enlargement in pulmonary hypertension; C, D-shape septum in transthoracic echocardiography; D, high tricuspid regurgitation pressure gradient

– Prostacyclin analogues and prostacyclin receptor agonists: Prostacyclin pathway dysregulate in PH patients, so the prostacyclin analogues (e.g. epoprostenol, iloprost and treprostinil) and prostacyclin receptor agonists (e.g. selexipag) can be suitable in these patients with anti-proliferative activity, vasodilation, and inhibition of platelet aggregation.

At the moment, there is no unique strategy for PAH treatment after HSCT. For severely ill patients, like those who are intubated or unstable, NO or prostacyclin analogues are recommended. One phosphodiesterase 5 inhibitor and one endothelin receptor antagonist are added to this therapy as well. If the patient is not getting enough responses to therapy, ECMO and lung transplantation should be considered.

Vasoreactivity test during RHC is recommended in patients with stable hemodynamics. In positive vasoreactivity test, calcium channel blockers may be useful. Phosphodiesterase 5 inhibitors and endothelin receptor antagonists should be considered in patients with negative test or non-responders to calcium channel blockers.

There is no definitive medical treatment for PVOD. Drugs used in PAH may be suitable for PVOD or worsen that. Lung transplantation is the only curative treatment for PVOD.

In addition to specific therapy for PH, supportive therapy is indicated. Diuretic therapy with loop diuretics (e.g. furosemide) and mineralocorticoid receptor antagonists (e.g. spironolactone) are necessary, especially in right-sided heart failure. Oxygen therapy is recommended in patients with lung disease and systemic O2 saturation < 60%. Anticoagulants are used in chronic thromboembolic patients. For multidisciplinary management with the oncology team, a PH center is recommended (Dandoy et al. 2013; Humbert et al. 2022; Kawashima et al. 2021; Levy et al. 2019; Lyon et al. 2022; Steward et al. 2004; Tanaka et al. 2021).

References

Cullivan S, Gaine S, Sitbon O. New trends in pulmonary hypertension. Euro Respir Rev. 2023;32(167).

Dandoy CE, Hirsch R, Chima R, Davies SM, Jodele S. Pulmonary hypertension after hematopoietic stem cell transplantation. Biol Blood Marrow Transplant. 2013;19(11):1546–56.

Frost A, Badesch D, Gibbs JSR, Gopalan D, Khanna D, Manes A, et al. Diagnosis of pulmonary hypertension. Euro Respir J. 2019;53(1).

Ghigna M-R, Dorfmüller P. Pulmonary vascular disease and pulmonary hypertension. Diagn Histopathol. 2019;25(8):304–12.

Humbert M, Kovacs G, Hoeper MM, Badagliacca R, Berger RM, Brida M, et al. 2022 ESC/ERS Guidelines for the diagnosis and treatment of pulmonary hypertension: developed by the task force for the diagnosis and treatment of pulmonary hypertension of the European society of cardiology (ESC) and the European respiratory society (ERS). Endorsed by the international society for heart and lung transplantation (ISHLT) and the European reference network on rare respiratory diseases (ERN-LUNG). Euro Heart J. 2022;43(38):3618–731.

Jodele S, Hirsch R, Laskin B, Davies S, Witte D, Chima R. Pulmonary arterial hypertension in pediatric patients with hematopoietic stem cell transplant–associated thrombotic microangiopathy. Biol Blood Marrow Transplant. 2013;19(2):202–7.

Kawashima N, Fukasawa Y, Nishikawa E, Ohta-Ogo K, Ishibashi-Ueda H, Hamada M, et al. Echocardiography monitoring of pulmonary hypertension after pediatric hematopoietic stem cell transplantation: pediatric pulmonary arterial hypertension and pulmonary veno-occlusive disease after hematopoietic stem cell transplantation. Transplant Cell Ther. 2021;27(9):786.e1–e8.

Levy M, Moshous D, Szezepanski I, Galmiche L, Castelle M, Lesage F, et al. Pulmonary hypertension after bone marrow transplantation in children. Euro Respir J. 2019;54(5).

Lyon AR, López-Fernández T, Couch LS, Asteggiano R, Aznar MC, Bergler-Klein J, et al. 2022 ESC guidelines on cardio-oncology developed in collaboration with the European hematology association (EHA), the European society for therapeutic radiology and oncology (ESTRO) and the international cardio-oncology society (IC-OS) developed by the task force on cardio-oncology of the European society of cardiology (ESC). Euro Heart J. 2022;43(41):4229–361.

Mandras SA, Mehta HS, Vaidya A, editors. Pulmonary hypertension: a brief guide for clinicians. In: Mayo clinic proceedings; 2020. Elsevier.

Peter Libby ROB, Mann DL, Tomaselli GF, Bhatt D, Solomon SD, Braunwald E. Braunwald's heart disease : a textbook of cardiovascular medicine. 12 ed. Philadelphia: Elsevier; 2021. 1322 p.

Simonneau G, Montani D, Celermajer DS, Denton CP, Gatzoulis MA, Krowka M, et al. Haemodynamic definitions and updated clinical classification of pulmonary hypertension. Euro Respir J. 2019;53(1).

Steward C, Pellier I, Mahajan A, Ashworth M, Stuart A, Fasth A, et al. Severe pulmonary hypertension: a frequent complication of stem cell transplantation for malignant infantile osteopetrosis. Br J Haematol. 2004;124(1):63–71.

Tanaka TD, Ishizawar DC, Saito T, Yoshii A, Yoshimura M. Successful treatment of pulmonary hypertension following hematopoietic stem cell transplant with a single oral tadalafil: a case report. Pulm Circ. 2021;11(3):20458940211027790.

Topyła-Putowska W, Tomaszewski M, Wysokiński A, Tomaszewski A. Echocardiography in pulmonary arterial hypertension: comprehensive evaluation and technical considerations. J Clin Med. 2021;10(15):3229.

Tuder RM, Abman SH, Braun T, Capron F, Stevens T, Thistlethwaite PA, et al. Development and pathology of pulmonary hypertension. J Am College Cardiol. 2009;54(1_Supplement_S):S3–S9.

Pericardial Disease in HSCT

Ardeshir Ghavamzadeh, Amir Hossein Emami, Kamran Roudini,
Kiara Rezaei Kalantari, Mina Mohseni, and Mehrdad Jafari Fesharaki

Abstract The pericardium has two layers: visceral and parietal. Pericardial diseases are common in cancer patients, which can be caused by underlying cancer or cancer treatments (e.g. chemotherapy, radiation therapy, and hematopoietic stem cell transplantation). The most common pericardial diseases are pericardial effusion, pericarditis and constrictive pericarditis. In this chapter, we examine pericardial diseases in patients who underwent hematopoietic stem cell transplantation (HSCT) (Pieri et al., in Eur Heart J Case Rep, 2020; Hoit, in Cardiol Clin 35:481–490, 2017).

A. Ghavamzadeh
Cancer and Cell Therapy Research Center, Tehran University of Medical Sciences, Tehran, Iran
e-mail: ghavamza@sina.tums.ac.ir

A. H. Emami
Hematology-Oncology Department, Cancer Institute, Imam Khomeini Hospital, School of medicine, Tehran University of Medical Sciences, Tehran, Iran
e-mail: emamiami@yahoo.com

K. Roudini
Department of Internal Medicine, Hematology and Medical Oncology Ward, Cancer Research Center, Cancer Institute, Imam Khomeini Hospital Complex, Tehran University of Medical Sciences, Tehran, Iran
e-mail: roudini.k@gmail.com

K. Rezaei Kalantari
Rajaie Cardiovascular Medical and Research Center, Iran University of Medical Sciences, Tehran, Iran
e-mail: rkkiara@gmail.com

M. Mohseni (✉)
Cardio-Oncology Research Center, Rajaie Cardiovascular Medical and Research Center, Iran University of Medical Sciences, Tehran, Iran
e-mail: mina.mohseni8694@gmail.com

M. Jafari Fesharaki
Department of Cardiology, School of Medicine, Shahid Beheshti University of Medical Sciences, Tehran, Iran
e-mail: Mehrjfmd@yahoo.com

Keywords Hematopoietic stem cell transplantation · Pericardial effusion · Pericarditis and constrictive pericarditis · Transplant-associated thrombotic microangiopathy · Graft versus host disease

Abbreviations

CMV	Cytomegalovirus
CP	Constrictive pericarditis
CRP	C-reactive protein
CT scan	Computerized tomography scan
CXR	Chest X-ray
ECG	Electrocardiogram
ESR	Erythrocyte sedimentation rate
GVHD	Graft versus host disease
HLA	Human leukocyte antigens
HSCT	Hematopoietic stem cell transplantation
ICU	Intensive care unit
IVC	Inferior vena cava
MRI	Magnetic resonance imaging
MV	Mitral valve
NSAID	Non-steroidal anti-inflammatory drugs
PE	Pericardial effusion
QTcD	Corrected QT dispersion
TA-TMA	Transplant-associated thrombotic microangiopathy
TBI	Total body irradiation
TV	Tricuspid valve

1 Pericardial Effusion

1.1 *Definition*

Hematopoietic stem cell transplantation (HSCT) is an effective and curative treatment for malignant and nonmalignant disorders such as cancers, hemoglobinopathies, and primary immunodeficiency syndromes in both adults and pediatrics. Although HSCT is an effective treatment, it is associated with significant morbidity and mortality (Versluys et al. 2018; Yanagisawa et al. 2015; Lyons et al. 2023; Kubo et al. 2021; Chen et al. 2016). Pericardial effusion (PE) is a known complication after HSCT. Significant pericardial effusion (PE) was defined as symptomatic moderate or large PE, with/without symptoms, is a rare complication following hematopoietic stem cell transplantation (HSCT). PE can result in significant morbidity and can progress to tamponade, which is life-threatening. As a result, early diagnosis and treatment and identification of risk factors for pericardial effusion are very important. Pericardial effusion can occur as early-onset (<100 days) or late-onset (>100 days) after HSCT. The development of significant pericardial effusion varies from 9 to 369 days in different studies (Yanagisawa et al. 2015; Chen et al. 2016; Liu et al. 2015, 2012; Diamond et al. 2018; Pfeiffer et al. 2017).

1.2   Diagnosis

Timely diagnosis of PE is very important because untreated PE can quickly turn into tamponade and require emergency drainage. Many times, PE is diagnosed incidentally during chest radiography, chest CT scan or echocardiography for other reasons, such as investigating the cause of infection or changing vital signs. This point can indicate that we need more frequent monitoring in these patients for timely diagnosis of PE. It is recommended that echocardiography be performed before HSCT and also upon admission to the intensive care unit (ICU), or during the occurrence of respiratory distress, hypoxia, shock, shortness of breath, and during the diagnosis of transplant-associated thrombotic microangiopathy (TA-TMA). Some studies suggest that in addition to performing echocardiography in the above conditions, patients should undergo echocardiography at specific intervals (for example, days +7, +30 and +100) for timely diagnosis of PE. However, the frequency of this monitoring is not well defined; depending on the patient's risk factors, these intervals can be shorter (Lyons et al. 2023; Pfeiffer et al. 2017).

ECG, chest radiography, M-mode and two-dimensional Doppler echocardiography, CT and MRI are helpful for diagnosing pericardial effusion and tamponade, but echocardiography is the best and most available evaluation tool. The changes that can be seen in the ECG in pericardial effusion are tachycardia, reduced voltage, electrical alternans, and non-specific changes. The CXR can be normal or the size of the cardiac shadow is increased. Evidence of tamponade in echocardiography includes early diastolic collapse of right ventricle, late diastolic collapse of right atrium, exaggerated respiratory variation on Mitral and tricuspid valves inflow, swinging heart and decreased IVC collapse. PEs are graded as trivial (only seen in systole), small (echo free space in diastole < 10 mm), moderate (10–20 mm), large (>20 mm), and very large (>25 mm). Evidence of tamponade in echocardiography are seen in a small number of PE cases. The occurrence of symptoms caused by pericardial effusion depends on the speed of fluid accumulation in the pericardial space and its amount. Echocardiography during follow-up should be repeated frequently to determine the effect of drugs prescribed for PE and to determine the optimal timing of pericardiocentesis (Versluys et al. 2018; Liu et al. 2015; Libby et al. 2021).

1.3 Symptoms

PE, especially in the early stages, can be completely asymptomatic or have non-specific symptoms, but there may be tamponade physiology in echocardiography. Non-specific symptoms such as tachypnea, dyspnea, vomiting and dizziness have been reported in patients with PE. The severity of clinical symptoms does not determine the severity of PE or patient outcome. Clinical and/or echocardiographic signs of tamponade are very important. Patients with significant PE can progress to tamponade. If the patient experiences tamponade, blood pressure drop and even shock can happen, which will be life-threatening, and timely diagnosis and treatment can save the patient's life and improve outcomes (Versluys et al. 2018; Chen et al. 2016; Pfeiffer et al. 2017).

1.4 Epidemiology

The incidence of post HSCT PE in pediatrics is significantly higher than in adults. Many studies have reported the incidence of PE in symptomatic or significant PE cases. If small or trivial PE cases are also considered, PE is definitely much higher in post HSCT cases (Versluys et al. 2018; Yanagisawa et al. 2015; Liu et al. 2015, 2012; Diamond et al. 2018; Pfeiffer et al. 2017; Tinianow et al. 2020). In one study incidence of Insignificant and significant PE after HSCT was 37.8% (Diamond et al. 2018). Whether or not PE affects survival is still controversial, but it seems that morbidity and mortality are higher in patients who develop PE after HSCT. It is still not clear that the reason for the lower survival in these patients is the presence of PE or other causes. Mortality from infection or graft versus host disease (GVHD) was higher in patients with PE than those without that. The exact incidence of PE is not known but the incidence of significant PE in children has been reported in different studies between 0.2 and 19% and in adults between 0.8 and 6.3% (Versluys et al. 2018; Yanagisawa et al. 2015; Lyons et al. 2023; Kubo et al. 2021; Chen et al. 2016; Liu et al. 2015, 2012; Diamond et al. 2018; Pfeiffer et al. 2017; Tinianow et al. 2020; Leong et al. 2020).

1.5 Etiology and Risk Factors

Many studies have investigated the risk factors of PE after HSCT. If PE occurs in patients after HSCT, the usual causes of PE should be investigated and ruled out first; including infection, heart failure, fluid overload, drug hypersensitivity, endocrine dysfunction and renal disorders. The possibility of recurrence of malignancy and primary disease relapse, must always be considered. GVHD both acute GVHD (in early onset PE < 100 days) and chronic GVHD (in late onset PE > 100 days) can cause PE. Cardiac injury by various mechanisms can cause PE. Acute drug toxicity from conditioning regimen can cause PE. Pre-transplant conditioning with radiation and high-dose cyclophosphamide can cause cardiac damage and PE. Myeloablative conditioning is a pre-HSCT risk factor associated with PE in pediatric patients. Iron overload caused by repeated transfusions may also cause cardiac damage. High dose cyclophosphamide and total body irradiation (TBI) are well-known causes of early onset post-transplant PE due to Immediate cardiac toxicity. Compared with autologous HSCT, allogenic HSCT (either with sibling or matched unrelated donor) is a risk factor for the occurrence of significant PE. Multiple transplant procedures seem to be associated with the incidence of PE after transplantation. Children are more prone to PE than adults. Age over 50 years in adults and younger age in children are possible risk factors for PE after HSCT. Delayed neutrophil engraftment and positive cytomegalovirus (CMV) serostatus of the recipient, Epstein Bar virus/cytomegalovirus viremia, systemic hypertension, toxicity of conditioning or post-transplant immune suppression, sinusoidal obstruction syndrome (venoocclusive disease), reaction to sirolimus, tacrolimus or any immunosuppressant, female sex, high-risk disease prior to HSCT, prolonged neutropenia, toxicity of conditioning, cord blood cell source, HLA mismatch, Diseases other than solid tumor and unrelated transplant donor are also other possible risk factors for PE after transplantation. Tapering of immunosuppressive agents for stable extensive chronic GVHD and Calcineurin inhibitor toxicity may cause PE after HSCT. transplant-associated thrombotic microangiopathy (TA-TMA) is an important risk factor for PE after HSCT. The occurrence of TA-TMA and PE after HSCT have many common risk factors; including advanced primary disease, unrelated donor, infection, conditioning regimen, HLA mismatch and GVHD. TA-TMA is also one of the side effects of calcineurin inhibitors. The occurrence of PE after TA-TMA can be due to third spacing and edema due to nephropathy, myocardial ischemia, congestive heart failure or myocarditis. Screening echocardiography is recommended in patients with TA-TMA to investigate post-HSCT PE. prolonged preoperative and the corrected QT dispersion (QTcD) may be a predictor of PE after HSCT. One of the causes of prolonged QTcD is previous chemotherapy, especially with an anthracycline-containing regimen (Versluys et al. 2018; Yanagisawa et al. 2015; Lyons et al. 2023; Kubo et al. 2021; Chen et al. 2016; Liu et al. 2015, 2012; Diamond et al. 2018; Pfeiffer et al. 2017; Tinianow et al. 2020; Leong et al. 2020) (Fig. 1).

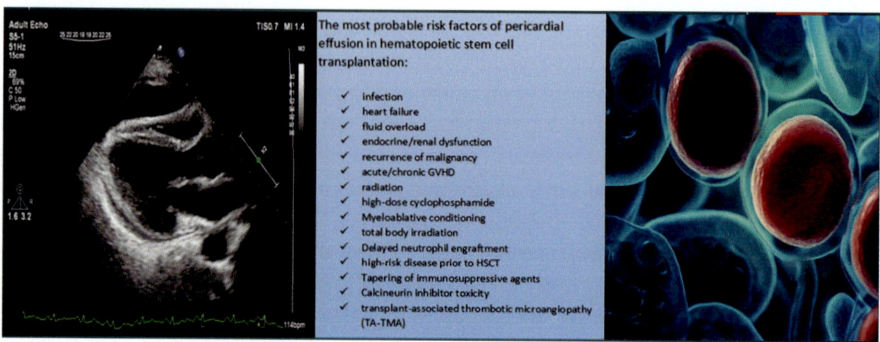

Fig. 1 The most probable risk factors of pericardial effusion in hematopoietic stem cell transplantation

1.6 Treatment

Significant PE can lead to cardiac tamponade, which is potentially life-threatening; As a result, careful management of PE is very important. One of the main treatments for significant PE and tamponade is pericardiocentesis, which is necessary in many cases of early-onset PE. Pericardiocentesis is an effective and safe procedure for the treatment of significant PE. enhanced immunosuppression should be considered for the treatment of late-onset PE caused by chronic GVHD, along with emergency pericardiocentesis or pericardial window placement. Because tapering immunosuppressive agents can cause PE, close monitoring of cardiac status and evaluation for recurrent PE is recommended. Follow-up echocardiography is recommended to check the incidence of recurrent PE. Discontinuation of calcineurin inhibitors is recommended as an effective treatment for PE in cases where the possible cause of PE is calcineurin inhibitor toxicity. If infection is suspected, appropriate antibiotic and antiviral treatments are recommended. If there is a doubt about drug side effects as the cause of PE, it is recommended to stop or switch the drug according to the patient's condition (Fig. 2) (Versluys et al. 2018; Yanagisawa et al. 2015; Kubo et al. 2021; Liu et al. 2015, 2012; Pfeiffer et al. 2017).

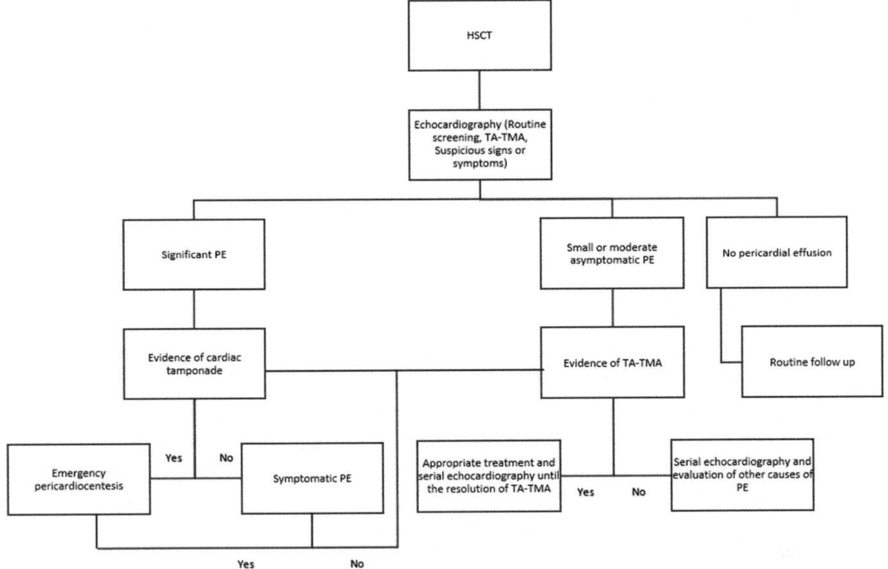

Fig. 2 Approach to PE after HSCT. HSCT: Hematopoietic stem cell transplantation; PE: Pericardial effusion; TA-TMA: Transplant-associated thrombotic microangiopathy

2 Pericarditis

2.1 Definition

The pericardium consists of two layers: visceral and parietal. Inflammation of the pericardial layers is called pericarditis. Pericarditis can be associated with PE. Pericarditis can occur as acute, chronic, recurrent. If pericarditis lasts more than 3 months, it is called chronic pericarditis. If the patient has pericarditis again after an asymptomatic period of 4 to 6 weeks, it is called recurrent pericarditis. After HSCT, serositis including pericarditis can occur for various reasons and can even cause pericardial effusion (Leonard et al. 2015; Chiabrando et al. 2020).

2.2 Diagnosis

Two of the following four criteria must be present to diagnose acute pericarditis. Criteria include chest pain, pericardial rub, electrocardiogram (ECG) changes, and development or worsening of pericardial effusion. Inflammatory markers such as ESR, CRP and evidence of inflammation in imaging such as cardiac MRI can also be helpful for diagnosis. Several diagnoses are proposed for patients who experience chest pain and ECG changes after HSCT, one of which is pericarditis (Chiabrando et al. 2020; Freyer et al. 2021; Shinohara et al. 2016).

2.3 Symptoms

Sharp chest pain that gets better with leaning forward and worse with breathing is the most important symptom of pericarditis. Other symptoms are non-specific and include restlessness, dyspnea, anxiety and low-grade fever (Chiabrando et al. 2020; Freyer et al. 2021; Shinohara et al. 2016).

2.4 Epidemiology

Pericarditis after HSCT is uncommon and the incidence of 0.9–5.7% has been reported in different studies. Pericarditis alone is not usually fatal unless it is myopericarditis. Although GVHD is more common in women, pericarditis is more common in men (Leonard et al. 2015; Freyer et al. 2021; Shinohara et al. 2016; Karakulska-Prystupiuk et al. 2018; Baker et al. 2020; Duléry et al. 2021).

2.5 Etiology and Risk Factors

In many cases, the exact etiology of pericarditis is not known. Pericarditis and pericardial effusion can occur simultaneously. Early onset pericarditis, especially those that occur before engraftment, can occur due to acute conditioning toxicity. Late onset pericarditis can occur due to acute or chronic GVHD, GVHD prophylaxis, or other immunologic events. Many cases of pericarditis occur during tapering of immunosuppression. Cyclophosphamide conditioning/GVHD prophylaxis is associated with both PE and pericarditis. High doses of cyclophosphamide before HSCT can cause myopericarditis. Cyclophosphamide-induced cardiac damage can occur anytime from a few days to about 20 days after administration of the drug. Cytarabine is one of the drugs that can cause pericarditis. Cardiac damage and pericarditis caused by cytarabine can be caused by an anaphylactic reaction, which usually occurs during drug administration. Cytarabine-induced pericarditis can occur even at low doses. Engraftment syndrome, mediastinal irradiation and TBI can also cause cardiac damage and pericarditis. Viral, bacterial or fungal Infections, should always be considered as a potential cause of pericarditis. human herpesvirus-6, Coxsackie B3, CMV infection, Legionella pneumophila and Mycobacterium tuberculosis can be mentioned among the infectious agents. Maintenance therapy with dasatinib or even nilotinib can cause pericarditis. Azacitidine is also among the drugs that can cause pericarditis (Leonard et al. 2015; Freyer et al. 2021; Shinohara et al. 2016; Baker et al. 2020; Duléry et al. 2021; Lopez-Fernandez et al. 2021; Yoshida et al. 2015; El-Asmar et al. 2016; Ataş et al. 2020; Tuzovic et al. 2019; Newman et al. 2016; Karamooz et al. 2017).

2.6   Treatment

Acute pericarditis can be a self-limiting disease. In the general population, the combination of NSAIDs and colchicine is recommended for the treatment of pericarditis. However, in patients after HSCT, the use of NSAID is limited due to the thrombocytopenia of the patient and the risk of bleeding and the possibility of NSAID-induced nephrotoxicity. The use of colchicine is also challenging due to its gastrointestinal side effects and drug interactions with calcineurin inhibitors and azole antifungals. Corticosteroids can be an alternative to NSAIDs for the treatment of pericarditis, but may increase the risk of recurrent pericarditis. Therefore, the treatment of pericarditis after HSCT is challenging and requires more studies. Long-term treatment with colchicine reduces the risk of recurrent pericarditis if the patient tolerates it. Treatment with colchicine for less than 3 months increases the risk of recurrent pericarditis. Corticosteroid therapy in patients with GVHD can also be effective for the treatment of pericarditis (Leonard et al. 2015; Freyer et al. 2021; Shinohara et al. 2016; Ataş et al. 2020).

3   Constrictive Pericarditis

3.1   Definition

Constrictive pericarditis (CP) occurs in the case of fibrosis, thickening or stiffness of the pericardium which reduces pericardial compliance and causes impaired ventricular filling (Pieri et al. 2020).

3.2   Diagnosis

Echocardiography is a good tool to diagnose CP. Findings in echocardiography favoring CP include septal bounce, annulus reversus, normal LV wall thickness, increased pericardial thickness, Inferior vena cava plethora and exaggerated respiratory variation on MV and TV inflows. ECG in CP can show nonspecific ST-segment changes, atrial fibrillation and sinus tachycardia. B-type natriuretic peptide (BNP) and N-terminal pro-B-type natriuretic peptide (Nt pro-BNP) levels are usually normal or mildly elevated. Pericardial calcification is uncommonly seen on CXR. Accurate diagnosis requires multimodality imaging including echocardiography, CT scan, cardiac MRI and even cardiac catheterization. In echocardiography, an increase in the thickness of the pericardium (>4 mm) may be seen, but CT scan and CMR are more accurate for examining the pericardium (Pieri et al. 2020; Chiabrando et al. 2020; Garcia 2016).

3.3 Symptoms

Common symptoms of CP include edema of the lower extremities, dyspnea on exertion, and fatigue. Patients with CP are at risk of pulmonary edema in case of excessive hydration (Garcia 2016; Farge et al. 2021).

It is difficult to differentiate constrictive pericarditis (CP) from restrictive cardiomyopathy. Signs of constrictive pericarditis include predominantly right heart failure, elevated JVP, prominent "x" and rapid "y" descent, Kussmaul's sign (failure to decrease in jugular venous pressure during inspiration) and pulsus paradoxus, hepatomegaly, splenomegaly, ascites, and edema. Pericardial knock is the specific heart sound of CP. Pleural effusion is more common than pulmonary edema in patients with CP (Garcia 2016; Cavet et al. 2000).

3.4 Etiology and Risk Factors

Radiation of the thorax and mediastinum can cause late onset CP and manifest even more than two decades after radiation exposure. Radiation affects all parts of the heart, including the myocardium, coronary arteries, heart valves, and pericardium. CP can be a complication of acute or chronic pericarditis, but it is uncommon. Leukemic infiltration can rarely manifest as CP. GVHD following HSCT is one of the rare causes of CP. Short-term use of methotrexate may lead to accumulation in the pericardium during acute pericarditis, which may contribute to the development of constrictive pericarditis (Pieri et al. 2020; Shinohara et al. 2016; Karamooz et al. 2017; Garcia 2016; Cavet et al. 2000; Acıbuca et al. 2019).

3.5 Treatment

In cases where significant degrees of pericardial constriction, pericardectomy is usually required. In cases where CP is caused by radiation, the prognosis is poor. In mild CP cases, sodium restriction and diuretic therapy can help control lower limb edema and hepatic congestion but surgery is needed in most cases. Long-term outcome in post-irradiation CP is worse due to the involvement of all parts of the heart, including the myocardium, coronary arteries and cardiac valves. Transient constrictive pericarditis may occur, especially in cases after heart surgery, which improves with anti-inflammatory treatments. In cases of CP caused by chronic GVHD, treatment with targeted immunotherapy can be helpful in addition to surgical treatment. Graft-versus-host disease is a rare but potentially reversible cause of CP, requiring surgery and appropriate immunomodulation (Pieri et al. 2020; Garcia 2016; Cavet et al. 2000).

4 Effusive-Constrictive Pericarditis

Effusive-constrictive pericarditis occurs in the form of pericardial effusion and constriction, so that right atrial pressure remains high even after pericardiocentesis. The presence of septal bounce in echocardiography indicates the presence of effusive-constrictive pericarditis. This condition occurs in about 10% of cases of tamponade, and its management is similar to CP cases (Karamooz et al. 2017; Garcia 2016).

References

Acıbuca A, Yeral M, Koçer NE, Koc Z, Güllü H. Relapsed acute myeloid leukemia presenting with myocardial hypertrophy and constrictive pericardial physiology. Anatol J Cardiol. 2019;21(5):287.

Ataş E, Kutluk MT, Akyüz C. Cardiac complications in patients who underwent to hematopoietic stem cell transplantation. J Cancer Res Ther. 2020;16(1):53–9.

Baker JK, Shank-Coviello J, Zhou B, Dixon J, McCorkle R, Sarpong D, et al. Cardiotoxicity in hematopoietic stem cell transplant: keeping the beat. Clin Lymphoma Myeloma Leukemia 2020;20(4):244–51. e4.

Cavet J, Lennard A, Gascoigne A, Finney R, Lucraft H, Richardson C, et al. Constrictive pericarditis post allogeneic bone marrow transplant for Philadelphia-positive acute lymphoblastic leukaemia. Bone Marrow Transplant. 2000;25(5):571–3.

Chen X, Zou Q, Yin J, Wang C, Xu J, Wei J, et al. Pericardial effusion post transplantation predicts inferior overall survival following allo-hematopoietic stem cell transplant. Bone Marrow Transplant. 2016;51(2):303–6.

Chiabrando JG, Bonaventura A, Vecchié A, Wohlford GF, Mauro AG, Jordan JH, et al. Management of acute and recurrent pericarditis: JACC state-of-the-art review. J Am Coll Cardiol. 2020;75(1):76–92.

Diamond M, Ruiz-Mesa C, Corrales-Medina FF, Tamariz LJ, Ziga E, Swaminathan S. Incidence and outcome of pericardial effusion in pediatric patients after hematopoietic stem cell transplant: a single-institution experience. J Pediatr Hematol Oncol. 2018;40(2):132–6.

Duléry R, Mohty R, Labopin M, Sestili S, Malard F, Brissot E, et al. Early cardiac toxicity associated with post-transplant cyclophosphamide in allogeneic stem cell transplantation. Cardio Oncol. 2021;3(2):250–9.

El-Asmar J, Kharfan-Dabaja MA, Ayala E. Acute pericarditis and tamponade from Coxsackie B3 in an adult hematopoietic-cell-allograft recipient: a rare but potentially serious complication. Hematol Oncol Stem Cell Ther. 2016;9(2):82–5.

Farge D, Abdallah NA, Marjanovic Z, Del Papa N. Autologous stem cell transplantation in scleroderma. La Presse Médicale. 2021;50(1):104065.

Freyer CW, Fradley M, Madnick D, Carver JR, Frey NV, Gill SI, et al. Characterization of pericarditis following allogeneic hematopoietic cell transplantation. Transplant Cellular Ther. 2021;27(11):934. e1–e6.

Garcia MJ. Constrictive pericarditis versus restrictive cardiomyopathy? J Am Coll Cardiol. 2016;67(17):2061–76.

Hoit BD. Anatomy and physiology of the pericardium. Cardiol Clin. 2017;35(4):481–90.

Karakulska-Prystupiuk E, Basak G, Dwilewicz-Trojaczek J, Paluszewska M, Boguradzki P, Jędrzejczak W, editors. Pericarditis in patients with chronic graft-vs-host disease. In: Transplantation proceedings; 2018. Elsevier.

Karamooz E, Nguyen N, Gold JA. Rapid onset of hypoxemia in a man with graft-versus-host disease after stem cell transplant leukemia. Ann Am Thorac Soc. 2017;14(2):283–7.

Kubo H, Imataki O, Fukumoto T, Oku M, Ishida T, Kubo YH, et al. Risk factors for and the prognostic impact of pericardial effusion after allogeneic hematopoietic stem cell transplantation. Transplant Cellular Therap. 2021;27(11):949. e1–e8.

Leonard J, Newell L, Meyers G, Hayes-Lattin B, Gajewski J, Heitner S, et al. Chronic GvHD-associated serositis and pericarditis. Bone Marrow Transplant. 2015;50(8):1098–104.

Leong K, Heal ME, Bass JL, Aggarwal V, Narasimhan S, Gupta A, et al. Effects of systemic steroid administration on recurrence of pericardial effusion in pediatric patients after hematopoietic stem cell transplantation. J Pediatr Hematol Oncol. 2020;42(4):256–60.

Libby P, Bonow RO, Mann DL, Tomaselli GF, Bhattt DL, Solomon S. Braunwald's heart disease: a textbook of cardiovascular medicine, 12th ed. Philadelphia: Elsevier; 2021. volumes cm p.

Liu Y-C, Gau J-P, Hong Y-C, Yu Y-B, Hsiao L-T, Liu J-H, et al. Large pericardial effusion as a life-threatening complication after hematopoietic stem cell transplantation—association with chronic GVHD in late-onset adult patients. Ann Hematol. 2012;91:1953–8.

Liu YC, Chien SH, Fan NW, Hu MH, Gau JP, Liu CJ, et al. Risk factors for pericardial effusion in adult patients receiving allogeneic haematopoietic stem cell transplantation. Br J Haematol. 2015;169(5):737–45.

Lopez-Fernandez T, Vadillo IS, de la Guia AL, Barbier KH. Cardiovascular issues in hematopoietic stem cell transplantation (HSCT). Curr Treat Options Oncol. 2021;22(6):51.

Lyons K, Dham N, Schwartz B, Dávila Saldaña BJ. Risk factors for symptomatic pericardial effusions posthematopoietic stem cell transplant. J. Transplant. 2023;2023.

Newman M, Malla M, Gojo I. Azacitidine-induced pericarditis: a case series. Pharmacother J Hum Pharmacol Drug Ther. 2016;36(4):443–8.

Pfeiffer T, Rotz S, Ryan T, Hirsch R, Taylor M, Chima R, et al. Pericardial effusion requiring surgical intervention after stem cell transplantation: a case series. Bone Marrow Transplant. 2017;52(4):630–3.

Pieri CA, Roberts N, Gribben J, Manisty C. Graft-versus-host disease: a case report of a rare but reversible cause of constrictive pericarditis. Eur Heart J. Case Rep. 2020.

Shinohara A, Honda A, Nukina A, Takaoka K, Tsukamoto A, Hangai S, et al. Acute pericarditis before engraftment in hematopoietic stem cell transplantation. Bone Marrow Transplant. 2016;51(3):459–61.

Tinianow A, Gay JC, Bearl DW, Connelly JA, Godown J, Kitko CL. Pericardial effusion following hematopoietic stem cell transplantation in children: Incidence, risk factors, and outcomes. Pediatr Transplant. 2020;24(5):e13748.

Tuzovic M, Mead M, Young PA, Schiller G, Yang EH. Cardiac complications in the adult bone marrow transplant patient. Curr Oncol Rep. 2019;21:1–16.

Versluys A, Grotenhuis H, Boelens M, Mavinkurve-Groothuis A, Breur J. Predictors and outcome of pericardial effusion after hematopoietic stem cell transplantation in children. Pediatr Cardiol. 2018;39:236–44.

Yanagisawa R, Ishii E, Motoki N, Yamazaki S, Morita D, Sakashita K, et al. Pretransplant-corrected QT dispersion as a predictor of pericardial effusion after pediatric hematopoietic stem cell transplantation. Transpl Int. 2015;28(5):565–74.

Yoshida M, Nakamae H, Okamura H, Nishimoto M, Hayashi Y, Koh H, et al. Pericarditis associated with human herpesvirus-6 reactivation in a patient after unrelated cord blood transplant. Exp Clin Transplant. 2015.

Metabolic Syndrome in HSCT

Zahra Ghaemmaghami, Mohammad Javad Alemzadeh-Ansari,
Alireza Rezvani, and Alireza Bari

Abstract Hematopoietic cell transplantation (HCT) increases metabolic syndrome risk that directly related to increased cardiovascular morbidity and mortality. The cause of this increase can be divided into direct or indirect injury. The direct injury is related to the cardiovascular effects of cancer treatment (cardiotoxicity of the chemotherapy regime and/or radiotherapy before HSCT and an indirect injury includes preexisting CVRF/CVD and CV lifestyle change (i.e., deconditioning, weight gain) during and after HSC all this factor leads to an increase in metabolic syndrome that has a higher prevalence in HSCT patients, it is important to have a strategy for prevention, early detection and appropriate treatment of metabolic syndrome component in this population. In this chapter, we discuss the importance, prevention, and therapeutic strategy of metabolic syndrome in HSCT patients.

Keywords Metabolic syndrome · Hyperlipidemia · Insulin resistance ·
Hypertriglyceridemia

Abbreviations

BMI	Body mass index
CV	Cardiovascular

Z. Ghaemmaghami
Rajaie Cardiovascular Medical and Research Center, Iran University of Medical Sciences, Tehran, Iran

M. J. Alemzadeh-Ansari
Cardiovascular Intervention Research Center, Rajaie Cardiovascular Medical and Research Center, Iran University of Medical Sciences, Tehran, Iran

A. Rezvani (✉)
Hematology and Oncology Department, Shiraz University of Medical Science, Shiraz, Iran
e-mail: ar.rezvani@hotmail.com

A. Bari
Hematology and Oncology Department, Faculty of Medicine, Mashhad University of Medical Sciences, Mashhad, Iran

© The Author(s), under exclusive license to Springer Nature Switzerland AG 2024
A. Alizadehasl et al. (eds.), *Cardiovascular Considerations in Hematopoietic Stem Cell Transplantation*, https://doi.org/10.1007/978-3-031-53659-5_17

CVD   Cardiovascular disease
CVRF   Cardiovascular risk factor
GVHD   Graft versus host disease
HSCT   Haematopoietic stem cell transplantation
Mets   Metabolic syndrome
PFM   Pure fat mass
TBI   Total body irradiation

1   Introduction

Life expectancy of HSCT recipient increased in recent years that related to significant decrease of treatment-related mortality due to advances in transplantation techniques, supportive care, prevention of GVHD and infections, and patient selection, it is estimated that the prevalence of HSCT survivors will reach more than half a million by 2030 (Passweg et al. 2020; Majhail et al. 2013).Furthermore, HSCT recipient have a higher chance of developing a number of metabolic and cardiovascular diseases, such as high blood pressure, Obesity, and dyslipidemia, abnormal glucose metabolism constitutes the metabolic syndrome component. cardiovascular diseases (CVD) are a major competing risk for morbidity and mortality, for up to 10 years, after transplantation (Majhail et al. 2013; Oliveira et al. 2021; Armenian et al. 2017). Long-term CV morbidity and mortality are exacerbated by metabolic syndrome and supportive medications (such as corticoids). This is especially true after allogeneic HSCT (Bielorai et al. 2017).

Metabolic syndrome (MetS), or syndrome X or Insulin resistance syndrome, is a combination of cardiovascular disease risk factors and, that described in 1999 for first time. The Three most known definitions for metabolic syndrome used for studies and health care plans are shown in (Table 1) (Alberti et al. 2009).

This chapter aims to provide HSCT recipients with a practical and multidisciplinary method for the avoidance, detection, and treatment of metabolic syndrome. Given the significance of diabetes and HTN, these two components of the metabolic syndrome are covered in different chapters.

2   Epidemiology and Importance of Metabolic Syndrome After HSCT

HSCT recipients experience mortality rates of 1.4 to 2.3 times higher than the general population (Bhatia et al. 2007, 2005) due to an increased risk of dyslipidemia, hypertension, and diabetes in these patients, which can result in non-fatal and fatal cardiovascular disease. Cardiovascular death in HSCT recipient is 3.6 per 1000 person-years, that more frequent than the overall population. The incidences of

Table 1 The most known definitions for metabolic syndrome

	NCEP ATP III 2005	IDF/AHA 2009
Definition	≥3 risk factors	≥ 3 risk factors
Risk factor		
Abdominal obesity	Waist circumference: >102 cm (>40 in) in men: >88 cm (>35 in) in women	Waist circumference population and country—specific definitions
Triglycerides	≥150 mg/dl (≥1.7 mmol/L) or drug treatment for elevated levels	≥ 150 mg/dl (≥ 1.7 mmol/L) or drug treatment for elevated levels
HDL cholesterol		
Men	<40 mg/dl (<1.0 mmol/L) for drug treatment for reduced levels	<40 mg/dl (<1.0 mmol/L) for drug treatment for reduced levels
Women	<50 mg/dl (<13 mmol/L) or drug treatment for reduced levels	<50 mg/dl (<13 mmol/L) or drug treatment for reduced levels
Blood pressure	≥130/≥85 mm Hg or drug treatment for HTN	≥130/≥85 mm Hg or drug treatment for HTN
Fasting glucose	≥100 mg/dl (≥ 6.11mmol/L) or drug treatment for DM	≥100 mg/dl (≥ 6.11mmol/L) or drug treatment for DM

stroke, rhythm problems, ischemic coronary heart disease, cardiomyopathy or heart failure, and vascular diseases rose among transplant recipients (Taskinen et al. 2000; Scott Baker et al. 2007).

In one study Diabetes and hypertension prevalence were reported 7.6% and 18.5% of HCT survivors, compared to 3.1% and 14.9% of their siblings. In another study, the prevalence of diabetes, dyslipidemia, and hypertension in HCST patients compared to the general population in the same range of age was 18.7 versus 8.5%, 43 versus 40%, 43 versus 34.6%, respectively significantly higher than the general population (Armenian et al. 2012).

There are no consistent variations in the rate and risks of other components after total body irradiation or receipt of an allogeneic versus an autologous graft, with the exception of an elevated rate of hypertension among allogeneic graft recipients.

In a retrospective cohort analysis, HSCT recipients had higher cardiovascular morbidity and mortality risks than the general population. Compared to those of the same age in the comparison group, the rates were highest for people under 60 years of age and those with several cardiovascular risk factors (diabetes, hypertension, dyslipidemia, and renal disease) (Chow et al. 2011). After the first two years after transplantation, the rates of hypercholesterolemia and hypertriglyceridemia have been found to be as high as 73.4% and 72.5%, respectively. After allogenic HCT, This the onset of dyslipidemia after HSCT is rapid as show hypertriglyceridemia and hypercholesterolemia typically appear 8 and 11 months after transplantation respectively (Marini et al. 2015).

Due to the increasing number of older people who are a candidate for HSCT and have comorbid conditions, manifestations of metabolic syndrome after HSCT are increased and detected earlier after HSCT (Scott et al. 2016; Chow et al. 2014).

With regard to the high prevalence of metabolic syndrome in HCST recipients, it is important to pay specific attention to this matter and develop a strategy for the prevention, early detection, and appropriate treatment of the metabolic syndrome component in these patients (Blaser et al. 2012; Griffith et al. 2010).

3 Pathogenesis

The cause of CVD acceleration in HCT recipients can be classified as direct or indirect injury. Direct injury is related to the cardiovascular side effects of cancer treatment (cardiotoxicity of chemotherapy and/or radiation therapy before HSCT), while indirect injury includes preexisting CVRF / CVD and CV lifestyle change (i.e., deconditioning, weight gain) during and after HSCT Patients referred for HSCT have an elevated CVD risk: among patients, 25% have hypertension, 5% have diabetes, and 32% have dyslipidemia.

Although, Immune reconstitution, elevated inflammatory cytokines, stem cell source, donor-recipient interactions, altered intestinal microbiome after HCT, immunosuppressive drugs, and endocrine dysfunction caused by drugs or radiation therapy have been evaluated in more studies and proposed as perpetrator factors, but the meticulous pathophysiology and etiology of metabolic and increased CVD mortality in HCT recipients is not yet fully understood (Turcotte et al. 2016).

In allogenic HSCT, body weight and composition change, and in the initial post-HSCT period, the BMI of the limb is lower (Ishikawa et al. 2019). Regardless of how comparable their BMIs were with their siblings, survivors of TBI with or without CNS radiation had a decrease in lean body mass (LBM), and PFM was higher in all groups of HCT survivors, according to a study that compares the body mass index (BMI) of HCT recipients with their siblings (Ketterl et al. 2018).

Increased protein catabolism, which causes muscular atrophy and a drop in lean body mass index in line with sarcopenic Obesity, is a significant contributor to the risk of metabolic syndrome (Lu et al. 2013; Skaarud et al. 2019) and are related to metabolic syndrome and heart disease (Shah et al. 2014). Obesity predisposes to a pro-inflammatory state by increasing levels of the completely anti-inflammatory protein adiponectin while decreasing levels of pro-inflammatory cytokines such as IL-6 and TNF-a (Ellulu et al. 2017). Early post-transplant changes in body composition are likely caused by stress-related factors such as infection and GVHD, while late effects on a lean body mass index are likely caused by hormonal and pituitary dysfunction, as well as metabolic stressors that can contribute to a reduced rate of muscle mass regeneration (Vantyghem et al. 2014).

TBI significantly increases the risk of heart failure, coronary artery disease, conduction problems, and pericardial effusion, as well as diabetes, dyslipidemia, metabolic syndrome, and hypertension Increased visceral Obesity and body composition alterations are associated with the prevalent finding of GH deficiency in HCT survivors exposed to TBI (Ratosa and Pantar 2019; Meacham et al. 2010; Attallah et al. 2006).

According to one study, HCT survivors who underwent TBI with or without CNS irradiation had higher levels of leptin and lower levels of adiponectin compared to their siblings. All HCT survivors' groups had higher levels of IL-6 than their siblings As we know, Obesity and increased adipose tissue as an endocrine organ lead to increased circulating levels of leptin and induce a state of chronic subclinical inflammation (Iikuni et al. 2008). Leptin plays a crucial role in metabolism and inflammatory reactions, so elevated leptin has a role in the pathogenesis of chronic inflammation. Leptin levels and chronic GVHD are related, as are an increase in plasma leptin level and leptin resistance, which are linked to the development of metabolic syndrome (Chait and Hartigh 2020; Tseng et al. 2017; Wu et al. 2016).

Early post-HSCT dyslipidemia, diabetes, hypertension, conditioning-related irradiation, and the use of cardiotoxic medications are associated with accelerated and premature atherosclerotic disease (Tichelli et al. 2008; Armenian and Bhatia 2008). Whole-body irradiation (TBI) conditioning has been shown to increase the incidence of dyslipidemia and diabetes. Hypertriglyceridemia and hypercholesterolemia are present in acute GVHD after allogeneic transplantation (OR 3.4) (Armenian et al. 2012; Rackley et al. 2005).

One explanation is that high-dose chemotherapy and radiation cause DNA damage, This damage is particularly severe within the mitochondria which results in mitochondrial dysfunction, that leads to fat storage and hyperlipidemia (Turcotte et al. 2016).

In other hand Dyslipidaemia has been associated with immunosuppressive treatments for graft versus host disease, including glucocorticoids, cyclosporine, sirolimus, and combination therapy with corticosteroids and tacrolimus. Corticosteroids are believed to contribute to weight gain Very low-density lipoprotein production from the liver increases along with total cholesterol and triglyceride levels when insulin resistance is present. By preventing the production of cholesterol bile acids and preventing its transit, cyclosporine increases LDL levels and affects the removal of VLDL and LDL cholesterol from the body Endocrinopathies such as uncontrolled diabetes, hypothyroidism, growth hormone deficiency, and hypogonadism that are seen as a result of HCT related treatment have been associated with dyslipidemia (Iannuzzo et al. 2022; Hulzebos et al. 2004; Savani and Tichelli 2021). Although treatment-related exposure in patients before, during, and after transplantation process is proposed as underlying mechanism in the evolution of metabolic syndrome after but is not fully explanatory and the literature supports the role of immune cells and cytokines in the pathophysiology of metabolic syndrome development and progression (Fahed et al. 2022). Chronic GVHD (grades II-IV) has been associated with an increased risk of hypertension, diabetes, and dyslipidemia (Rackley et al. 2005) (Table 2) shows a reasonable etiology of dyslipidemia in metabolic syndrome.

Low-grade systemic inflammation and adipokines. Although we do not yet know whether the inflammatory status of the stem cell donor can affect the graft inflammatory responses, it has been suggested that the source of stem cells may be another factor that contributes to the development of metabolic syndrome due to differences in the inflammatory response between stem cell sources, including bone marrow,

Table 2 Cause of dyslipidemia in allo-HSCT recipient

	Decrease activity of LDL-C enzyme TGs metabolism In adipose tissue at high glucose concentration	VLDL TGs, and TC↑
Immunosuppressant's	**Glucocorticoids** Reduced expression of GLUT2 and glucokinase, B-cell apoptosis, increased insulin resistance in skeletal muscle, increased production of triglycerides and VLDL-C as a result of insulin resistance, and decreased clearance of triglyceride-rich lipoproteins like chylomicron and VLDL-C by corticosteroid-induced LPL are all effects of this. lowers LDL receptors and enhances esterase activity. result in weight increase	VLDL TGs, and TC↑
	CNI Inhibition of cholesterol conversion into bile acid by Inhibit steroid 26 hydroxylase, LDL receptor blocking leading to elevation of serum LDL, change in lipase activity affecting VLDL and LEL clcarancc. Incrcase hepatic lipase activity and reduce LPL via interfering with the binding of LDL-C to LDL receptors. Stop pancreatic B cells from secreting insulin by attaching to cyclin-D and triggering b cell death. Interfere with LDL-C to LDL receptors binding increase hepatic lipase activity, decrease of LPL. Inhibit insulin secretion of pancreatic B cells by binding to	VLDL, and LDL↑
	mTOR inhibitors Boost hepatic VLDL production and boost the FFA pool. Enhance lipolysis through enhancing hormone-sensitive lipase (raising circulating FFA), interfering with triglyceride metabolism, reducing triglyceride storage, and interfering with the insulin signaling system	VLDL, and TGs ↑
GVHD	Produce nephritic syndrome and intrahepatic cholestasis. To address the high metabolic needs of rapid clonal growth, T cells involved in GVHD rely on lipid biosynthesis pathways. gut microbiota disturbance	TC and TGs↑
Intestinal micro flora	The enzymes involved in the metabolism of three substances can change according to changes in the gastrointestinal micro flora	Multiple metabolic disorders

(continued)

Table 2 (continued)

	Decrease activity of LDL-C enzyme TGs metabolism In adipose tissue at high glucose concentration	VLDL TGs, and TC↑
Nutrition intake	Excessive intake of nutrients that are heavy in sugar and fat	TC and TGs↑
Others	Hypogonadism, hyperthyroidism, hypothyroidism, and a lack of growth hormone. The majority of them are connected to diseases like TBI or cytotoxic agents	Multiple metabolic disorders

Abbreviations GLUT2 Glucose transporter 2, IR Insulin resistance, VLDL Very-low-density lipoprotein, LDL Low-density lipoprotein, LPL Lipoprotein lipase, FFA Free fatty acid, TBI Total body irradiation (Lu et al. 2022)

peripheral blood and umbilical cord blood. Alterations may be involved in the acceleration of metabolic and CVD complications According to certain studies, recurrence of the original pretransplantation disease in survivors who underwent HCT for a peripheral blood source of stem cells mobilised with primary growth factor was associated with an increased risk of several cardiovascular and related diseases compared to controls (Chow et al. 2011).

Another factor that may have a role in the improvement of metabolic syndrome after HCT is the alteration of the intestinal microbial composition in HCT recipients that due to prolonged antibiotic use and specific complications after HCT such as GVHD, with regard to the emerging evidence for the role of the gut microbiome in changing metabolic health, it is likely that this alteration contributes to a higher prevalence of metabolic syndrome in HCT recipients (Vantyghem et al. 2014).

4 Prevention, Treatment, and Management of Metabolic Syndrome Component After HSCT

The baseline CV assessment aims to manage the CV toxicity risk and diagnose and cure subclinical illnesses. People without overt CVD or a history of cardiotoxicity should pursue primary prevention measures, whereas those with known CVDs or signs of CV toxicity should consider secondary prevention therapy (Lyon et al. 2020). Complete medical history and physical examination, body mass index calculation, heart rate and blood pressure assessment, fasting blood glucose, lipid profile and HbA1c test should all be part of CV evaluation (Piepoli et al. 2020). Clinically obvious CVDs can be delayed or prevented by early detection, alteration of CVRF and subclinical CVDs, and appropriate therapies (Piepoli et al. 2020).

It is recommended to aggressively manage CVRF during the HSCT process using best practices According to current recommendations for HCT recipients, similar screening procedures for dyslipidemia are recommended in the general population because the development of dyslipidemia after HSCT can occur up to three months after the procedure (Marini et al. 2015; Bravo-Jaimes et al. 2020; Majhail et al. 2012).

Lifestyle changes, including quitting smoking, eating a balanced diet, quitting drinking too much alcohol, and participating in regular aerobic activity, are essential for survivorship care in the treatment and prevention of CVRF and CVD. The best CVRF objectives and treatment plans for HSCT survivors are not determined by prospective, randomised studies.

Treatment requirement and drug selection are like the general population in each specific condition, but the in hsct recipient it is crucial to pay specific attention to comorbidity and drug-drug interactions.

To treat dyslipidemia Before HSCT, check the lipid profile at baseline. Three key measures to prevent and treat dyslipidemia include a good diet, lifestyle changes, and medication. These should be implemented at 3 months, 12 months, and then annually following HSCT, as well as every 3–6 months in high-risk individuals. Lifestyle modifications are important to keep LDL less than 70mg /dl in high-risk patients and less than 55 mg/dl in very high-risk patients.

Is if TG is more than 500 mg/dL initiation of fibrates is recommended for general people (Lopez-Fernandez et al. 2021).

Evaluation of body mass index with each visit for Obesity at least once every year—A change in lifestyle In view of the existence of GH shortage and insulin resistance in HCT patients as well as the involvement of GH and insulin-like growth factor 1 in the formation and function of visceral adipose tissue, it is advised to reduce caloric intake and increase moderate-intensity exercise. It may be worthwhile to evaluate GH levels and investigate GH replacement treatment in individuals with GH deficits to reduce insulin resistance and central Obesity, since GH insufficiency may play a role in the establishment of a sarcopenic obese phenotype that is predisposed to metabolic and CVD risks (Lewitt 2017).

Beneficiaries of hrct A nutritionist should supervise overall calorie intake to ensure that patients receive the daily nutrients they need.

Lipid-lowering drugs:

Statins, ezetimibe, fibrates, and niacin are the lipid-lowering treatments for hyperlipidemia that are used the most frequently. Some more recent lipid-lowering medications, including PCSK9 inhibitors, have progressively made their way into clinical practice and may have both lipid-lowering and anti-GVHD influence; as a result, they may be employed as a good anti-GVHD option (Armenian et al. 2012)though their effect on lipid levels and CV outcome are promising, but we do not have any trial for their safety and efficacy in HSCT recipients. (Table 3) summarised the lipid-lowering drugs that were used in HSCT recipients.

With progress in supportive medical care and improvement in patient outcome survival in HSCT recipient, on the other hand, some complication and drugs increased CV risk factors in these patients. Therefore, appropriate preventive and therapeutic strategies are an important issue and require special attention.

Table 3 Lipid-lowering drugs after allo- HSCT

Lipid-lowering drugs	Mechanism	Side effect	Drug interacting
Statins	(1) Inhibit competitively the conversion of substances that appeared to be more severe to HMGCOA reeducates (2) Block the process that produces cholesterol via key isoprenoid intermediates (3) Reduce immune activity and maintain it. Reduce MHC-II to lower T cell activating and co-stimulatory chemicals on APCs. Boost the production of interleukin 10, a TH2 subtype anti-inflammatory cytokine (1) Inhibit competitively the conversion of substances that appeared to be more severe to HMGCOA reeducates (2) Block the process that produces cholesterol via key isoprenoid intermediates (3) Reduce immune activity and maintain it. Reduce MHC-II to lower T cell activating and co-stimulatory chemicals on APCs. Boost the production of interleukin 10, a TH2 subtype anti-inflammatory cytokine (4) Other effects include decreasing oxidative stress and inflammation, increasing atherosclerotic plaques, inhibiting the thrombosis response, and enhancing endothelial function	Elevated transaminases, myositis, and rhabdomyolysis	Through its actions on the cell membrane transporter mulidrugresistant protein2, cyclosporine can raise the blood level of statins Statins are more likely to be harmful Detrimental when used with CYP3A4 inhibitors such as azole antifungals, non-dihydropyridine calcium channel blockers (such as verapamil and diltiazem), and macrolide antibiotics

(continued)

Table 3 (continued)

Lipid-lowering drugs	Mechanism	Side effect	Drug interacting
Niacin	(1) inhibits glycerin esterase activity in adipose tissue (2) Enhances LPL activity and promotes the hydrolysis of plasma TGs	Flushing, gastrointestinal intolerance, hyperglycemia, and hyperuricemia are made worse, and the blood pressure is raised	Lower effect of ganglion blockers, calcium channel bookers, and adenoid inhibitors
PCSK9 inhibitor	Binds to LDLR, facilitates LDLR entry into liver cells, and is lysosomally destroyed	Respiratory tract infection	

Abbreviations HMG-COA 3—hydroxyl 3- methylglutaryl-coenzyme A, APC—Antigen-presenting cells, MHC-II Major histocompatibility complex II expression. TH2 T helper 2, LPL lipoprotein lipase, LDLR LDL receptor

References

Alberti KG, Eckel RH, Grundy SM, Zimmet PZ, Cleeman JI, Donato KA, et al. Harmonizing the metabolic syndrome: a joint interim statement of the international diabetes federation task force on epidemiology and prevention; national heart, lung, and blood institute; American heart association; world heart federation; international atherosclerosis society; and international association for the study of obesity. Circulation. 2009;120(16):1640–5.

Armenian SH, Bhatia S. Cardiovascular disease after hematopoietic cell transplantation–lessons learned. Haematologica. 2008;93(8):1132–6.

Armenian SH, Sun C-L, Vase T, Ness KK, Blum E, Francisco L, et al. Cardiovascular risk factors in hematopoietic cell transplantation survivors: role in development of subsequent cardiovascular disease. Blood J Am Soc Hematol. 2012;120(23):4505–12.

Armenian SH, Horak D, Scott JM, Mills G, Siyahian A, Teh JB, et al. Cardiovascular function in long-term hematopoietic cell transplantation survivors. Biol Blood Marrow Transplant. 2017;23(4):700–5.

Attallah H, Friedlander AL, Hoffman AR. Visceral obesity, impaired glucose tolerance, metabolic syndrome, and growth hormone therapy. Growth Hormon IGF Res. 2006;16:62–7.

Bhatia S, Robison LL, Francisco L, Carter A, Liu Y, Grant M, et al. Late mortality in survivors of autologous hematopoietic-cell transplantation: report from the bone marrow transplant survivor study. Blood. 2005;105(11):4215–22.

Bhatia S, Francisco L, Carter A, Sun C-L, Baker KS, Gurney JG, et al. Late mortality after allogeneic hematopoietic cell transplantation and functional status of long-term survivors: report from the bone marrow transplant survivor study. Blood J Am Soc Hematol. 2007;110(10):3784–92.

Bielorai B, Weintraub Y, Hutt D, Hemi R, Kanety H, Modan-Moses D, et al. The metabolic syndrome and its components in pediatric survivors of allogeneic hematopoietic stem cell transplantation. Clin Transplant. 2017;31(3):e12903.

Blaser BW, Kim HT, Alyea EP III, Ho VT, Cutler C, Armand P, et al. Hyperlipidemia and statin use after allogeneic hematopoietic stem cell transplantation. Biol Blood Marrow Transplant. 2012;18(4):575–83.

Bravo-Jaimes K, Marcellon R, Varanitskaya L, Kim PY, Iliescu C, Gilchrist SC, et al. Opportunities for improved cardiovascular disease prevention in oncology patients. Curr Opin Cardiol. 2020;35(5):531–7.

Chait A, Den Hartigh LJ. Adipose tissue distribution, inflammation and its metabolic consequences, including diabetes and cardiovascular disease. Front Cardiovasc Med. 2020;7:22.

Chow EJ, Mueller BA, Baker KS, Cushing-Haugen KL, Flowers ME, Martin PJ, et al. Cardiovascular hospitalizations and mortality among recipients of hematopoietic stem cell transplantation. Ann Intern Med. 2011;155(1):21–32.

Chow EJ, Baker KS, Lee SJ, Flowers ME, Cushing-Haugen KL, Inamoto Y, et al. Influence of conventional cardiovascular risk factors and lifestyle characteristics on cardiovascular disease after hematopoietic cell transplantation. J Clin Oncol. 2014;32(3):191.

Ellulu MS, Patimah I, Khaza'ai H, Rahmat A, Abed Y. Obesity and inflammation: the linking mechanism and the complications. Arch Med Sci. 2017;13(4):851–63.

Fahed G, Aoun L, Bou Zerdan M, Allam S, Bou Zerdan M, Bouferraa Y, et al. Metabolic syndrome: updates on pathophysiology and management in 2021. Int J Mol Sci. 2022;23(2):786.

Griffith ML, Savani BN, Boord JB. Dyslipidemia after allogeneic hematopoietic stem cell transplantation: evaluation and management. Blood J Am Soc Hematol. 2010;116(8):1197–204.

Hulzebos CV, Bijleveld CM, Stellaard F, Kuipers F, Fidler V, Slooff MJ, et al. Cyclosporine A-Induced reduction of bile salt synthesis associated with increased plasma lipids in children after liver transplantation. Liver Transpl. 2004;10(7):872–80.

Iannuzzo G, Cuomo G, Di Lorenzo A, Tripaldella M, Mallardo V, Iaccarino Idelson P, et al. Dyslipidemia in transplant patients: which therapy? J Clin Med. 2022;11(14):4080.

Iikuni N, Kwan Lam QL, Lu L, Matarese G, Cava AL. Leptin and inflammation. Curr Immunol Rev. 2008;4(2):70–9.

Ishikawa A, Otaka Y, Kamisako M, Suzuki T, Miyata C, Tsuji T, et al. Factors affecting lower limb muscle strength and cardiopulmonary fitness after allogeneic hematopoietic stem cell transplantation. Support Care Cancer. 2019;27:1793–800.

Ketterl TG, Chow EJ, Leisenring WM, Goodman P, Koves IH, Petryk A, et al. Adipokines, inflammation, and adiposity in hematopoietic cell transplantation survivors. Biol Blood Marrow Transplant. 2018;24(3):622–6.

Lewitt MS. The role of the growth hormone/insulin-like growth factor system in visceral adiposity. Biochem Insights. 2017;10:1178626417703995.

Lopez-Fernandez T, Vadillo IS, de la Guia AL, Barbier KH. Cardiovascular issues in hematopoietic stem cell transplantation (HSCT). Curr Treat Options Oncol. 2021;22(6):51.

Lu C-W, Yang K-C, Chang H-H, Lee L-T, Chen C-Y, Huang K-C. Sarcopenic obesity is closely associated with metabolic syndrome. Obes Res Clin Pract. 2013;7(4):e301–7.

Lu Y, Ma X, Pan J, Ma R, Jiang Y. Management of dyslipidemia after allogeneic hematopoietic stem cell transplantation. Lipids Health Dis. 2022;21(1):1–13.

Lyon AR, Dent S, Stanway S, Earl H, Brezden-Masley C, Cohen-Solal A, et al. Baseline cardiovascular risk assessment in cancer patients scheduled to receive cardiotoxic cancer therapies: a position statement and new risk assessment tools from the Cardio-oncology study group of the heart failure association of the European society of cardiology in collaboration with the international cardio-oncology society. Eur J Heart Fail. 2020;22(11):1945–60.

Majhail NS, Rizzo JD, Lee SJ, Aljurf M, Atsuta Y, Bonfim C, et al. Recommended screening and preventive practices for long-term survivors after hematopoietic cell transplantation. Rev Bras Hematol Hemoter. 2012;34:109–33.

Majhail NS, Tao L, Bredeson C, Davies S, Dehn J, Gajewski JL, et al. Prevalence of hematopoietic cell transplant survivors in the United States. Biol Blood Marrow Transplant. 2013;19(10):1498–501.

Marini BL, Choi SW, Byersdorfer CA, Cronin S, Frame DG. Treatment of dyslipidemia in allogeneic hematopoietic stem cell transplant patients. Biol Blood Marrow Transplant. 2015;21(5):809–20.

Meacham LR, Chow EJ, Ness KK, Kamdar KY, Chen Y, Yasui Y, et al. Cardiovascular risk factors in adult survivors of pediatric cancer—a report from the childhood cancer survivor study. Cancer Epidemiol Biomark Prev. 2010;19(1):170–81.

Oliveira GH, Al-Kindi SG, Guha A, Dey AK, Rhea IB, deLima MJ. Cardiovascular risk assessment and management of patients undergoing hematopoietic cell transplantation. Bone Marrow Transplant. 2021;56(3):544–51.

Passweg JR, Baldomero H, Chabannon C, Basak GW, Corbacioglu S, Duarte R, et al. The EBMT activity survey on hematopoietic-cell transplantation and cellular therapy 2018: CAR-T's come into focus. Bone Marrow Transplant. 2020;55(8):1604–13.

Piepoli MF, Abreu A, Albus C, Ambrosetti M, Brotons C, Catapano AL, et al. Update on cardiovascular prevention in clinical practice: a position paper of the European association of preventive cardiology of the European society of cardiology. Eur J Prev Cardiol. 2020;27(2):181–205.

Rackley C, Schultz KR, Goldman FD, Chan KW, Serrano A, Hulse JE, et al. Cardiac manifestations of graft-versus-host disease. Biol Blood Marrow Transplant. 2005;11(10):773–80.

Ratosa I, Pantar MI. Cardiotoxicity of mediastinal radiotherapy. Rep Pract Oncol Radiother. 2019;24(6):629–43.

Savani BN, Tichelli A. Blood and marrow transplantation long term management: survivorship after transplant. Wiley Online Library; 2021.

Scott JM, Armenian S, Giralt S, Moslehi J, Wang T, Jones LW. Cardiovascular disease following hematopoietic stem cell transplantation: pathogenesis, detection, and the cardioprotective role of aerobic training. Crit Rev Oncol Hematol. 2016;98:222 34.

Scott Baker K, Ness KK, Steinberger J, Carter A, Francisco L, Burns LJ, et al. Diabetes, hypertension, and cardiovascular events in survivors of hematopoietic cell transplantation: a report from the bone marrow transplantation survivor study. Blood. 2007;109(4):1765–72.

Shah RV, Murthy VL, Abbasi SA, Blankstein R, Kwong RY, Goldfine AB, et al. Visceral adiposity and the risk of metabolic syndrome across body mass index: the MESA study. JACC Cardiovasc Imag. 2014;7(12):1221–35.

Skaarud K, Veierød MB, Lergenmuller S, Bye A, Iversen PO, Tjønnfjord GE. Body weight, body composition and survival after 1 year: follow-up of a nutritional intervention trial in allo-HSCT recipients. Bone Marrow Transplant. 2019;54(12):2102–9.

Taskinen M, Saarinen-Pihkala UM, Hovi L, Lipsanen-Nyman M. Impaired glucose tolerance and dyslipidaemia as late effects after bone-marrow transplantation in childhood. The Lancet. 2000;356(9234):993–7.

Tichelli A, Bhatia S, Socié G. Cardiac and cardiovascular consequences after haematopoietic stem cell transplantation. Br J Haematol. 2008;142(1):11–26.

Tseng P-W, Wu D-A, Hou J-S, Hsu B-G. Leptin is an independent marker of metabolic syndrome in elderly adults with type 2 diabetes. Tzu-Chi Med J. 2017;29(2):109.

Turcotte LM, Yingst A, Verneris MR. Metabolic syndrome after hematopoietic cell transplantation: at the intersection of treatment toxicity and immune dysfunction. Biol Blood Marrow Transplant. 2016;22(7):1159–66.

Vantyghem M, Cornillon J, Decanter C, Defrance F, Karrouz W, Leroy C. Le 617 Mapihan K, Couturier MA, De Berranger E, Hermet E, Maillard N, Marcais A, 618 Francois S, Tabrizi R, Yakoub-Agha I, Société Française de Thérapie Cellulaire. 619 Management of endocrino-metabolic dysfunctions after allogeneic hematopoietic stem 620 cell transplantation. Orphanet J Rare Dis. 2014;9(162):621.

Wu D, Dawson N, Levings M. Obesity-associated adipose tissue inflammation and transplantation. Am J Transplant. 2016;16(3):743–50.

HSCT in Patients with Cardiac Amyloidosis

Marjan Hajahmadi and Soroush Rad

Abstract Cardiac amyloidosis characterize as a syndrome due to deposition of amyloid in the extra cardiomyocyte space. It divides in two most common subtypes: AL and ATTR that are the reason of more than 95% of cases. Because of the heterogeneity in clinical presentation delay in diagnosis is common. In recent years with effective therapies, the outcomes improved and the importance of early detection and prevention of profound organ failure is further enhanced. Treatment of AL amyloidosis is complicated and The backbone of treatment is mailnly multi anti-plasma cell therapies. A growing number of studies on this syndrome are ongoing, to make treatment strategies more and more evidence-based. One of the main part of AL Amyloidosis management is supportive care. When inherent organ dysfunction (specially heart failure),occurs effective therapy is going to have limited efficiency. High dose of Melphalan is indicated for eligible patients for stem cell transplantation, followed by stem cell rescue. ASCT is done for selected AL amyloidosis patients when therapy-related mortality is acceptable. This review brings an update summery on the diagnostic approach and treatment of cardiac amyloidosis AL type.

Abbreviations

ASCT	Autologous stem cell transplantation
CCB	Calcium channel blocker
CM	Cardiomyopathy

M. Hajahmadi
Cardiovascular Department, Rasoul Akram General Hospital Iran University of Medical Sciences, Tehran, Iran
e-mail: hajahmadipour.m@iums.ac.ir

S. Rad (✉)
Internal Medicine Department, Hematology and Oncology Ward, Tehran University of Medical Science Shariati General Hospital, Tehran, Iran
e-mail: srad@sina.tums.ac.ir

Hematology, Oncology and Stem Cell Transplantation Research Center, Hematology, Oncology and Stem Cell Transplantation Research Institute, Tehran, Iran

A. Alizadehasl et al. (eds.), *Cardiovascular Considerations in Hematopoietic Stem Cell Transplantation*, https://doi.org/10.1007/978-3-031-53659-5_18

CMR	Cardiac MRI
FLC	Free light chain
GLS	Global longitudinal strain
HSCT	Hematopoietic stem cell transplantation
hATTR	Hereditary transthyretin amyloidosis
HFpEF	Heart failure preserved ejection fraction
IFE	Immunofixation electrophoresis
ICD	Implantable cardioverter defibrillator
IHC	Immunohistochemistry
LVH	Left ventricular hypertrophy
MGUS	Monoclonal gammopathy of undetermined significance
NT-proBNP	N-terminal prohormone of brain natriuretic peptide
TDI	Tissue Doppler echocardiography

1 Introduction

When amyloid fibril deposit in the extra myocyte space a syndrome called Cardiac amyloidosis develop. It can cause signs or symptoms or may be diagnosed by screening in patients with extra cardiac signs of amyloidosis. Nonspecific presentation make diagnosis delay happen so many patients are diagnosed when advanced organ failure has occurred.

High dose chemotherapy and stem cell rescue (Auto-HSCT) is the best optional treatment for AL-amyloidosis. It is important that we know different subtypes of amyloidosis that cause cardiac manifestations and distinct the AL type between them to plan an effective treatment as soon as possible.

In this chapter we focus on diagnostic approach to cardiac amyloidosis and recognizing the AL subtype and treatment options for AL-amyloidosis including HSCT.

2 Types of Cardiac Amyloidosis

The most common types of cardiac amyloidosis are (Fontana et al. 2019):

1. Light chain (AL amyloidosis)—is known also as primary systemic amyloidosis. AL amyloidosis caused by deposition of misfolded immunoglobulin light chains from a plasma cell dyscrasia. Diagnostic criteria and suspicious patients evaluation algorithm are in Table 1 and Fig. 1 (Finsterer et al. 2019; Fontana 2023).

Table 1 Systemic Al Amyloidosis (all criteria are needed)

Amyloid related systemic syndrome
– Nephrotic proteinuria-non diabetic
– Restrictive CMP or non specified CHF
– Increase in NT-proBNP level, without known heart disease
– Carpal tunnel disease, indescribable edema, hepatosplenomegaly
– Indescribable neck and facial purpura
– Macroglossia

Verification of a monoclonal plasma-cell proliferative disorder
– Serum&Urine IF (Immunofixation)
– Imbalance in kappa/lambda ratio
– Bone marrow clonal plasma cells

Evidence of amyloid deposition by positive Congo Red in involved organ or any other tissue
– Bone marrow (65%)
– Fat pad (77%)
– Biopsy of involved organ

Evidence of Light chain amyloid deposition
– Mass spectrometry
– Immunohistochemistry (IHC)

2. Transthyretin amyloidosis (ATTR amyloidosis)—the deposited amyloid's percursor protein is transthyretin (TTR or Prealbumin), TTR is a protein synthesized by the liver responsible for thyroid hormone and retinol (vitamin A) transport. Two subtypes for ATTR amyloidosis is known:

   - Wild-type amyloidosis (wtATTR)—formerly called senile systemic amyloidosis is due to misfolded or wild-type transthyretin deposition. vATTR or AL Amyloidosis are more common and more severe than this subtype (Connors et al. 2016).
   - Hereditary amyloidosis (hATTR)—occurs by TTR gene mutations, these mutations make instability of the tetrameric complex of transthyretin,,make it misfold then deposit. The hATTR is usually transmitted as autosomal dominant with variable penetrance. Over 120 different mutations linked with hATTR amyloidosis is known (Finsterer et al. 2019).

3. Other types: Other types of amyloidosis including AA amyloidosis (serum amyloid-A), AApoA-1 (hereditary ApolipoproteinA-1), AApoA-4 (ApolipoproteinA-4 amyloidosis) and IAA (Isolated Atrial Amyloidosis) over producing of ANP involves primarily atrium which manifest commonly with AF, rarely causes cardiac amyloidosis (Fontana 2023).

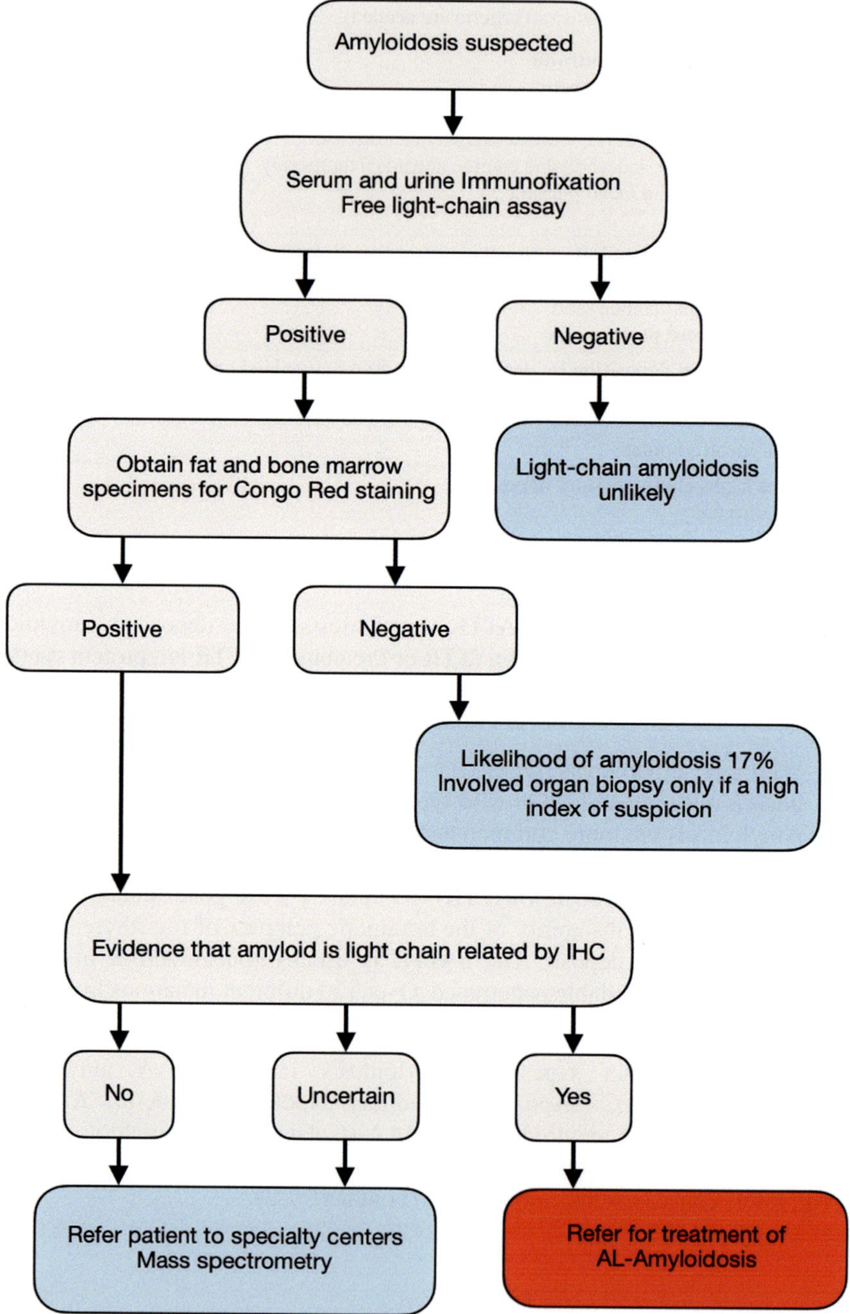

Fig. 1 Diagnostic algorithm for AL-amyloidosis suspected patients

3 Epidemiology

Contrary to what was thought Cardiac Amyloidosis is relatively common in certain patient groups however since it is under diagnosed. The real incidence of cardiac amyloidosis is not well known, but probably estimated between 5 and 13 per million per year.

AL amyloidosis is the most common subtype of amyloidosis rate as 65% in UK and for 93% of cases in China. In 50–75% of AL amyloidosis there is evidence of heart involvement (Nienhuis et al. 2016).

The tendency of men above women to develop AL amyloidosis is mostly in China compared to Western countries (2:1 & 1.3:1). Incidence increase in both developing and developed countries, by age (Nienhuis et al. 2016).

In John Hopkins hospital 108 patients with HFpEF, 14% were diagnosed with cardiac amyloidosis after endomyocardial biopsy(15 patient, 7 wATTR, 4 with vATTR, 3 with AL) (Hahn et al. 2020). Untreated survival of AL CMP is on average 1.5 years, vATTR CMP 2.5 years and for wATTR the survival is 3.6 years (Kittleson et al. 2020). However with early detection and novel therapies of disease, this malignant nature would change (Sukhacheva et al. 2016).

4 Clinical Manifestation (When to Suspect)

Progressive deposit of amyloid fibers in the heart leads to restrictive cardiomyopathy and due to elevated diastolic pressures, heart failure syndrome presents. Symptoms can be non specified and mimic HFpEF syndrome. Atrial fibrillation is more common in AL and ATTR and increasing with advanced age and comorbidities. AF associates with more clinical symptoms but has no impact on survival. In AL amyloidosis sudden cardiac death can be happen commonly due to electromechanical dissociation rather than ventricular arrhythmia (Fuster et al. 2022).

Cardiac amyloidosis should be suspected in these specific scenarios:

- When a patient presents sign and symptoms of HFpEF without hypertension.
- Beta-blockers and non-dihydropyridine calcium channel blockers intolerance developing blunted cardiac output due to lowering heart rate in state of small LV size and fixed cardiac output.
- Presence of LVH, but reduced tolerance of antihypertensive medication.
- Orthostatic hypotension or syncope after taking angiotensin converting enzyme inhibitors reflections neurohormonal insufficiency.
- Autonomic dysfunction and heart failure syndrome.
- Neurological signs history of neuropathy, carpal tunnel syndrome, spinal cord stenosis associated with heart failure presentation.
- Inappropriate ECG voltage with the severity of LVH seen in imaging modalities.

5 Diagnostic Evaluation

A schematic approach to patients with suspicion of cardiac amyloidosis is in Fig. 2 (Kittleson et al. 2020; Castano et al. 2016; Gillmore et al. 2016; Knight et al. 2019; Carvalho et al. 2019).

The key for achieving success in early detection and hence timely treatment of cardiac amyloidosis is having appropriate suspicion and following the diagnostic work up for suspected patients. The presence of non cardiac manifestation is always

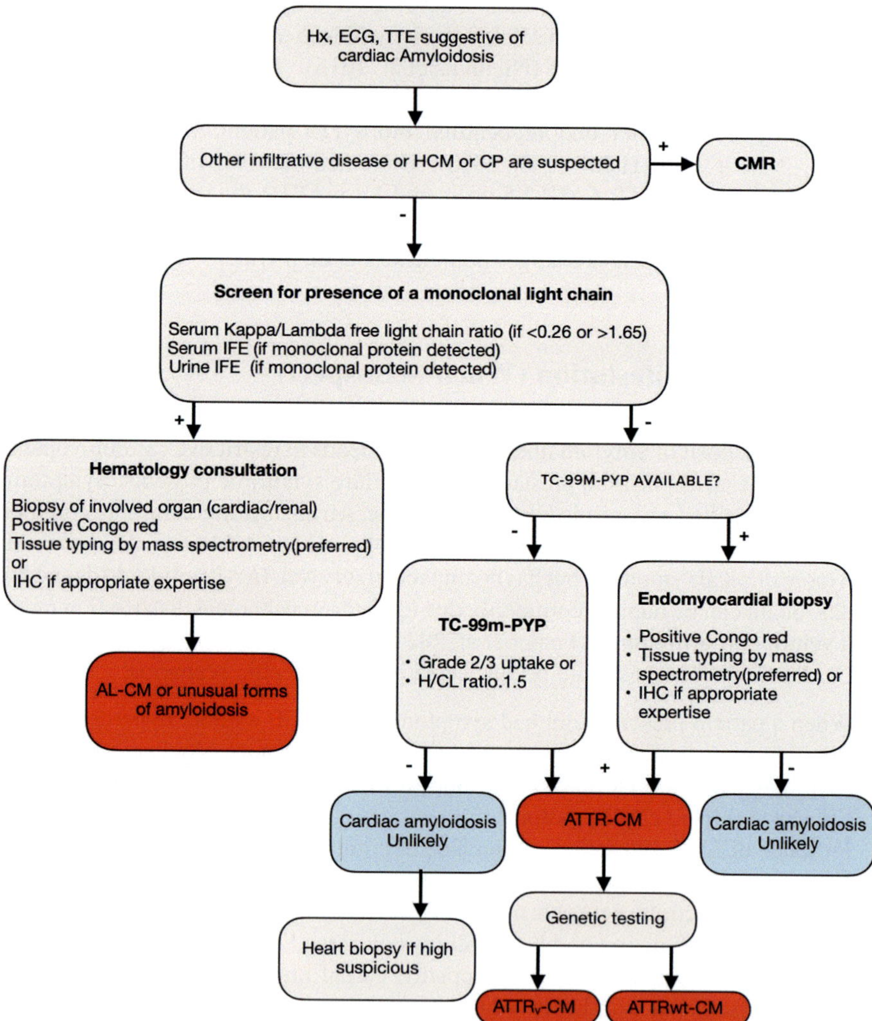

Fig. 2 Diagnostic algorithm for patients suspected to cardiac amyloidosis

a clue to diagnosis, however, some patients have only sign and symptoms of cardiac involvement.

6 ECG

Low voltage ECG although traditionally named a classic sign for cardiac amyloidosis however, studies showed less than 40% of patients have this sign on their ECG. Therefore it is not longer a sensitive finding. The best ECG signs are pseudo-infarct pattern and discrepancy between QRS voltage and wall thickness on ECG.

7 Echocardiography

The most standard and available tool for diagnostic imaging in patients with cardiac amyloidosis is echocardiography. As the first line study 2-D echocardiography reveals concentric LVH (more than 12 mm) Fig. 3. The texture of wall thickness is brighter than other forms of wall thickness, however this characteristic is not useful when using harmonic imaging.

Valvular thickness, biatrial enlargement, right ventricular hypertrophy, Aortic stenosis (mostly low flow-low gradient AS) and mild pericardial effusion are other findings which are associated with cardiac amyloidosis.

The most important echocardiography measure in these patients is evaluation of diastolic function. Severely decreased tissue Doppler velocities of mitral annular TDI with exaggerated E/e' ratio reflects a restrictive pattern for diastolic dysfunction in these patients. Speckle tracking with global longitudinal strain echocardiography reveals very low GLS numbers in presence of preserved ejection fraction, there are regional deformation in strain reduction, base and mid walls strain reduction is more prominent with apical sparing pattern result in (Cherry-on-top) pattern in bull's eye display. (Fig. 3).

Although all above findings should raise the level of suspicion for cardiac amyloidosis however, these findings are not sensitive enough to rule out or rule in disease and echocardiographic per se is not validated enough to confirm the diagnosis, or initiate treatment.

8 Cardiac Magnetic Resonance Imaging

CMR have an important role in evaluating amyloidosis and other infiltrative cardiomyopathies.MRI provides us data about chambers volume and function as well as tissue characterization subendocardial delayed enhancement of GAD which can progress to full thickness with disease advancement is a diagnostic sign, however

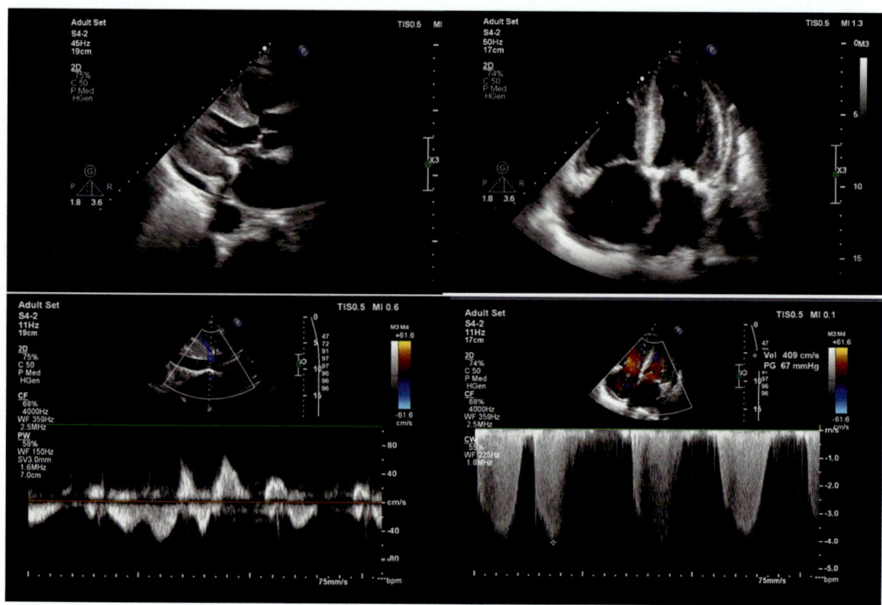

Fig. 3 65 Y/O lady with AL-Amyloidosis her TTE images shows significant concentric LVH associated with valvular thickness, bi-atrial enlargement, high TR gradients and mild pericardial effusion.very small e' in TDI and diastolic flow reversal in hepatic vein Doppler indicates restrictive pattern diastolic dysfunction. All above data suggest cardiac amyloidosis

again CMR can't differentiate between AL and ATTR disease and the findings are not enough to be used for treatment initiation by itself. These following findings are suggestive of amyloidosis (Knight et al. 2019; Carvalho et al. 2019):

- Diffuse subendocardial delayed enhancement
- Concomitant RV wall thickness with delayed enhancement
- Biatrial enlargement and delayed enhancement

CMR can provide prognostic data. GAD enhancement is a strong predictor of mortality in all types of cardiac amyloidosis. In addition by T1 mapping extra cellular volume and amyloid burden can be estimated and used for diagnostic and prognostic value and also can identify response to therapy in both AL and ATTR amyloidosis.

9 Nuclear Imaging

Nuclear imaging is an important tool in diagnosing ATTR cardiomyopathy. In some case nuclear imaging preclude need for biopsy. Tc-99 m PYP (US) Tc-99 m DPD and Tc-99 m HMDP (Europe) are commonly used. Uptake of these bone tracer on planar images visually grades ranging from 0 to 3: 0- No uptake, 1- uptake less than rib, 2- equal to rib uptake and 3- more than rib uptake.

In all positive scans to differentiate between myocardial uptake and blood uptake SPECT should be used by planar imaging. In ATTR patients nuclear imaging has more than 99% sensitivity and 86% specified in detection of cardiac evolvement. Only about 21% of patients with AL amyloidosis shows visual grade 2 or 3 of uptake when 94% of ATTR involvement proved by EMB shows this level of bone tracer uptake. So if monoclonal detecting test failed and there was moderate to high uptake confirmed with SPECT, ATTR diagnosis is prompted and genetic testing for distinguish between Wt ATTR and vATTR cardiomyopathy could be done. False positive results can happen in severe renal disease, significant coronary artery disease, hydroxychloroquine intake and APOA4 cardiomyopathy. False negative results also can occur. In case of strong clinical suspicion, endomyocardial biopsy is needed to confirm the diagnosis (Castano et al. 2016; Gillmore et al. 2016).

10 Cardiac CT

Extra cellular volume expansion can provide an index of suspicion to amyloidosis (specifically ATTR) in patients who are going to underwent TAVR for their aortic stenosis. Although if monoclonal protein is absent, Cardiac CT is not sufficient enough to substitute Tc-99 m scintigraphy and myocardial biopsy.

11 Cardiac Biomarkers

Cardiac Biomarkers (high-sensitive Troponin T and NT-proBNP) provide information for risk stratification of patients and helps prognostic estimation. There are some scoring system. See cardiac staging (Muchtar et al. 2021).

12 Monoclonal Protein

Identification of monoclonal protein is an important part of the diagnosis of AL-amyloidosis (Fig. 1). Three diagnostic tests (Urine and serum protein immunofixation and serum free light chain ratio analysis) have an accuracy 99.3% and if all of them become negative, then the diagnosis of AL-amyloidosis is unlikely (Muchtar et al. 2021; Garcia et al. 2018).

13 Tissue Biopsy

Tissue biopsy is required for patients undergoing diagnostic evaluation of AL-cardiac amyloidosis (Figs. 1 and 2). It is important that differentiate AL-amyloidosis from an unrelated monoclonal gammopathy of undetermined significance (MGUS) with positive Congo Red staining and determination of light chains origin (70% Lambda origin) for the amyloid deposits by IHC or Mass spectrometry.

The initial site for biopsy is bone marrow biopsy and abdominal fat pad aspiration biopsy because of their ease, convenience, and high yield (83%). If both sites are negative and AL amyloidosis is still believed to exist then the affected organ should be biopsied (Garcia et al. 2018; Rajkumar et al. 2014).

14 Chromosomal Changes

When a myeloma FISH panel is applied to patients with AL amyloidosis, t(11;14)(q13;q32) is the most common abnormality, seen in nearly 50% of patients.

15 Prognosis

The AL amyloidosis prognosis is resting on two factors the involved organ and the basal plasma cell clone. The severity of cardiac involvement is the single most important prognostic index for short and longterm survival. There are several criterion systems that evaluate the prognosis of AL-amyloidosis. We use the revised Mayo 2012 criteria (Muchtar et al. 2021).

The system can be used to evaluate prognosis at diagnosis. The revised Mayo system can also be used for re-staging at three and six months from treatment initiation and at the time of second line therapy (Hwa et al. 2019; Abdallah et al. 2021).

In AL amyloidosis, changes in NT-proBNP have also been used to predict response to treatment and disease progression, with a decrease in NT-proBNP of >30% and >300 ng/L from a baseline value $\geq$ 650 ng/L associated with better prognosis (Palladini et al. 2012).

16 Other Organs Involvement Evaluation

For treatment purposes, organ involvement by amyloidosis is defined 0 (Dispenzieri 2014):

Kidney—Biopsy verification of amyloid deposition plus clinical evidence of renal dysfunction or laboratory data, if amyloid deposits have been confirmed at another site, 24-h urine protein > 0.5 g/day, predominantly albumin.

Liver—Biopsy verification of amyloid deposition plus laboratory evidence of liver dysfunction, or, if amyloid deposits have been established at another site, total liver span more than 15 cm without heart failure, or alkaline phosphatase > 1.5 times.

Nerve—Clinical symmetric sensorimotor peripheral polyneuropathy in lower extremities or gastric-emptying disorder, intestinal pseudo obstruction, voiding dysfunction not directly linked to organ infiltration.

Gastrointestinal tract—Confirmation of amyloid deposition by biopsy plus GI symptoms (e.g., diarrhea, motility disturbances, and weight loss).

17 Treatment

Generally all patients with systemic AL amyloidosis require treatment at the time of diagnosis. Patients with amyloid cardiomyopathy due to AL amyloidosis need careful management of cardiac complications, including heart failure (HF), atrial fibrillation, and conduction disease.

18 Cardiac Failure

The main manifestation of cardiac Amyloidosis is congestion and volume management is the cornerstone of the therapy diuretics are used commonly however, their usage in this subgroup of heart failure patients should be with cautious because they have small LV size and fixed stroke volume so volume management in these patients is a state of art and diuretics with more bioavailability are preferable like torsemide over furosemide (Kittleson et al. 2020).

Low blood pressure is an important issue in these patients so unlike other forms of heart failure ACEI and ARB or ARNis have no role in cardiac amyloidosis and can even make symptoms worse.

Since these patients have fixed stroke volume, lowering heart rate can cause low cardiac out put symptoms and worsens fatigue, so Beta-Blockers and nondihydropydine CCB should not be used in them. Orthostatic hypotension and low blood pressure symptoms can be alleviated by midodrine, pressure stocks, waist high with minimum 30 mmHg ankle pressure and also tilt training can provide.

19 Arrhythmia

Atrial fibrillation occurs frequently in cardiac amyloidosis. AF can aggravate diastolic dysfunction however since these patients have fixed stroke volume, heart rate control can lower cardiac out put even more.

The dosage of beta-blockers should be as low as possible and digoxin usage should be with cautious as toxicity is seen more often in these group of patients. However in selected case with very extremely elevated heart rate digoxin can be used safely with cautious (Donnelly et al. 2020). In case of resistance heart rate or beta blocker intolerance Amiodarone can be used for rate control in cardiac amyloidosis. Rhythm control can be done by cardio version or ablation in early stage of atrial fibrillation since sinus rhythm can provide partial alleviation in diastolic dysfunction symptoms,

20 Anticoagulation

Cardiac amyloidosis are frequently associated with thrombus formation and cardiac source of emboli so regardless of $CHADS_2$-VAS_C warfarin or DOAC should be considered. Of note Intra cardiac thrombi formation can occur even in sinus rhythm and very low A velocity in mitral inflow Doppler can be considered as a risk factor in these patients.

21 Device Therapy

Heart block due to conduction system involvement and some times severe chronotropic insufficiency (specifically in ATTR amyloidosis) which is seen more frequently in patients with cardiac amyloidosis needs pacemaker implantation. Choosing between biventricular pacing and RV pacing can be made by percentage of RV pacing. In ATTR patients who needs more than 40% of pacing biventricular pacing might be preferred over RV pacing if patient's survival is acceptable. However, these recommendation needs more investigation.

Survival benefits of ICD implantation for primary prevention in cardiac amyloidosis is controversial and depends on estimated survival and whether our patient is appropriate candidate for cardiac transplantation or not sudden cardiac death happens in these patients more often due to electromechanical dissociation rather than ventricular arrhythmia so ICD implantation should be reserved for secondary prevention based on information gathered to date (Varr et al. 2014).

22   Treatment of Al-Amyloidosis

All newly diagnosed AL amyloidosis have to be assessed for autologous hematopoietic stem cell transplantation (HSCT) eligibility (Dispenzieri et al. 2013; Sanchorawala et al. 2022; Szalat et al. 2021). Unfortunately only, 20% of newly diagnosed patients will be eligible for transplant.

AL amyloidosis patients should meet all of the following criteria in order to be eligible for autologous HSCT:

- Age $\leq$ 70 years
- ECOG performance status $\leq$ 2
- Troponin T < 0.06 ng/mL
- New York Heart Association (NYHA) functional status class I or II
- Systolic blood pressure $\geq$ 90 mmHg
- Creatinine clearance $\geq$ 30 mL/min (unless on chronic stable dialysis)
- No more than two seriously involved organ (heart, kidney, liver or autonomic nerve).

23   Patients Eligible for Transplant

Review of data suggest better outcomes for patients who receive two to four cycles of bortezomib-based induction therapy prior to stem cell mobilization and transplantation. The preferred induction therapy is Daratumumab plus Cyclophosphamide, Bortezomib and Dexamethasone (CyBorD). If daratumumab is not available, induction with CyBorD alone is acceptable option (Fig. 4) (Palladini et al. 2020; Manwani et al. 2019).

Treatment regimens at first line that contain Lenalidomide/Thalidomide are avoided as these drugs are usually less well-tolerated in patients with AL amyloidosis (Kastritis et al. 2019).

The dose-intensive regimens with melphalan 200 mg/m^2 used to treat patients with AL-amyloidosis. Currently majority of transplant centers use, G-CSF mobilization And An optimal collection defined as $\geq$ 5 $\times$ 10^6 CD34 + cells from peripheral blood apheresis (Sanchorawala et al. 2022; Skinner et al. 2004; Szalat et al. 2021).

Patients achieving a complete or more than partial response, had significantly better survival than those with less response. Autologous HSCT allows for the delivery of myeloablative doses of Melphalan aimed at the underlying plasma cell dyscrasia. HSCT stops amyloid production, amyloid deposits are slowly resorbed, and organ function, performance status, and quality of life improve (Sanchorawala et al. 2022; Szalat et al. 2021).

Maintenance therapy is offered after HSCT to all patients with overt multiple myeloma and in patients with complete response that doesn't accept the HSCT.

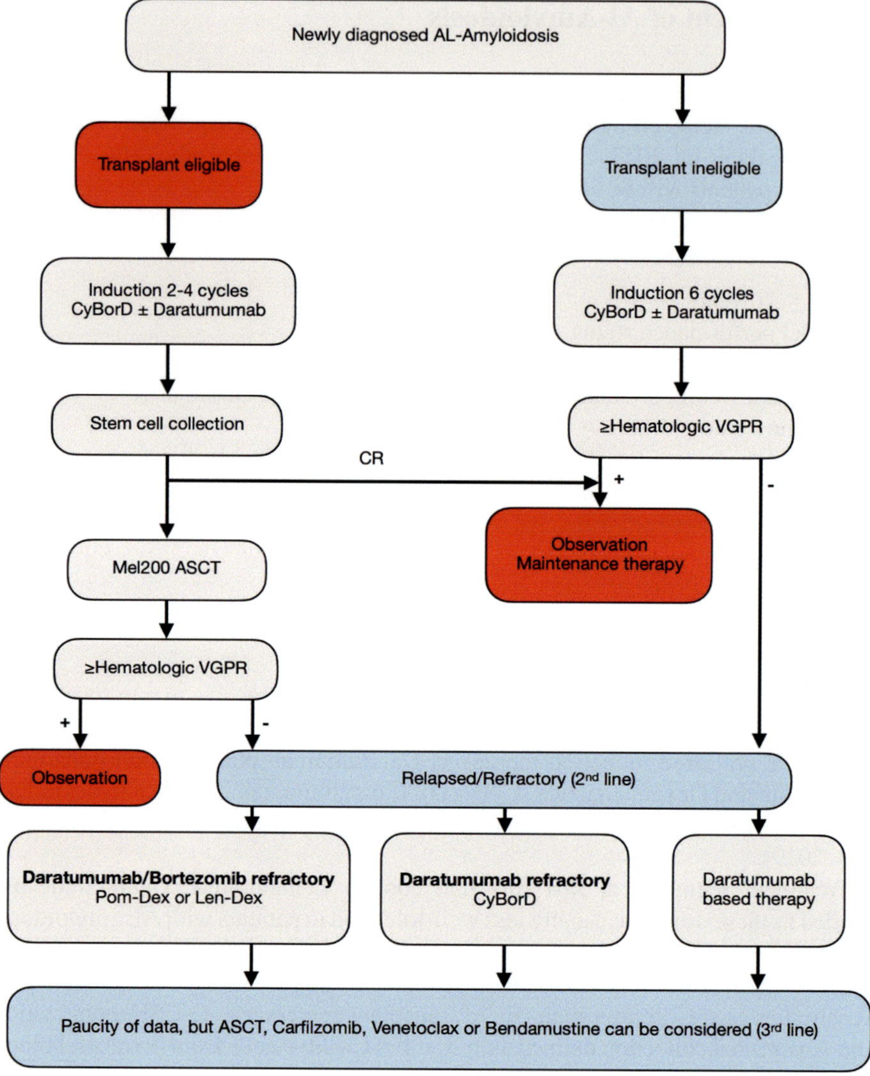

Fig. 4 Treatment algorithm in AL-amyloidosis patients (HSCT eligible and ineligible, Relapsed/ Refractory)

24 Patients Ineligible for Transplant

Best management for patients who are not eligible for HCT is combination therapy with Bortezomib, Cyclophosphamide and Dexamethasone with or without Daratumumab (CyBorD ± Dara), or bortezomib, Melphalan and dexamethasone (Fig. 4) (Palladini et al. 2020; Manwani et al. 2019; Kastritis et al. 2020).

25 Relapsed/refractory Patients

For patients who relapsed/refractory to melphalan, bortezomib-based regimen, and dexamethasone, or hematopoietic cell transplantation HSCT, other options include, proteasome inhibitor-based regimens, daratumumab and immunomodulatory-based regimens (Roussel et al. 2020; Sanchorawala et al. 2020; Lecumberri et al. 2020).

For patients who relapsed/refractory to Daratumumab, proteasome inhibitor-based regimens, and immunomodulatory-based regimens are the choices of treatment (Fig. 4) (Cohen et al. 2020; Dispenzieri et al. 2021; Basset et al. 2021; Palladini et al. 2017; Milani et al. 2018).

There aren't enough evidence to decide which of these regimens are most effective but the preferred option will be dictated by physician and patient preferences, drug availability, prior therapy, expected toxicity, and insurance coverage.

26 Survival of Patients

ASCT for selected patients is supported by several studies, showing high response rate, more response durability and also better long-term survival. A large report of ASCT in AL amyloidosis, hematologic CR was achieved in 43–56%, overall response in 56–69%, median Overall Survival was 6.3–10.9 years (Muchtar et al. 2021; Sanchorawala et al. 2022; Szalat et al. 2021).

Because of the complexity in the management and rarity of this disease, it is important to refer the suspected patients to an expert center with cardiac amyloidosis to enhance diagnostic aspects and to optimize therapy.

References

Abdallah N, Dispenzieri A, Muchtar E, et al. Prognostic restaging after treatment initiation in patients with AL amyloidosis. Blood Adv. 2021;5:1029.

Basset M, Kimmich CR, Schreck N, et al. Lenalidomide and dexamethasone in relapsed/refractory immunoglobulin light chain (AL) amyloidosis: results from a large cohort of patients with long follow-up. Br J Haematol. 2021;195:230.

Carvalho FP, Erthal F, Azevedo CF. The role of cardiac MR imaging in the assessment of patients with cardiac amyloidosis. Magn Reson Imaging Clin N Am. 2019;27:453.

Castano A, Haq M, Narotsky DL, et al. Multicenter study of planar technetium 99m pyrophosphate cardiac imaging: predicting survival for patients with ATTR cardiac amyloidosis. JAMA Cardiol. 2016;1(8):880–9.

Cohen OC, Sharpley F, Gillmore JD, et al. Use of ixazomib, lenalidomide and dexamethasone in patients with relapsed amyloid light-chain amyloidosis. Br J Haematol. 2020;189:643.

Connors LH, Sam F, Skinner M, et al. Heart failure resulting from age-related cardiac amyloid disease associated with wild-type transthyretin: a prospective, observational cohort study. Circulation. 2016;133(3):282–90.

Dispenzieri A. Renal risk and response in amyloidosis. Blood. 2014;124:2315.

Dispenzieri A, Seenithamby K, Lacy MQ, et al. Patients with immunoglobulin light chain amyloidosis undergoing autologous stem cell transplantation have superior outcomes compared with patients with multiple myeloma: a retrospective review from a tertiary referral center. Bone Marrow Transplant. 2013;48:1302.

Dispenzieri A, Kastritis E, Wechalekar AD, et al. A randomized phase 3 study of ixazomib-dexamethasone versus physician's choice in relapsed or refractory AL amyloidosis. Leukemia. 2021.

Donnelly JP, Sperry BW, Gabrovsek A, et al. Digoxin use in cardiac amyloidosis. Am J Cardiol. 2020;133:134–138.

Finsterer J, Iglseder S, Wanschitz J, et al. Hereditary transthyretin-related amyloidosis. Acta Neurol Scand. 2019;139:92.

Fontana M, Ćorović A, Scully P, Moon JC. Myocardial amyloidosis: the exemplar interstitial disease. JACC Cardiovasc Imaging. 2019;12:2345.

Fontana MD, Cardiac amyloidosis: epidemiology, clinical manifestations, and diagnosis. Uptodate 2023.

Fuster V, Narula J, Vaishnava P, Leon MB, Callans DJ, Rumsfeld J, Poppas A, editors. Fuster and hurst's the heart, 15e. McGraw Hill; 2022.

Garcia Y, Collins AB, Stone JR. Abdominal fat pad excisional biopsy for the diagnosis and typing of systemic amyloidosis. Hum Pathol. 2018;72:71.

Gillmore JD, Maurer MS, Falk RH, et al. Nonbiopsy diagnosis of cardiac transthyretin amyloidosis. Circulation. 2016;133(24):2404–12.

Hahn VS, Yanek LR, Vaishnav J, et al. Endomyocardial biopsy characterization of heart failure with preserved ejection fraction and prevalence of cardiac amyloidosis. JACC Heart Fail. 2020;8(9):712–24.

Hwa YL, Gertz MA, Kumar SK, et al. Prognostic restaging at the time of second-line therapy in patients with AL amyloidosis. Leukemia. 2019;33:1268.

Kastritis E, Dialoupi I, Gavriatopoulou M, et al. Primary treatment of light-chain amyloidosis with bortezomib, lenalidomide, and dexamethasone. Blood Adv. 2019;3:3002.

Kastritis E, Leleu X, Arnulf B, et al. Bortezomib, melphalan, and dexamethasone for light-chain amyloidosis. J Clin Oncol. 2020;38:3252.

Kittleson MM, Maurer MS, Ambardekar AV, et al, American heart association heart failure and transplantation committee of the council on clinical cardiology. Cardiac amyloidosis: evolving diagnosis and management: a scientific statement from the American heart association. Circulation. 2020;142(1):e7–e22.

Knight DS, Zumbo G, Barcella W, et al. Cardiac structural and functional consequences of amyloid deposition by cardiac magnetic resonance and echocardiography and their prognostic roles. JACC Cardiovasc Imaging. 2019;12:823.

Lecumberri R, Krsnik I, Askari E, et al. Treatment with daratumumab in patients with relapsed/refractory AL amyloidosis: a multicentric retrospective study and review of the literature. Amyloid. 2020;27:163.

Manwani R, Cohen O, Sharpley F, et al. A prospective observational study of 915 patients with systemic AL amyloidosis treated with upfront bortezomib. Blood. 2019;134(25):2271–80.

Milani P, Schönland S, Merlini G, et al. Treatment of AL amyloidosis with bendamustine: a study of 122 patients. Blood. 2018;132:1988.

Muchtar E, Dispenzieri A, Gertz MA, et al. Treatment of AL amyloidosis: mayo stratification of myeloma and risk-adapted therapy (mSMART) consensus statement 2020 update. Mayo Clin Proc. 2021;96(6):1546–77.

Nienhuis HL, Bijzet J, Hazenberg BP. The prevalence and management of systemic amyloidosis in western countries. Kidney Dis (basel). 2016;2(1):10–9.

Palladini G, Dispenzieri A, Gertz MA, et al. New criteria for response to treatment in immunoglobulin light chain amyloidosis based on free light chain measurement and cardiac biomarkers: impact on survival outcomes. J Clin Oncol. 2012;30:4541.

Palladini G, Milani P, Foli A, et al. A phase 2 trial of pomalidomide and dexamethasone rescue treatment in patients with AL amyloidosis. Blood. 2017;129:2120.

Palladini G, Kastritis E, Maurer MS, et al. Daratumumab plus CyBorD for patients with newly diagnosed AL amyloidosis: safety run-in results of ANDROMEDA. Blood. 2020;136:71.

Rajkumar SV, Dimopoulos MA, Palumbo A, et al. International myeloma working group updated criteria for the diagnosis of multiple myeloma. Lancet Oncol. 2014;15:e538.

Roussel M, Merlini G, Chevret S, et al. A prospective phase 2 trial of daratumumab in patients with previously treated systemic light-chain amyloidosis. Blood. 2020;135:1531.

Sanchorawala V, Sarosiek S, Schulman A, et al. Safety, tolerability, and response rates of daratumumab in relapsed AL amyloidosis: results of a phase 2 study. Blood. 2020;135:1541.

Sanchorawala V, Boccadoro M, Gertz M, et al. Guidelines for high dose chemotherapy and stem cell transplantation for systemic AL amyloidosis: EHA-ISA working group guidelines. Amyloid. 2022;29:1.

Skinner M, Sanchorawala V, Seldin DC, et al. High-dose melphalan and autologous stem-cell transplantation in patients with AL amyloidosis: an 8-year study. Ann Intern Med. 2004;140:85–93.

Sukhacheva TV, Eremeeva MV, Ibragimova AG, et al. Isolated atrial amyloidosis in patients with various types of atrial fibrillation. Bull Exp Biol Med. 2016;160(6):844–9.

Szalat R, Sarosiek S, Havasi A, et al. Organ responses after high dose melphalan and stem cell transplantation in AL amyloidosis. Leukemia. 2021;35:916.

Varr BC, Zarafshar S, Coakley T, et al. Implantable cardioverter-defibrillator placement in patients with cardiac amyloidosis. Heart Rhythm. 2014;11(1):158–62.

HSCT in Patients with Hemoglobinopathies

Tahereh Rostami, Azadeh Kiumarsi, and Mina Mohseni

Abstract Hemoglobinopathies, are a group of monogenic inherited disorders that affect the structure or production of the globin chains. Improvements in red blood cell transfusion management and its potential complications have influenced the prognosis of patients. However cardiovascular disease represents the leading etiology of morbidity and mortality in thalassemia, as well as a main determinant of prognosis in SCD. In spite of noteworthy advances in medical care for severe hemoglobinopathies, allogeneic HSCT remains the only definitive treatment available. This chapter aims to discuss the cardiovascular considerations of patients with hemoglobinopathies in the setting of hematopoietic stem cell transplantation.

Keywords Hemoglobinopathies · Thalassemia · Sickle cell disease · Cardiovascular · Heart failure · Anemia

Abbreviations

AF	Atrial fibrillation
BM	Bone marrow
Bu	Busulfan
Bu-Cy	Busulfan and cyclophosphamide
CMR	Cardiovascular magnetic resonance

T. Rostami (✉)
Hematologic Malignancies Research Center, Research Institute for Oncology, Hematology and Cell Therapy, Tehran University of Medical Science, Tehran, Iran
e-mail: trostami@sina.tums.ac.ir

A. Kiumarsi
Children's Medical Center Hospital, School of Medicine, Tehran University of Medical Sciences, Tehran, Iran
e-mail: Akiumarsi@sina.tums.ac.ir

M. Mohseni
Cardio-Oncology Research Center, Rajaie Cardiovascular Medical and Research Center, Iran University of Medical Sciences, Tehran, Iran

A. Alizadehasl et al. (eds.), *Cardiovascular Considerations in Hematopoietic Stem Cell Transplantation*, https://doi.org/10.1007/978-3-031-53659-5_19

CMV	Cytomegalovirus
CV	Cardiovascular
Cy	Cyclophosphamide
EBV	Epstein-barr virus
EFS	Event-free survival
GVHD	Graft-versus-host disease
Hb	Hemoglobin
HPFH	Hereditary persistence of fetal hemoglobin
HSCs	Hematopoietic stem cells
HSCT	Hematopoietic stem cell transplantation
LV	Left ventricular
LVEF	Left ventricular ejection fraction
MAC	Myeloablative conditioning
MR	Magnetic resonance
NMA	Non-myeloablative
NO	Nitric oxide
NTDT	Non-transfusion-dependent thalassemia
OS	Overall survival
PH	Pulmonary hypertension
RBC	Red blood cell
RIC	Reduced-intensity conditioning
RV	Right ventricular
SCD	Sickle cell disease
TBI	Total body irradiation
TDT	Transfusion-dependent thalassemia
TDT	Transfusion-dependent thalassemia
TFS	Thalassemia free survival
TI	Thalassemia intermedia
TNF	Tumor necrosis factor
VOC	Vaso-occlusive crises

1   An Introduction to Hemoglobinopathies

Hemoglobinopathies are the most frequent inherited monogenic disorders globally, and it is estimated that 5–7% of the world's population carry a defective hemoglobin (Hb) trait.[1]

Hemoglobinopathies are divided in to two categories according to the causal mutations, that result in changes in the the structure of globin chains (abnormal Hb

---

[1] Modell B, Darlison M. Global epidemiology of haemoglobin disorders and derived service indicators. Bull World Health Organ. 2008;86(6):480–487.

Piel FB. The present and future global burden of the inherited disorders of hemoglobin. Hematology/Oncology Clinics. 2016 Apr 1;30(2):327–41.

or Hb variants e.g., sickle cell anemia syndrome) or defects in the synthesis process of globin chains (thalassemia).[2]

The clinical phenotypes are variable in severity, depending on the type and the zygosity of the mutation and the probability of genetic modifiers co-inheritance.[3]

β-thalassemia and sickle cell disease (SCD) are the severe clinical cases of hemoglobinopathies. Patients with β-thalassemia who have mutations in the β-globin genes resulting in adult hemoglobin (Hb A) reduction or absence and so transfusion dependent anemia.[4] A mutant β-globin chain which incorporates in an Hb tetramer (Hb S) characterizes SCD. This pathogenic sickle hemoglobin (Hb S), has a propensity to polymerize under deoxygenated conditions. Hb S polymerization leads to impaired red blood cell (RBC) survival due to RBC sickling and hemolysis, acute chest syndrome, vaso-occlusive crises (VOC), long-term organ damage, and early mortality.[5]

Although improvements in both red blood cell transfusion and iron overload prevention and treatment by using iron chelating agents have influenced the prognosis of the patients, cardiovascular disease represents the major cause of morbidity and mortality in thalassemia patients, as well as a main prognosis determinant in SCD.[6]

---

[2] Farmakis D, Triposkiadis F, Lekakis J, Parissis J. Heart failure in haemoglobinopathies: pathophysiology, clinical phenotypes, and management. European Journal of Heart Failure. 2017 Apr;19(4):479–89.

Lucarelli G, Isgrò A, Sodani P, Gaziev J. Hematopoietic stem cell transplantation in thalassemia and sickle cell anemia. Cold Spring Harbor perspectives in medicine. 2012 May 1;2(5):a011825.

Trent RJ. Diagnosis of the haemoglobinopathies. Clinical Biochemist Reviews. 2006 Feb;27(1):27.

Weatherall DJ, Clegg JB. The thalassaemia syndromes. (4th ed), Blackwell Scientific; 2001.

Angastiniotis M, Modell B. Global epidemiology of hemoglobin disorders. Ann N Y Acad Sci 1998;850:251–69.

Kohne E. Hemoglobinopathies: clinical manifestations, diagnosis, and treatment. Deutsches Ärzteblatt International. 2011 Aug;108(31–32):532.

[3] Zittersteijn HA, Harteveld CL, Klaver-Flores S, Lankester AC, Hoeben RC, Staal FJ, Gonçalves MA. A small key for a heavy door: genetic therapies for the treatment of hemoglobinopathies. Frontiers in Genome Editing. 2021 Feb 4;2:617780.

[4] Musallam KM, Rivella S, Vichinsky E, Rachmilewitz EA. Non-transfusion-dependent thalassemias. Haematologica. 2013;98(6):833–844.

Rivella S. The role of ineffective erythropoiesis in non-transfusion-dependent thalassemia. Blood Rev. 2012;26(suppl 1):S12–S15.

[5] Telen MJ, Malik P, Vercellotti GM. Therapeutic strategies for sickle cell disease: towards a multi-agent approach. Nat Rev Drug Discov. 2019;18(2):139–158.

Guilcher GM, Truong TH, Saraf SL, Joseph JJ, Rondelli D, Hsieh MM. Curative therapies: allogeneic hematopoietic cell transplantation from matched related donors using myeloablative, reduced intensity, and nonmyeloablative conditioning in sickle cell disease. In: Seminars in hematology 2018 Apr 1 (Vol. 55, No. 2, pp. 87–93). WB Saunders.

[6] Romano MM. Haemoglobinopathies from the cardiac point of view. Hematology, Transfusion and Cell Therapy. 2019 Oct 10;41:195–6.

Akiki N, Hodroj MH, Bou-Fakhredin R, Matli K, Taher AT. Cardiovascular Complications in β-Thalassemia: Getting to the Heart of It. Thalassemia Reports. 2023 Jan 30;13(1):38–50.

In the most severe forms, hematopoietic stem cell transplantation (HSCT) is the only curative treatment, but its applicability is limited due to lack of suitable donors and the risks of this procedure.[7]

2   Stem Cell Transplantation for Hemoglobinopathies

In spite of progressive improvements in management of patients with hemoglobinopathies, allogeneic HSCT is the only available definitive therapeutic option.[8] Hematopoietic stem cells (HSCs) are located within the niches of bone marrow (BM) and construct the peripheral blood cells, lifelong.[9] So, replacing pathogenic HSCs with healthy donor HSCs in a hereditary hematologic disease such as β-thalassemia and SCD could lead to a one-time cure.[10]

In thalassemia, the principal indication for HSCT is considered transfusion dependence and so HSCT should be offered to children with thalassemia who have an available healthy, HLA-identical sibling, as early as possible.[11] Despite improvement in conventional treatment of thalassemia patients, and although β-globin gene therapy has been recently used in non- $\beta^0\beta^0$ thalassemia; however in clinical practice

Ali, S.; Mumtaz, S.; Shakir, H.A.; Khan, M.; Tahir, H.M.; Mumtaz, S.; Mughal, T.A.; Hassan, A.; Kazmi, S.A.R.; Sadia; et al. Current status of beta-thalassemia and its treatment strategies. Mol. Genet. Genom. Med. 2021, 9, e1788.

Barbero, U.; Fornari, F.; Guarguagli, S.; Gaglioti, C.M.; Longo, F.; Doronzo, B.; Anselmino, M.; Piga, A. Atrial fibrillation in beta -thalassemia Major Patients: Diagnosis, Management and Therapeutic Options. Hemoglobin 2018, 42, 189–193.

Lopes A, Dantas MT, Ladeia AM. Prevalence of Cardiovascular Complications in Individuals with Sickle Cell Anemia and Other Hemoglobinopathies: A Systematic Review. Arquivos Brasileiros de Cardiologia. 2022 Nov 21.

Santarone S, Angelini S, Natale A, Vaddinelli D, Spadano R, Casciani P, Papola F, Di Lembo E, Iannetti G, Di Bartolomeo P. Survival and late effects of hematopoietic cell transplantation in patients with thalassemia major. Bone Marrow Transplantation. 2022 Nov;57(11):1689–97.

[7] Wienert, B., Martyn, G. E., Funnell, A. P. W., Quinlan, K. G. R., and Crossley, M. (2018). Wake-up Sleepy Gene: Reactivating Fetal Globin for β-Hemoglobinopathies. Trends Genet. 34, 927–940. https://doi.org/10.1016/j.tig.2018.09.004.

[8] Cavazzana M, Antoniani C, Miccio A. Gene therapy for β-hemoglobinopathies. Molecular Therapy. 2017 May 3;25(5):1142–54.

[9] Hoggatt J, Pelus LM. Mobilization of hematopoietic stem cells from the bone marrow niche to the blood compartment. Stem cell research and therapy. 2011 Apr;2(2):1–8.

[10] Bhatia M, Walters MC. Hematopoietic cell transplantation for thalassemia and sickle cell disease: past, present and future. Bone marrow transplantation. 2008 Jan;41(2):109–17.

[11] Leonard A, Bertaina A, Bonfim C, Cohen S, Prockop S, Purtill D, Russell A, Boelens JJ, Wynn R, Ruggeri A, Abraham A. Curative therapy for hemoglobinopathies: an International Society for Cell and Gene Therapy Stem Cell Engineering Committee review comparing outcomes, accessibility and cost of ex vivo stem cell gene therapy versus allogeneic hematopoietic stem cell transplantation. Cytotherapy. 2022 Mar 1;24(3):249–61.

allogeneic hematopoietic stem cell transplantation (HSCT) remains the only curative approach.[12]

Over the last four decades, HSCT has been successfully performed in patients with transfusion-dependent thalassemia (TDT), with a cure rate of 80–90%.[13] However, outcomes of HSCT correlate with severity of iron overload and duration of exposure; thus, HSCT results are best in individuals in whom the risks of organ damage from excess iron stores have been minimized.[14]

To predict the outcome, a prognostic score has been developed by Pesaro group in patients younger than 17 years and stratified them into 3 risk groups considering the adequacy of iron chelation and the presence of hepatomegaly and portal fibrosis.[15]

Another influential factor on HSCT outcome in TDT patients is the type of donor. HLA-genoidentical donors are only available to < 20% of patients. Although improvements in transplantation protocols and management of transplantation-related complications have permitted moving towards matched unrelated donors, haploidentical related donors, and alternative stem cell sources such as umbilical-cord blood, they could be linked to potential morbidity and mortality. Outcomes differ geographically, and the high risk alternative approaches should only be considered

---

[12] Cavazzana M, Antoniani C, Miccio A. Gene therapy for β-hemoglobinopathies. Molecular Therapy. 2017 May 3;25(5):1142–54.

Sabloff M, Chandy M, Wang Z, et al. HLA matched sibling bone marrow transplantation for beta-thalassemia major. Blood. 2011;117(5):1745–1750.

Rahal I, Galambrun C, Bertrand Y, Garnier N, Paillard C, Frange P, Pondarré C, Dalle JH, de Latour RP, Michallet M, Steschenko D. Late effects after hematopoietic stem cell transplantation for β-thalassemia major: the French national experience. Haematologica. 2018 Jul;103(7):1143.

Baronciani D, Angelucci E, Potschger U, et al. Hematopoietic stem cell transplantation in thalassemia: a report from the European Society for Blood and Bone Marrow Transplantation Hemoglobinopathy Registry, 2000–2010. Bone Marrow Transplant. 2016;51(4):536–541.

Angelucci E. Hematopoietic stem cell transplantation in thalassemia. Hematology Am Soc Hematol Educ Program. 2010;2010:456–462.

Lucarelli G, Galimberti M, Polchi P, et al. Marrow Transplantation in Patients with Thalassemia Responsive to Iron Chelation Therapy. N Engl J Med. 1993;329(12):840–844.

Angelucci E, Barosi G, Camaschella C, et al. Italian Society of Hematology Practice Guidelines for the Management of Iron Overload in Thalassemia Major and Related Disorders. Haematologica. 2008;93(5):741–752.

Rachmilewitz EA, Giardina PJ. How I treat thalassemia. Blood. 2011;118(13):3479–3488.

Thomas ED, Buckner CD, Sanders JE, Papayannopoulou T, et al. Marrow Transplantation for Thalassaemia. Lancet. 1982;2(8292):227–229.

Lucarelli G, Polchi P, Galimberti M, et al. Marrow Transplantation for Thalassaemia Following Busulphan and Cyclophosphamide. Lancet. 1985;1(8442): 1355–1357.

[13] Cappellini. MD., Farmakis. D., Porter. J., Taher. A. 4th edition, 2022 Guidelines for the Management of Transfusion Dependent Thalassaemia (TDT), Thalassaemia Intertational Federation.

[14] Angelucci E, Pilo F, Coates TD. Transplantation in thalassemia: revisiting the Pesaro risk factors 25 years later. American Journal of Hematology. 2017 May;92(5):411–3.

[15] Strocchio L, Locatelli F. Hematopoietic stem cell transplantation in thalassemia. Hematology/ Oncology Clinics. 2018 Apr 1;32(2):317–28.

Lucarelli G, Galimberti M, Polchi P, et al. Bone marrow transplantation in patients with thalassemia. N Engl J Med 1990;322(7):417–21.

Giardini C, Lucarelli G. Bone marrow transplantation for beta-thalassemia. Hematol Oncol Clin North Am 1999;13(5):1059–64.

Table 1 Outcomes of HSCT from different donor types in β-thalassemia

Donor type	Patient age	Pesaro class		OS (%)	TFS (%)
Matched sibling donor (MSD)	< 17 years	Class I/II	Age < 7 years	98	> 94
			Age ≥ 7 years	86	83
		Class III		92	92
	Adult (> 17 years)	Class III		65	65
Matched unrelated donor (MUD)	< 17 years	Class I/II		96.7	80
		Class III		65.2	54.5
	Adult (> 17 years)	Class III		70.4	70.4
Haploidentical donor	1–28 years	Class I/II/III		96	96

OS: Overall survival; TFS: Thalassemia free survival

at expert centers. Table 1 summarizes the HSCT outcomes according to different donor types.[16]

SCD could lead to considerable morbidity, affecting the patients' quality of life. Better-quality surveillance and preventive strategies, together with treatment with hydroyurea have contributed to significant improvements of children survival in the last two decades. Yet, mortality is substantial in adult patients.

HSCT is the only available approach with curative intent for SCD.[17] As the clinical phenotypes of SCD are very variable, unlike TDT, indications for HSCT in SCD patients are not universally defined.[18] One of the generally accepted indications for HSCT in SCD is the presence of central nervous system disease despite adequate transfusion therapy due the risk of recurrent or progressive cerebral infarcts.[19]

---

[16] Gaziev J, Sodani P, Lucarelli G. Hematopoietic stem cell transplantation in thalassemia. Bone marrow transplantation. 2008 Aug;42(1):S41.

Anurathapan U, Hongeng S, Pakakasama S, Songdej D, Sirachainan N, Pongphitcha P, Chuansumrit A, Charoenkwan P, Jetsrisuparb A, Sanpakit K, Rujkijyanont P. Hematopoietic stem cell transplantation for severe thalassemia patients from haploidentical donors using a novel conditioning regimen. Biology of Blood and Marrow Transplantation. 2020 Jun 1;26(6):1106–12.

[17] Lucarelli G, Isgrò A, Sodani P, Gaziev J. Hematopoietic stem cell transplantation in thalassemia and sickle cell anemia. Cold Spring Harbor perspectives in medicine. 2012 May 1;2(5):a011825.

Nouraie M, Lee JS, Zhang Y, Kanias T, Zhao X, Xiong Z, Oriss TB, Zeng Q, Kato GJ, Gibbs JS, Hildesheim ME. The relationship between the severity of hemolysis, clinical manifestations and risk of death in 415 patients with sickle cell anemia in the US and Europe. Haematologica. 2013 Mar;98(3):464.

[18] Leonard A, Bertaina A, Bonfim C, Cohen S, Prockop S, Purtill D, Russell A, Boelens JJ, Wynn R, Ruggeri A, Abraham A. Curative therapy for hemoglobinopathies: an International Society for Cell and Gene Therapy Stem Cell Engineering Committee review comparing outcomes, accessibility and cost of ex vivo stem cell gene therapy versus allogeneic hematopoietic stem cell transplantation. Cytotherapy. 2022 Mar 1;24(3):249–61.

[19] Hulbert ML, McKinstry RC, Lacey JL, Moran CJ, Panepinto JA, Thompson AA, Sarnaik SA, Woods GM, Casella JF, Inusa B, Howard J. Silent cerebral infarcts occur despite regular blood transfusion therapy after first strokes in children with sickle cell disease. Blood, The Journal of the American Society of Hematology. 2011 Jan 20;117(3):772–9.

Other severe complications including osteonecrosis, pulmonary hypertension, sickle nephropathy, recurrent VOC, acute chest syndrome and splenic sequestration could also be considered as indications for HSCT.[20]

A large international survey has reported the results of HLA-identical sibling HSCT in over 1000 SCD patients with SCD, as 5-year EFS and OS of 91.4 and 92.9%, respectively.[21] However, experience of HSCT from unrelated and haploidentical donors for SCD is scant, with high rates of reported infectious complications, graft-versus-host disease (GvHD) and graft rejection.[22]

3 Cardiac Complications in Hemohlobinopathies

The clinical outcomes of HSCT has significantly improved but it is linked to substantial acute and long term complications, including graft failure, GvHD, infections, neurologic, hormonal and cardiovascular complications.[23] Although cardiovascular

---

Scothorn DJ, Price C, Schwartz D, Terrill C, Buchanan GR, Shurney W, Sarniak I, Fallon R, Chu JY, Pegelow CH, Wang W. Risk of recurrent stroke in children with sickle cell disease receiving blood transfusion therapy for at least 5 years after initial stroke. The Journal of pediatrics. 2002 Mar 1;140(3):348–54.

[20] Kassim AA, Sharma D. Hematopoietic stem cell transplantation for sickle cell disease: The changing landscape. Hematology/oncology and stem cell therapy. 2017 Dec 1;10(4):259–66.

Leonard A, Bertaina A, Bonfim C, Cohen S, Prockop S, Purtill D, Russell A, Boelens JJ, Wynn R, Ruggeri A, Abraham A. Curative therapy for hemoglobinopathies: an International Society for Cell and Gene Therapy Stem Cell Engineering Committee review comparing outcomes, accessibility and cost of ex vivo stem cell gene therapy versus allogeneic hematopoietic stem cell transplantation. Cytotherapy. 2022 Mar 1;24(3):249–61.

Shenoy S. Hematopoietic stem-cell transplantation for sickle cell disease: current evidence and opinions. Therapeutic advances in hematology. 2013 Oct;4(5):335–44.

Kassim AA, Sharma D. Hematopoietic stem cell transplantation for sickle cell disease: The changing landscape. Hematology/oncology and stem cell therapy. 2017 Dec 1;10(4):259–66.

[21] Gluckman E, Cappelli B, Bernaudin F, Labopin M, Volt F, Carreras J, Pinto Simões B, Ferster A, Dupont S, De La Fuente J, Dalle JH. Sickle cell disease: an international survey of results of HLA-identical sibling hematopoietic stem cell transplantation. Blood, The Journal of the American Society of Hematology. 2017 Mar 16;129(11):1548–56.

[22] Shenoy S, Eapen M, Panepinto JA, Logan BR, Wu J, Abraham A, Brochstein J, Chaudhury S, Godder K, Haight AE, Kasow KA. A trial of unrelated donor marrow transplantation for children with severe sickle cell disease. Blood, The Journal of the American Society of Hematology. 2016 Nov 24;128(21):2561–7.

Gluckman E, Cappelli B, Scigliuolo GM, De la Fuente J, Corbacioglu S. Alternative donor hematopoietic stem cell transplantation for sickle cell disease in Europe. Hematology/Oncology and Stem Cell Therapy. 2020 Dec 1;13(4):181–8.

Bolaños-Meade J, Cooke KR, Gamper CJ, Ali SA, Ambinder RF, Borrello IM, Fuchs EJ, Gladstone DE, Gocke CB, Huff CA, Luznik L. Effect of increased dose of total body irradiation on graft failure associated with HLA-haploidentical transplantation in patients with severe haemoglobinopathies: a prospective clinical trial. The Lancet Haematology. 2019 Apr 1;6(4):e183–93.

[23] Bhatia, S. Long-term health impacts of hematopoietic stem cell transplantation inform recommendations for follow-up. Expert Rev. Hematol. 2011, 4, 437–452.

problems are not very common among all the other HSCT-related morbidities, they could cause a high mortality rate and a significant reduction in the long-term survivors' quality of life.[24] Hence, there is a necessity for amplified alertness and better understanding of the probable cardiovascular problems so that their diagnosis and treatment take place in a proper and timely manner. This review aims to focus on the possible cardiac complications that may be faced in patients with hemoglobinopathies which go through HSCT.

3.1 Underlying Disease Associated Cardiac Complications

The main pathophysiologic mechanism of hemoglobinopathies is chronic hemolysis, which fallouts to anemia and compensatory elevation of cardiac output. Moreover, anemia treatment with regular blood transfusions leads to iron overload in the myocard. Chronic hemolysis is also linked to vasculopathy which encompasses depletion of nitric oxide, endothelial dysfunction and abnormalities in elastic tissue. In SCD, the abnormal Hb S could root in to recurrent microvascular obstruction, followed by ischemia and reperfusion injury. Ultimately, left ventricular (LV) and right ventricular (RV) dysfunction and pulmonary hypertension (PH) could occur.[25] Figure 1 shows An overview of the pathophysiology of cardiomyopathy in patients with hemoglobinopathies.

Cardiac Diseases in Patients with Hemoglobinopathies

1. Cardiomyopathies

    1.1 Iron overload cardiomyopathy

    Iron overload cardiomyopathy, as the main cardiac disease in TDT patients, has two different manifestations including restrictive phenotype (a diastolic LV dysfunction with restricted filling, preserved contractility and LV ejection fraction, with or without pulmunary hypertension and RV dilatation) and dilated phenotype or hypokinetic cardiomyopathy (LV remodeling with dilatation and condensed LVEF).[26] Iron overload cardiomyopathy is diagnosed by the existence of iron overload (transferrin saturation > 55%, serum ferritin > 300 ng/mL) together with the following issues: cardiac siderosis (cardiac iron on $T2^*$ < 20 ms) and evidence of cardiac

---

[24] Ohmoto, A.; Fuji, S. Cardiac complications associated with hematopoietic stem-cell transplantation. Bone Marrow Transplant. 2021, 56, 2637–2643.

[25] Farmakis D. Cardiac complications in haemoglobinopathies. Proceedings and abstract book. 2017 Apr 1;102(1):39–40.

[26] Farmakis D. Cardiac complications in haemoglobinopathies. Proceedings and abstract book. 2017 Apr 1;102(1):39–40.

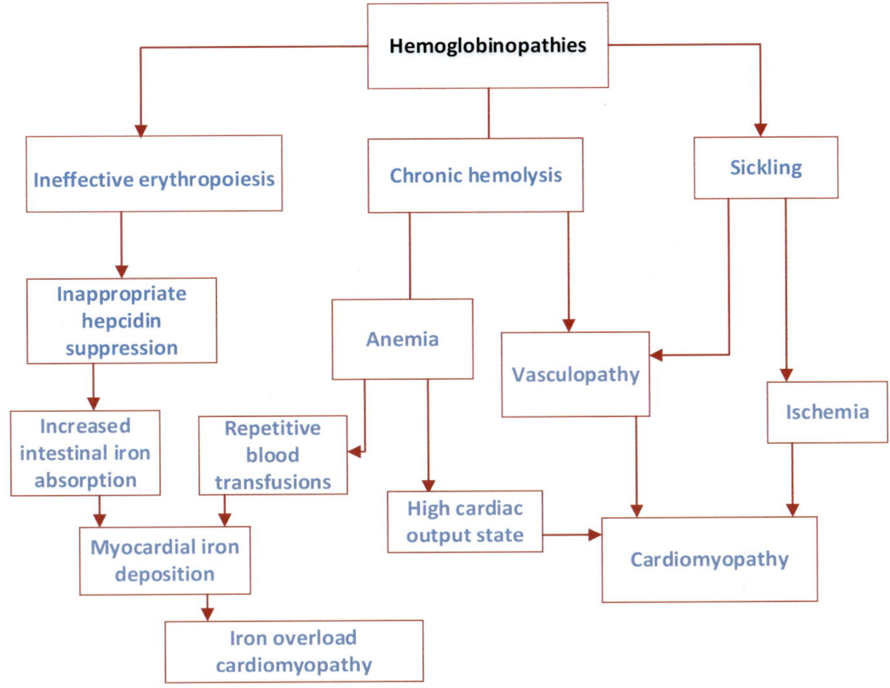

Fig. 1 An overview of the pathophysiology of cardiomyopathy in hemoglobinopathies

disease (impaired ventricular geometry or/and function) and evidence of heart disease (impaired ventricular geometry or/and function).[27]

Assessment of myocardial iron deposition by measurement of T2* magnetic resonance (MR) relaxation has become the most widely used.[28] The T2* threshold of 10 ms could predict heart failure with an acceptable sensitivity and specificity. The risk of heart failure associated with cardiac T2* values < 10 ms, in comparison with cardiac T2* values 10 ms, is significantly higher.[29]

In TDT patients, the incidence of iron overload cardiomyopathy is reported to be 11.4–15.1%, but in SCD, it only occur in 2–5% of the patients. Later onset of chronic transfusion, the efficient erythropoiesis consuming

---

[27] Romano MM. Haemoglobinopathies from the cardiac point of view. Hematology, Transfusion and Cell Therapy. 2019 Oct 10;41:195–6.

Kremastinos DT, Farmakis D. Iron overload cardiomyopathy in clinical practice. Circulation. 2011 Nov 15;124(20):2253–63.

[28] Romano MM. Haemoglobinopathies from the cardiac point of view. Hematology, Transfusion and Cell Therapy. 2019 Oct 10;41:195–6.

[29] Kirk P, Roughton M, Porter JB, Walker JM, Tanner MA, Patel J, Wu D, Taylor J, Westwood MA, Anderson LJ, Pennell DJ. Cardiac T2* magnetic resonance for prediction of cardiac complications in thalassemia major. Circulation. 2009 Nov 17;120(20):1961–8.

the excess iron together with the long-lasting inflammation sequestering iron in reticuloendothelial cells, lowers the chance of myocardial iron deposition in SCD compared with TDT. Myocardial iron deposition can cause LV systolic dysfunction, arrhythmias and sudden death.[30]

It is noteworthy that in TDT patients, iron overload in myocytes might enhance their susceptibilty to developing viral infections associated myocarditis that seems to be of particular relevance in causing dilated cardiomyopathy.[31]

1.2 High cardiac output cardiomyopathy

The high cardiac output cardiomyopathy has a complex pathophysiology and is developed as a consequence of several factors including:

- Continued tissue hypoxia due to chronic anemia, decreased level of 2,3-bisphosphoglycerate in the transfused blood and also the high oxygen affinity of the abnormal hemoglobin types.
- Compensatory bone marrow expansion and extramedullary hematopoiesis (splenomegaly) resulting in the intrasplenic shunts.
- Iron-induced liver damage and also transfusion-associated viral infections, resulting hepatic cirrhosis, which contributes to high cardiac output state.[32]

---

[30] Barbero, U.; Fornari, F.; Guarguagli, S.; Gaglioti, C.M.; Longo, F.; Doronzo, B.; Anselmino, M.; Piga, A. Atrial fibrillation in beta -thalassemia Major Patients: Diagnosis, Management and Therapeutic Options. Hemoglobin 2018, 42, 189–193.

Kaur H, Aurif F, Kittaneh M, Chio JP, Malik BH. Cardiomyopathy in sickle cell disease. Cureus. 2020 Aug 8;12(8).

Meloni A, Puliyel M, Pepe A, et al.: Cardiac iron overload in sickle-cell disease. Am J Hematol. 2014, 89:678–683.

de Montalembert M, Ribeil JA, Brousse V, et al.: Cardiac iron overload in chronically transfused patients with thalassemia, sickle cell anemia, or myelodysplastic syndrome. PLoS One. 2017, 12:e0172147.

[31] Kremastinos, D.T.; Farmakis, D.; Aessopos, A.; Hahalis, G.; Hamodraka, E.; Tsiapras, D.; Keren, A. Beta-thalassemia cardiomyopathy: History, present considerations, and future perspectives. Circ. Heart Fail. 2010, 3, 451–458.

Du, Z.D.; Roguin, N.; Milgram, E.; Saab, K.; Koren, A. Pulmonary hypertension in patients with thalassemia major. Am. Heart J. 1997, 134, 532–537.

Jabbar, D.A.; Davison, G.; Muslin, A.J. Getting the iron out: Preventing and treating heart failure in transfusion-dependent thalassemia. Cleve. Clin. J. Med. 2007, 74, 807–810, 813–816.

[32] Cappellini. MD., Farmakis. D., Porter. J., Taher. A. 4th edition, 2022 Guidelines for the Management of Transfusion Dependent Thalassaemia (TDT), Thalassaemia Intertational Federation.

Hahalis G, Manolis AS, Apostolopoulos D, Alexopoulos D, Vagenakis AG, Zoumbos NC. Right ventricular cardiomyopathy in beta-thalassaemia major. Eur Heart J. 2002; 23:147–156.

Kremastinos, D.T.; Farmakis, D.; Aessopos, A.; Hahalis, G.; Hamodraka, E.; Tsiapras, D.; Keren, A. Beta-thalassemia cardiomyopathy: History, present considerations, and future perspectives. Circ. Heart Fail. 2010, 3, 451–458.

Rund D, Rachmilewitz E. Beta-thalassemia. N Engl J Med. 2005; 353:1135–1146.

Modell B, Khan M, Darlison M. Survival in beta-thalassaemia major in the UK: data from the UK Thalassaemia Register. Lancet. 2000; 355:2051–2052.

2. Pulmonary hypertension

Hemoglobinopathies are among the most common etiologies of pulmonary hypertension (PH), worldwide.[33] Although PH is prevalent in both TDT and also patients with non-transfusion-dependent thalassemia (NTDT), it has been the foremost cause of right-sided heart failure in thalassemia intermedia patients and its occurrence is five times more freequent in patients with NTDT.[34] Around one-third of homozygous SCA patients experience PH, which is associated with adverse prognosis in them.[35]

Catheterization of right heart is the gold standard method for PH diagnosis but is invasive. Screening by echocardiography and pro-BNP measurements should be considered in all patients, especially those with NTDT.[36] Figure 2 illustrates the pathophysiology of pulmonary hypertension in hemoglobinopathies.

3. Acute pericarditis and myocarditis

Kremastinos DT, Toutouzas PK, Vyssoulis GP, Venetis CA, Vretou HP, Avgoustakis DG. Global and segmental left ventricular function in beta-thalassemia. Cardiology. 1985; 72:129–139.

Aessopos A, Farmakis D, Hatziliami A, Fragodimitri C, Karabatsos F, Joussef J, Mitilineou E, Diamanti-Kandaraki E, Meletis J, Karagiorga M. Cardiac status in well-treated patients with thalassemia major. Eur J Haematol. 2004; 73:359–366.

Spirito P, Lupi G, Melevendi C, Vecchio C. Restrictive diastolic abnormalities identified by Doppler echocardiography in patients with thalassemia major. Circulation. 1990; 82:88–94.

[33] Barnett CF, Hsue PY, Machado RF. Pulmonary hypertension: an increasingly recognized complication of hereditary hemolytic anemias and HIV infection. JAMA. 2008; 299:324–331.

[34] Farmakis D, Aessopos A. Pulmonary hypertension associated with hemoglobinopathies: prevalent but overlooked. Circulation. 2011 Mar 22;123(11):1227–32.

Taher, A.T.; Cappellini, M.D. How I manage medical complications of _-thalassemia in adults. Blood 2018, 132, 1781–1791.

Musallam, K.M.; Taher, A.T.; Rachmilewitz, E.A. beta-thalassemia intermedia: A clinical perspective. Cold Spring Harb. Perspect. Med. 2012, 2, a013482.

Derchi, G.; Galanello, R.; Bina, P.; Cappellini, M.D.; Piga, A.; Lai, M.E.; Quarta, A.; Casu, G.; Perrotta, S.; Pinto, V.; et al. Prevalence and risk factors for pulmonary arterial hypertension in a large group of beta-thalassemia patients using right heart catheterization: AWebthal study. Circulation 2014, 129, 338–345.

Taher, A.; Vichinsky, E.; Musallam, K.; Cappellini, M.D.; Viprakasit, V. Guidelines for the Management of Non Transfusion Dependent Thalassaemia (NTDT); Weatherall, D., Ed.; Thalassaemia International Federation: Nicosia, Cyprus, 2013.

[35] Covitz W, Espeland M, Gallagher D, Hellenbrand W, Leff S, Talner N. The heart in sickle cell anemia. The Cooperative Study of Sickle Cell Disease (CSSCD). Chest 1995;108:1214–1219.

Gladwin MT, Sachdev V, Jison ML, Shizukuda Y, Plehn JF, Minter K, Brown B, Coles WA, Nichols JS, Ernst I, Hunter LA, Blackwelder WC, Schechter AN, Rodgers GP, Castro O, Ognibene FP. Pulmonary hypertension as a risk factor for death in patients with sickle cell disease. N Engl J Med 2004;350:886–895.

[36] Akiki N, Hodroj MH, Bou-Fakhredin R, Matli K, Taher AT. Cardiovascular Complications in β-Thalassemia: Getting to the Heart of It. Thalassemia Reports. 2023 Jan 30;13(1):38–50.

Motta, I.; Mancarella, M.; Marcon, A.; Vicenzi, M.; Cappellini, M.D. Management of age-associated medical complications in patients with beta-thalassemia. Expert Rev. Hematol. 2020, 13, 85–94.

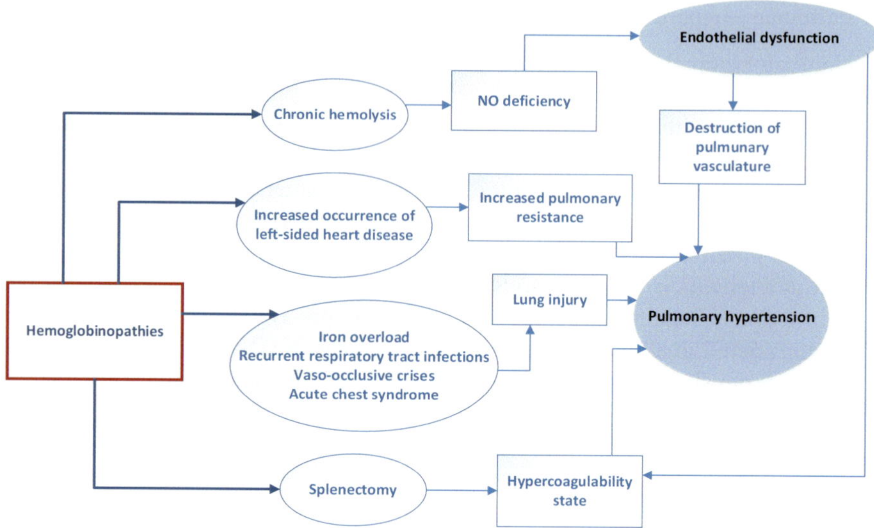

Fig. 2 Pathophysiology of pulmonary hypertension in hemoglobinopathies

Before the era of regular blood transfusion and iron chelation therapy, at least half of the thalassemia major patients experienced acute pericarditis as a potential cause of heart failure during their first two decades of life.[37] With advances in anemia and iron overload management, which made patients more vulnerable to infections, and improved quality of blood products, a dramatic reduction of pericarditis and myocarditis has been reported in transfusion-dependent patients.[38]

4. Myocardial ischemia

Vascular complications and parenchymal damage of myocardium are also found in hemoglobinopathies. Iron overload, reduced nitric oxide and immune dysregulation due to red blood cell transfusions, infections, and the high cytokine levels in the stored blood products could lead to endothelial damage and arterial stiffness in TDT patients.[39] In patients with SCD, myocardial ischemia leads to

---

[37] Engle MA, Erlandson M, Smith CH. Late cardiac complications of chronic,refractory anemia with hemochromatosis.Circulation1964;30:698–705.

[38] Jessopos A, Farmakis D, Hatziliami A, Fragodimitri C, Karabatsos F, Joussef J,Mitilineou E, Diamanti-Kandaraki E, Meletis J, Karagiorga M. Cardiac status inwell-treated patients with thalassemia major.Eur J Haematol2004;73:359–366.

[39] Kremastinos, D.T.; Farmakis, D.; Aessopos, A.; Hahalis, G.; Hamodraka, E.; Tsiapras, D.; Keren, A. Beta-thalassemia cardiomyopathy: History, present considerations, and future perspectives. Circ. Heart Fail. 2010, 3, 451–458.

Hahalis, G.; Kremastinos, D.T.; Terzis, G.; Kalogeropoulos, A.P.; Chrysanthopoulou, A.; Karakantza, M.; Kourakli, A.; Adamopoulos, S.; Tselepis, A.D.; Grapsas, N.; et al. Global vasomotor dysfunction and accelerated vascular aging in _-thalassemia major. Atherosclerosis 2008, 198, 448–457.

LV dysfunction.[40] It seems that small-vessel disease following sickling process in these patients plays the key role in parenchymal damage of myocardium in SCD.[41]

5. Arrhythmia

An inclusive range of ventricular and supraventricular arrhythmias are frequently seen in patients with cardiac dysfunction.[42] In a study by Kirk et al., the occurrence of arrhythmia was reported to be closely correlated to the indicators of cardiac iron overload, i.e. the annual relative risk for arrhythmias was 4.6 in cardiac $T2* < 20$ ms and 8.8 in $T2* < 6$ ms.[43] Even though cardiac siderosis is considered to be the main etiology, arrhythmias like atrial fibrillation which is a common arrhythmia in TDT, it could also develop in patients without iron overload. This may be due to fibrosis as a consequence of previous resolved iron deposition.[44] Moreover, iron chelation could reverse iron-induced atrial fibrillation if heart remodeling would have not occured.[45,46]

Diagnosis of Cardiovascular Complications in Hemoglobinopathies

Cheung, Y.F.; Chan, G.C.; Ha, S.Y. Arterial stiffness and endothelial function in patients with beta-thalassemia major. Circulation 2002, 106, 2561–2566.

Stoyanova, E.; Trudel, M.; Felfly, H.; Lemsaddek, W.; Garcia, D.; Cloutier, G. Vascular endothelial dysfunction in beta-thalassemia occurs despite increased eNOS expression and preserved vascular smooth muscle cell reactivity to NO. PLoS ONE 2012, 7, e38089.

[40] Damy T, Bodez D, Habibi A, Guellich A, Rappeneau S, Inamo J, Guendouz S,Gellen-Dautremer J, Pissard S, Loric S, Wagner-Ballon O, Godeau B, AdnotS, Dubois-Randé JL, Hittinger L, Galactéros F, Bartolucci P. Haematologicaldeterminants of cardiac involvement in adults with sickle cell disease.Eur Heart J2016;37:1158–1167.

[41] Aessopos A, Farmakis D, Trompoukis C, Tsironi M, Moyssakis I, TsaftaridesP, Karagiorga M. Cardiac involvement in sickle beta-thalassemia.Ann Hematol2009;88:557–564.

Aessopos A, Tsironi M, Vassiliadis I, Farmakis D, Fountos A, Voskaridou E,Perakis A, Defteraios S, Loutradi A, Loukopoulos D. Exercise-induced myocardialperfusion abnormalities in sickle beta-thalassemia: Tc-99m tetrofosmin gatedSPECT imaging study. Am J Med2001;111:355–360.

[42] Farmakis D. Cardiac complications in haemoglobinopathies. Proceedings and abstract book. 2017 Apr 1;102(1):39–40.

Engle MA, Erlandson M, Smith CH. Late cardiac complications of chronic,refractory anemia with hemochromatosis.Circulation1964;30:698–705.

Jessup M, Manno CS. Diagnosis and management of iron-induced heart diseasein Cooley's anemia.Ann N Y Acad Sci1998;850:242–250.

[43] Kirk P, Roughton M, Porter JB, Walker JM, Tanner MA, Patel J, Wu D, Taylor J, Westwood MA, Anderson LJ, Pennell DJ. Cardiac T2* magnetic resonance for prediction of cardiac complications in thalassemia major. Circulation. 2009 Nov 17;120(20):1961–8.

[44] Taher, A.T.; Cappellini, M.D. How I manage medical complications of _-thalassemia in adults. Blood 2018, 132, 1781–1791.

Malagù, M.; Marchini, F.; Fiorio, A.; Sirugo, P.; Clò, S.; Mari, E.; Gamberini, M.R.; Rapezzi, C.; Bertini, M. Atrial Fibrillation in β-Thalassemia: Overview of Mechanism, Significance and Clinical Management. Biology 2022, 11, 148.

[45] Nomani, H.; Bayat, G.; Sahebkar, A.; Fazelifar, A.F.; Vakilian, F.; Jomezade, V.; Johnston, T.P.; Mohammadpour, A.H. Atrial fibrillation in _-thalassemia patients with a focus on the role of iron-overload and oxidative stress: A review. J. Cell. Physiol. 2019, 234, 12249–12266.

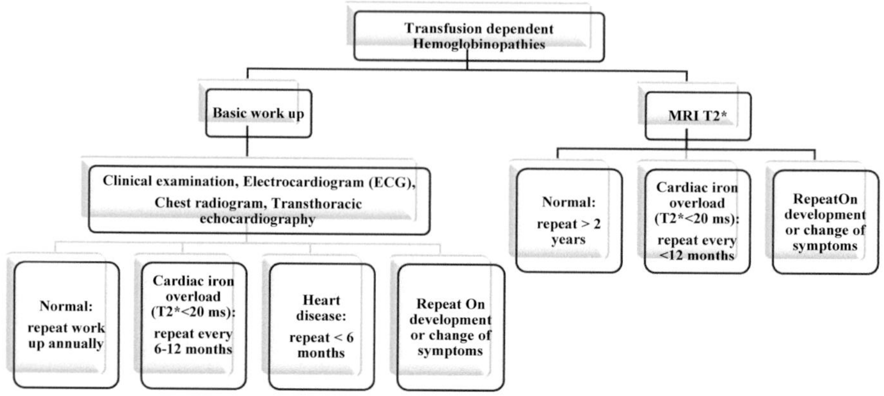

Fig. 3 Basic algorithm for the cardiac evaluation of patients with thalassemia

Figure 3 presents an algorithm indicative of the pathway for cardiac evaluation for thalassemia patients.

Prevention and Treatment of Cardiovascular Disease

Prevention and treatment of cardiovascular disease in hemoglobinopathies have been founded on the basis of disease-related principles including blood transfusions, iron chelation and hydroxyurea, together with cardio-active therapeutic strategies.[46]

With currently available iron chelators and effective chelation regimens guided by CMR (cardiovascular magnetic resonance) imaging, reversal of iron overload cardiomyopathy will be possible. Particularly in severe myocardial siderosis and LV dysfunction, combination therapy with two different chelators could decrease myocardial iron and recover cardiac function. However, very long time of treatment with iron chelators considering repeated myocardial T2* scans would be necessary to achieve it.[47] Figure 4 presents the therapeutic algorithm in patients on regular transfusions with or at risk of iron overload cardiomyopathy.

---

[46] Cappellini. MD., Farmakis. D., Porter. J., Taher. A. 4th edition, 2022 Guidelines for the Management of Transfusion Dependent Thalassaemia (TDT), Thalassaemia Intertational Federation.

Aessopos A, Kati M, Farmakis D. Heart disease in thalassemia intermedia: a review of the underlying pathophysiology. Haematologica. 2007; 92:658–665.

[47] Kremastinos DT, Farmakis D. Iron overload cardiomyopathy in clinical practice. Circulation. 2011 Nov 15;124(20):2253–63.

Gujja P, Rosing DR, Tripodi DJ, Shizukuda Y. Iron overload cardiomyopathy: better understanding of an increasing disorder. J Am Coll Cardiol 2010;56:1001–1012.

Tanner MA, Galanello R, Dessi C, Smith GC, Westwood MA, Agus A, Pibiri M, Nair SV, Walker JM, Pennell DJ. Combined chelation therapy in thalassemia major for the treatment of severe myocardial siderosis with left ventricular dysfunction. J Cardiovasc Magn Reson 2008;10:12.

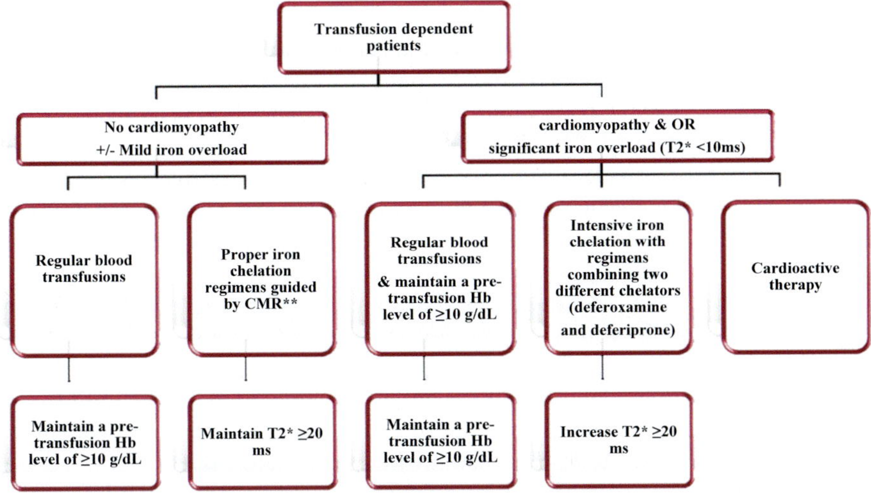

Fig. 4 Therapeutic algorithm in patients on regular transfusions with or at risk of iron overload cardiomyopathy

3.2 Transplant-Related Cardiac Complications

Myocardial dysfunction and heart failure, vascular and structural heart abnormalities are considered the main HSCT-related cardiovascular complications.[48]

During or until 100 days post-transplant, cardiovascular problems including acute heart failure, arrhythmias, and pericardial tamponade may occur. In long-term survivors, cardiomyopathy, ischemic heart disease and vascular disorder together with cardiac-related comorbidities such as diabetes mellitus and hypertension have been reported.[49] Transplant-survivors show a higher risk of evolving cardiovascular disorders as well as cardiovascular event-related death.[50]

---

[48] Roziakova, L.; Mistrik, M.; Batorova, A.; Kruzliak, P.; Bojtarova, E.; Dubrava, J.; Gergel, J.; Mladosievicova, B. Can we predict clinical cardiotoxicity with cardiac biomarkers in patients after haematopoietic stem cell transplantation? Cardiovasc Toxicol. 2015, 15, 210–216.

[49] Tuzovic, M.; Mead, M.; Young, P.A.; Schiller, G.; Yang, E.H. Cardiac Complications in the Adult Bone Marrow Transplant Patient. Curr. Oncol. Rep. 2019, 21, 28.

[50] Leger, K.J.; Baker, K.S.; Cushing-Haugen, K.L.; Flowers, M.E.D.; Leisenring, W.M.; Martin, P.J.; Mendoza, J.A.; Reding, K.W.; Syrjala, K.L.; Lee, S.J.; et al. Lifestyle factors and subsequent ischemic heart disease risk after hematopoietic cell transplantation. Cancer 2018, 124, 1507–1515.

Leger, K.J.; Cushing-Haugen, K.; Hansen, J.A.; Fan, W.; Leisenring, W.M.; Martin, P.J.; Zhao, L.P.; Chow, E.J. Clinical and Genetic Determinants of Cardiomyopathy Risk among Hematopoietic Cell Transplantation Survivors. Biol. Blood Marrow Transplant. 2016, 22, 1094–1101.

Chow, E.J.; Mueller, B.A.; Baker, K.S.; Cushing-Haugen, K.L.; Flowers, M.E.; Martin, P.J.; Friedman, D.L.; Lee, S.J. Cardiovascular hospitalizations and mortality among recipients of hematopoietic stem cell transplantation. Ann. Intern. Med. 2011, 155, 21–32.

Conditioning Regimens

Pre-HSCT preparative regimen is an essential part of transplantation to eradicate the patient's bone marrow cells, and provide immunosuppressive effect needed for the donor cells engraftment. Conditioning protocols are categorized in to myeloablative (MAC), reduced-intensity (RIC) or non-myeloablative.[51]

Considering the particular features of β-thalassemia, i.e. bone marrow hyperplasia and alloimmunisation due to multiple pre-transplant blood transfusions, a myeloablative conditioning regimen seems necessary in ordert to reduce the risk of graft failure.[52]

Most myeloablative conditioning regimes are comprised of busulfan (Bu) and cyclophosphamide (Cy) which is an alkylating agent with immunosuppressive properties. Several studies have shown that high dose of Cy (i.e. > 100 mg/kg), could cause severe cardiovascular problems such as arrhythmia, severe pericarditis and myocarditis and heart failure.[53] Cy-induced cardiac toxicity is dose-dependent and its incidence is increased to 8.5% in administration of high dose Cy.[54]

Considering Pesaro risk classification for predicting outcome of HSCT in TDT patients, classical dosage of Cy (50 mg/kg/day) in patients classified as class III has been associated with worse outcome compared to class I patients.[55] So dose reduction of Cy in patients classified as clas III yielded encouraging results with improvement

---

[51] Tuzovic, M.; Mead, M.; Young, P.A.; Schiller, G.; Yang, E.H. Cardiac Complications in the Adult Bone Marrow Transplant Patient. Curr. Oncol. Rep. 2019, 21, 28.

Zhao Y, He R, Oerther S, Zhou W, Vosough M, Hassan M. Cardiovascular Complications in Hematopoietic Stem Cell Transplanted Patients. Journal of Personalized Medicine. 2022 Oct 31;12(11):1797.

Bacigalupo, A.; Ballen, K.; Rizzo, D.; Giralt, S.; Lazarus, H.; Ho, V.; Apperley, J.; Slavin, S.; Pasquini, M.; Sandmaier, B.M.; et al. Defining the intensity of conditioning regimens: Working definitions. Biol. Blood Marrow Transplant. 2009, 15, 1628–1633.

[52] Mulas O, Mola B, Caocci G, La Nasa G. Conditioning regimens in patients with β-thalassemia who underwent hematopoietic stem cell transplantation: a scoping review. Journal of Clinical Medicine. 2022 Feb 9;11(4):907.

Lucarelli, G.; Gaziev, J. Advances in the allogeneic transplantation for thalassemia. Blood Rev. 2008, 22, 53–63.

[53] Shanholtz, C. Acute life-threatening toxicity of cancer treatment. Crit. Care Clin. 2001, 17, 483–502.

Goldberg, M.A.; Antin, J.H.; Guinan, E.C.; Rappeport, J.M. Cyclophosphamide cardiotoxicity: An analysis of dosing as a risk factor. Blood 1986, 68, 1114–1118.

[54] Ishida, S.; Doki, N.; Shingai, N.; Yoshioka, K.; Kakihana, K.; Sakamaki, H.; Ohashi, K. The clinical features of fatal cyclophosphamide-induced cardiotoxicity in a conditioning regimen for allogeneic hematopoietic stem cell transplantation (allo-HSCT). Ann. Hematol. 2016, 95, 1145–1150.

[55] Lucarelli, G.; Galimberti, M.; Polchi, P.; Angelucci, E.; Baronciani, D.; Giardini, C.; Politi, P.; Durazzi, S.M.; Muretto, P.; Albertini, F. Bone marrow transplantation in patients with thalassemia. N. Engl. J. Med. 1990, 322, 417–421.

of overall survival and also reduced mortality. The same imrovement in outcome of HSCT has been achieved by replacing Cy with fludarabine.[56]

MAC regimen based on Bu-Cy has been the most common approach used in matched related donor HSCT for patients with SCD. However, as MAC regimen is associated with increased toxicity, RIC and reduced-toxicity conditioning regimens including substitution of Cy with fludarabine or substitution of Bu with treosulfan and the addition of thiotepa have been evaluated and used recently.[57]

Although NMA approach with TBI 200 cGy and fludarabine was associated with graft failure, TBI with alemtuzumab has been linked with improved survival.[58]

Infections

HSCT recipients are at vulnerable to developing opportunistic infections. Recieving several immunosuppressive agents, having indwelling catheters, and the skin and mucosal integrity disruption due to chemoradiotherapy and GVHD predispose them to differnt kinds of infections. Infections may also cause cardiac manifestations.

Infectious endocarditis is an uncommon transplant-related complication.[59] In HSCT recipients who had developed infectious endocarditis, presence of right-sided

---

[56] Lucarelli, G.; Galimberti, M.; Giardini, C.; Polchi, P.; Angelucci, E.; Baronciani, D.; Erer, B.; Gaziev, D. Bone marrow transplantation in thalassemia: The experience of Pesaro. Ann. N. Y. Acad. Sci. 1998, 850, 270–275.

Sauer, M.; Bettoni, C.; Lauten, M.; Ghosh, A.; Rehe, K.; Grigull, L.; Beilken, A.; Welte, K.; Sykora, K.W. Complete substitution of cyclophosphamide by fludarabine and ATG in a busulfan-based preparative regimen for children and adolescents with β-thalassemia. Bone Marrow Transplant. 2005, 36, 383–387.

[57] Arnold SD, Bhatia M, Horan J, Krishnamurti L. Haematopoietic stem cell transplantation for sickle cell disease–current practice and new approaches. British journal of haematology. 2016 Aug;174(4):515–25.

Krishnamurti L. Hematopoietic cell transplantation for sickle cell disease. Frontiers in Pediatrics. 2021 Jan 5;8:551170.

[58] Hsieh, M.M., Fitzhugh, C.D., Weitzel, R.P., Link, M.E., Coles, W.A., Zhao, X., Rodgers, G.P., Powell, J.D. & Tisdale, J.F. (2014) Nonmyeloablative HLA-matched sibling allogeneic hematopoietic stem cell transplantation for severe sickle cell phenotype. JAMA, 312, 48–56.

Saraf SL, Oh AL, Patel PR, et al. Nonmyeloablative stem cell transplantation with alemtuzumab/low-dose irradiation to cure and improve the quality of life of adults with sickle cell disease. Biol Blood Marrow Transplant. 2016;22:441–448.

Guilcher GM, Monagel DA, Nettel-Aguirre A, Truong TH, Desai SJ, Bruce A, Shah RM, Leaker MT, Lewis VA. Nonmyeloablative matched sibling donor hematopoietic cell transplantation in children and adolescents with sickle cell disease. Biology of Blood and Marrow Transplantation. 2019 Jun 1;25(6):1179–86.

Guilcher GM, Monagel D, Leaker M, Truong TH, Nettel-Aguirre A, Bruce A. Alemtuzumab/low dose tbi conditioning facilitates stable long-term donor hematopoietic cell engraftment from sibling donors in children with sickle cell disease. Blood. (2017) 130(Suppl. 1):4592.

[59] Safdar A, Childs BH, Keefe D, Sepkowitz K. Acute bacterial endocarditis during granulocytopenia in an allogeneic marrow transplant recipient. Am J Med 2000; 109: 514–515.

Cancelas JA, Lopez J, Cabezudo E et al. Native valve endocarditis due to Candida parapsilosis: a late complication after bone marrow transplantation-related fungemia. Bone Marrow Transplant 1994; 13: 33–34.

Rao K, Sah V. Medical management of Aspergillus flavus endocarditis. Pediatr Hematol Oncol 2000; 17: 425–427.

valvular lesions suggest that indwelling catheter could be the main source of the infection.[60] On the other hand, gastrointestinal and cutaneous GVHD may predispose patients to septicemia due to the mucosal or skin integrity being compromised. This could justify the reports indicating that prior evidence of acute GVHD existed in most of the patients who had developed infectious endocarditis post-HSCT.[61]

Fungal infections, also could cause coronary artery stenosis and arrhythmias.[62] Viral infections have been reported to cause pericardial effusions and arrhythmias. Infection by adenovirus, CMV, and enterovirus has been reported to be associated with coronary artery disease.[63] Pericarditis linked with EBV reactivation after allogeneic HSCT has also been described.[64]

Graft Versus Host Disease (GVHD)

After allogeneic HSCT, GVHD could cause considerable morbidity and mortality, however, cardiac complications associated with GVHD are uncommon.[65] GVHD is caused by immune recognition of recipient organs by donor T cells, leading to secretion of proinflammatory cytokines and chemokines and host tissue damage.[66] The heart has infrequently been reported as a direct focus tissue of GVHD, so the cardiac manifestations linked to GVHD are possibly cytokine-induced.[67] Increased levels of tumor necrosis factor (TNF)-α and IL-2 in patients with GVHD contribute to myocardial injury. TNF could harm muscle electrical activity, leading to reduced myocardial contractility. arrhythmias and third-degree atrioventricular block could be induced by IL-2.[68]

---

[60] Martino P, Micozzi A, Venditti M et al. Catheter-related rightsided endocarditis in bone marrow transplant recipients. Rev Infect Dis 1990; 15: 500–502.

[61] Kuruvilla J, Forrest DL, Lavoie JC, Nantel SH, Shepherd JD, Song KW, Sutherland HJ, Toze CL, Hogge DE, Nevill TJ. Characteristics and outcome of patients developing endocarditis following hematopoietic stem cell transplantation. Bone marrow transplantation. 2004 Dec;34(11):969–73.

[62] B.S. Andersson, M.A. Luna, K.B. McCredie Systemic aspergillosis as cause of myocardial infarction. Cancer, 58 (1986), pp. 2146–2150.

[63] G.S. Shirali, J. Ni, R. Chinnock, et al. Association of viral genome with graft loss in children after cardiac transplantation N Engl J Med, 344 (2001), pp. 1498–1503.

[64] Aoyama Y, Nakao Y, Ohta K, Sakai T, Nakamae H, Yamamura R, Yamane T, Hino M. Pericarditis associated with Epstein-Barr virus reactivation in a patient following allogeneic peripheral blood stem cell transplantation from an HLA genotypic 1-locus mismatched sibling donor. Leukemia and lymphoma. 2004 Feb 1;45(2):393–5.

[65] Kim DH, You JJ, Im HJ, Ko JK. Coronary artery involvement in chronic graft-versus-host disease presenting as sudden cardiac arrest. Pediatric Transplantation. 2019 Aug;23(5):e13474.

[66] Mohty B, Mohty M. Long-term complications and side effects after allogeneic hematopoietic stem cell transplantation: an update. Blood Cancer J. 2011;1:e16.

[67] Rackley C, Schultz KR, Goldman FD, Chan KW, Serrano A, Hulse JE, Gilman AL. Cardiac manifestations of graft-versus-host disease. Biology of Blood and Marrow Transplantation. 2005 Oct 1;11(10):773–80.

[68] D.L. Mann. Inflammatory mediators and the failing heart: past, present, and the foreseeable future Circ Res, 91 (2002), pp. 988–998.

S.D. Prabhu. Cytokine-induced modulation of cardiac function Circ Res, 95 (2004), pp. 1140–1153.

The cardiac manifestations of GVHD variy from asymptomatic to serious and lethal. Pericardial effusion and arrhythmias such as complete atrioventricular block or sinus node dysfunction have been related to heart involvement by GVHD.[69] Coronary artery complications related to chronic GVHD after HSCT has been rarely reported.[70] Coronary endothelium could be a potential target for the graft-versus-host immune response in chronic GVHD.[71]

Chronic GVHD-associated serositis (pleural/pericardial effusion) together with pericarditis usually occurs while tapering immunosuppression.[72] In cases with large pericardial effusion and cardiac tamponade, emergent pericardiocentesis together with enhanced immunosuppression can effectively control this life-threatening complication.[73] Pericardial constriction and restrictive cardiomyopathy may also occur late after HSCT.[74] A successful response to infliximab in recurrent pericardial GVHD has been reported.[75]

Increased awareness towards the mentioned uncommon manifestations is necessary. Patients with bradycardias in association with GVHD could improve with imuunosuppressive drugs. Electrophysiological studies and endomyocardial biopsy looking for lymphocytic infiltrates are applied as diagnostic tools for cardiac GVHD. A noninvasive method could also be a therapeutic test of steroid therapy or increasing immunosuppressive drugs.[76]

P.T. Vaitkus, D. Grossman, K.R. Fox, M.D. McEvoy, J.U. Doherty. Complete heart block due to interleukin-2 therapy. Am Heart J, 119 (1990), pp. 978–990.

J. Zhang, Z. Yu, S.L. Hilbert, et al. Cardiotoxicity of human recombinant interleukin-2 in rats: a morphological study Circulation, 87 (1993), pp. 1340–1353.

[69] Tichelli A, Bhatia S, Socié G. Cardiac and cardiovascular consequences after haematopoietic stem cell transplantation. Br J Haematol. 2008;142:11–26.

[70] Kim DH, You JJ, Im HJ, Ko JK. Coronary artery involvement in chronic graft-versus-host disease presenting as sudden cardiac arrest. Pediatric Transplantation. 2019 Aug;23(5):e13474.

[71] Tichelli A, Gratwohl A. Vascular endothelium as 'novel' target of graft-versus-host disease. Best Pract Res Clin Haematol. 2008;21:139–148.

[72] Jenner MJ, Micallef IN, Rohatiner AZ, Kelsey SM, Newland AC, Cavenagh JD. Successful therapy of transplant-associated venoocclusive disease with a combination of tissue plasminogen activator DQGGH¿EURWLGHMed Oncol. 2000; 17(4):333–336.

Jessica and Leonard, Chronic Gvhd-Associated Serositis andPericarditis, Internal Medicine, Oregon Health and Science University.

[73] Rhodes M, Lautz T, Kavanaugh-Mchugh A, Manes B, Calder C, Koyama T, Liske M, Parra D, Frangoul H. Pericardial effusion and cardiac tamponade in pediatric stem cell transplant recipients. Bone marrow transplantation. 2005 Jul;36(2):139–44.

[74] Cavet J, Lennard A, Gascoigne A, Finney RD, Lucraft HH, Richardson C, et al. Constrictive pericarditis post allogenic Bone Marrow Transplantation for Philadelphia positive acute lymphoblastic leukemia. Bone Marrow Transplantat. 2000; 25(5): 571–573.

[75] Rubio PM, Labrandero C, Riesco S, Plaza D, Gonzalez B, Munoz GM, versus host disease in a pediatric patient. Bone Marrow Transplant. 2013; 48 (1): 144–145.

[76] Rackley C, Schultz KR, Goldman FD, Chan KW, Serrano A, Hulse JE, Gilman AL. Cardiac manifestations of graft-versus-host disease. Biology of Blood and Marrow Transplantation. 2005 Oct 1;11(10):773–80.

Rehabilitation in HSCT

Maryam Barkhordar, Iraj Nazeri, Majid Maleki, Ghasem Janbabai, Azin Alizadehasl, Amir Ghaffari Jolfayi, Amir Askarinejad, Erfan Kohansal, Rasoul Azarfarin, and Sara Adimi

Abstract Rehabilitative interventions play a crucial role in the care of patients who have undergone Hematopoietic Stem Cell Transplantation (HSCT), and a multidisciplinary rehabilitation program can help manage psychological challenges, improve physical and functional abilities and overall Quality Of Life (QOL) in phases of HSCT's patients' treatment that associated with the transplant process. Historically, long-term Cardiovascular Disease (CVD) and CVD risk factors such as Myocardial Infarction (MI), Heart Failure (HF), and arrhythmias, are a major cause of morbidity and mortality among HSCT survivors; Cardiotoxicity has been associated with chemotherapeutic agents used as part of primary treatment or preparative regimens in addition to cardiovascular issues related to HSCT.

Keywords Hematopoietic stem cell transplantation · Quality of life · Multidisciplinary rehabilitation · Rehabilitation

Abbreviations

CVD   Cardiovascular disease
CBT   Cognitive behavioral therapy

M. Barkhordar
Hematology Oncology and Stem Cell Transplantation Research Center, Tehran University of Medical Sciences, Tehran, Iran

I. Nazeri
Department of Cardiology, Day General Hospital, Tehran, Iran

M. Maleki · A. Alizadehasl · A. G. Jolfayi · A. Askarinejad · E. Kohansal · R. Azarfarin ·
S. Adimi (✉)
Cardio-Oncology Research Center, Rajaie Cardiovascular Medical and Research Center, Iran University of Medical Sciences, Tehran, Iran
e-mail: sara_adimi@yahoo.com

G. Janbabai
Hematologic Malignancies Research Center, Tehran University of Medical Sciences, Tehran, Iran
e-mail: janbabai@yahoo.com

© The Author(s), under exclusive license to Springer Nature Switzerland AG 2024   273
A. Alizadehasl et al. (eds.), *Cardiovascular Considerations in Hematopoietic Stem Cell Transplantation*, https://doi.org/10.1007/978-3-031-53659-5_20

ET   Exercise training
GVHD   Graft-versus-host disease
HF   Heart failure
HSCT   Hematopoietic stem cell transplantation
MI   Myocardial infarction
OT   Occupational therapy
PT   Physical therapy
QOL   Quality of life
ROM   Range of motion

1   Rehabilitation Approach for HSCT

The rehabilitation approach for HSCT typically involves several components, including Physical Therapy (PT), Occupational Therapy (OT), speech therapy, and psychological support. The specific goals of rehabilitation depend on the individual HSCT treatment process (Steinberg et al. 2015; Mohammed et al. 2019a).

PT, a key component of HSCT rehabilitation, focuses on restoring physical function, strength, and endurance which may involve exercise therapy to improve mobility, balance, and coordination. Furthermore, specialized treatments for the management of lymphedema to prevent or reduce swelling.

The other important component of HSCT rehabilitation is OT, it may help the patients to regain their independence and improve their ability to manage their daily activities through adaptive training and techniques, also How to use assistive devices, such as wheelchairs or prosthetics.

Speech therapy may be required for HSCT patients who experience aphasia, dysarthria, or difficulty swallowing. However, this therapy can help to improve the patient's speech and language skills.

Also, mental health support is a crucial part of rehabilitation for HSCT patients with psychological challenges during and after treatment (Mohammed et al. 2019a, 2019b; Mohananey et al. 2021).

Additionally, Rehabilitation goals and interventions may based on the individual's specific HSCT treatment and Evaluation progress.

Here are some rehabilitation areas, goals, and interventions for HSCT (Table 1).

1.1   Rehabilitation Evaluation Assessment

The rehabilitation process begins with evaluating the physical, psychological, and social status of patients undergoing HSCT. This procedure is critical in developing individualized rehabilitation plans that address their needs. The following are some examples of assessment tools that can be used to evaluate HSCT patients:

Table 1 Rehabilitation area, goals and Interventions for HSCT

Rehabilitation area	Goals	Interventions
Physical	Increase strength, endurance, flexibility, mobility, gross and fine motor skills, promote bone health and prevent fractures	Physical therapy, occupational Therapy and exercise programs, pain and fatigue management, sensory integration therapy, lymphedema management and provide adaptive equipment to individual needs
Activities of daily living	Increase independence in ADLs and Improve self-care skills	Physical therapy, occupational therapy, skin and wound care education, pacing and energy conservation, leisure education, and social skills training and provide adaptive equipment as needed
Speech and language	Improve speech function and communication skills	Speech therapy and cognitive deficits
Psychological	Address anxiety, depression, and Improve memory, attention, and executive functioning	Psychotherapy, cognitive rehabilitation, stress and pain management, sleep hygiene education and mental health interventions as needed
Nutritional	Improve nutritional status, promote a healthy diet and weight management	Nutritional counseling education to promote healthy eating habits and weight management
Cardiopulmonary	Improve cardiovascular and respiratory function	Pulmonary rehabilitation and exercise therapy to improve Lung function, cardiovascular health, and education on managing symptoms and cardiovascular risk factors
Complementary and alternative medical therapies (CAM)	Promote holistic healing and well-being	CAM therapies, such as acupuncture, massage, homeopathy, herbal supplements and mindfulness meditation
Financial and social support	Address financial concerns, Improve social support and employment issues	Financial counseling, employment support, education on disability benefits, Social skills training, family and couples therapy, and support group participation

1. **Functional Assessment**: A functional assessment should be performed to evaluate the patient's ability to perform activities of daily living, such as bathing, dressing, and grooming. This assessment can help determine the level of assistance and support the patient may require during rehabilitation (Silva et al. 2021; Ishikawa et al. 2019).

2. **Range of Motion Assessment**: A range of motion (ROM) assessment should be performed to evaluate the patient's ability to move their limbs and joints. This assessment can help identify any limitations or weaknesses requiring PT or other rehabilitation interventions (Schmidt et al. 2016; Kvinge et al. 2022).

3. **Strength Assessment**: A strength assessment should evaluate the patient's muscle strength and endurance. This assessment can help determine the patient's ability to perform physical activities and identify any weakness that may require PT or other rehabilitation interventions (Ishikawa et al. 2019; Brotelle et al. 2018).

4. **Balance and Coordination Assessment**: A balance and coordination assessment should evaluate the patient's ability to maintain and coordinate their movements. This assessment can help identify deficits requiring PT or other rehabilitation interventions (Mohammed et al. 2019b; Kanter et al. 2021).

5. **Pain Assessment**: A pain assessment should be performed to evaluate the patient's level of pain and discomfort. This assessment can help determine the need for pain management interventions during rehabilitation (Stiff et al. 2006; Esser et al. 2017a).

6. **Cardiovascular and Cardiopulmonary Assessment**: A cardiovascular assessment should evaluate the patient's heart function and endurance. This assessment can help identify any limitations or risks related to cardiovascular exercise and determine the appropriate level of physical activity for the patient during rehabilitation. Pulmonary function testing may be performed to assess the patient's lung function and to identify any pre-existing respiratory conditions that may impact the transplant procedure. The patient's Cardiovascular and Cardiopulmonary status can be assessed using various tools, such as a cardiopulmonary exercise test, electrocardiogram, echocardiogram, stress test, 6-min walk test (Silva et al. 2021; Ishikawa et al. 2019)

7. **Psychological Assessment**: A psychological assessment should evaluate the patient's emotional and psychological well-being. This assessment can help identify any psychosocial issues that may impact the patient's rehabilitation and determine the appropriate level of support and intervention (Bjelland et al. 2002; Osborn et al. 2006; Amonoo et al. 2019).

Overall, a comprehensive rehabilitation evaluation is essential to ensure that HSCT patients receive the appropriate level of support and intervention during the rehabilitation period to help them recover and regain their strength and function.

1.2   Development of Individualized Rehabilitation Plans

Overall, the evaluation of the patient's physical, psychological, and social status is crucial in developing comprehensive and individualized rehabilitation programs for HSCT patients. Using validated assessment tools can help ensure that the patient's needs are accurately identified and addressed and ultimately optimize their recovery, QOL and overall outcomes (Mohammed et al. 2019b; Cheville et al. 2021; John and Chandy 2020).

It is essential to develop individualized rehabilitation plans for HSCT patients to improve their physical activity, functional capacity, muscle strength, and cognitive function to decrease fatigue and psychological distress. These plans should address the patient's specific needs and may include strategies such as rehabilitative exercise and psychological, nutritional, and social support (Bertz 2021; Manettas et al. 2022; Syrjala et al. 2012).

1.3   Patient Education on the Transplant Process and Potential Complications

Structured educational program for the patient's on the transplant process and potential complications are essential for reducing the status of depression, anxiety, stress, and risk of post-transplant complications, such as Graft-Versus-Host Disease (GVHD) and infections, while improving patient's understanding of the transplant process, potential complications and satisfaction, so leading to better outcomes (Wu et al. 2022; Jessop et al. 2019; Thygesen et al. 2012; Roberts et al. 2020).

1.4   Preparation for Physical Therapy and Exercise

Preparation for PT and exercise treatment is an essential component of the rehabilitation process for HSCT patients. The feasibility and effectiveness of a rehabilitation program for HSCT patients are highlighted by improved physical function and quality, The results of studies showed that patients who participated in rehabilitation programs had significantly better physical function and lower risk of post-transplant complications such as fatigue, compared to those who did not participate (Mohammed et al. 2019a; Mohananey et al. 2021; Bertz 2021; Roberts et al. 2020; Takekiyo et al. 2015; Rupnik et al. 2020; Hamada et al. 2021).

Overall, preparation for PT and exercise is crucial in the rehabilitation process for HSCT patients. Exercise and physical activity can improve physical function, reduce fatigue and other symptoms, and potentially reduce the risk of post-transplant complications.

2 Post-transplantation Rehabilitation

2.1 Timing of Rehabilitation Interventions

The timing of rehabilitation interventions is a challenging factor in the rehabilitation process for HSCT patients (Bertz 2021; Lans et al. 2019).

A study investigated the impact of early rehabilitation interventions on the outcomes of HSCT patients. The study found that early rehabilitation interventions, initiated during hospitalization for HSCT, led to better functional outcomes and improved QOL compared to delayed interventions (Mohammed et al. 2019a). Another study evaluated the timing of PT interventions in HSCT patients with acute GVHD. The study found that early PT interventions started within two weeks of HSCT led to significantly better outcomes in terms of pain, joint range of motion, and overall physical function compared to delayed interventions (Diep et al. 2023).

These studies emphasize the importance of early initiation of rehabilitation interventions in HSCT patients, particularly those with acute GVHD. Early rehabilitation interventions may include PT, OT, psychological support, and nutritional interventions and should be tailored to the individual patient's needs and abilities.

2.2 Rehabilitation Phases in HSCT

The rehabilitation of HSCT patients involves a multidisciplinary approach that spans several phases (Santa Mina et al. 2020; Fiuza-Luces et al. 2016; Kuehl et al. 2016; Hacker and Mjukian 2014; Wood et al. 2020; Stein et al. 2008) (Fig. 1); here are the details of each phase:

1. **Acute Phase**: The acute phase begins immediately after the HSCT procedure and lasts several weeks. During this phase, the primary goal of rehabilitation is to manage the patient's symptoms and prevent complications. The rehabilitation team will work closely with the medical team to monitor the patient's vital signs, manage pain, prevent infection, and maintain nutrition.
2. **Subacute Phase**: The subacute phase starts after the acute phase and lasts several months. During this phase, the primary goal of rehabilitation is to restore the patient's physical function and improve their QOL. The rehabilitation team will work with the patient to develop an exercise program that addresses their needs and goals. The program may include a combination of aerobic exercise, strength training, and flexibility exercises. The rehabilitation team will also provide education and counseling to help the patient manage any psychological or social issues that may arise.
3. **Long-term Phase**: The long-term phase begins after the patient completes the subacute phase and may last several years. The primary goal of rehabilitation

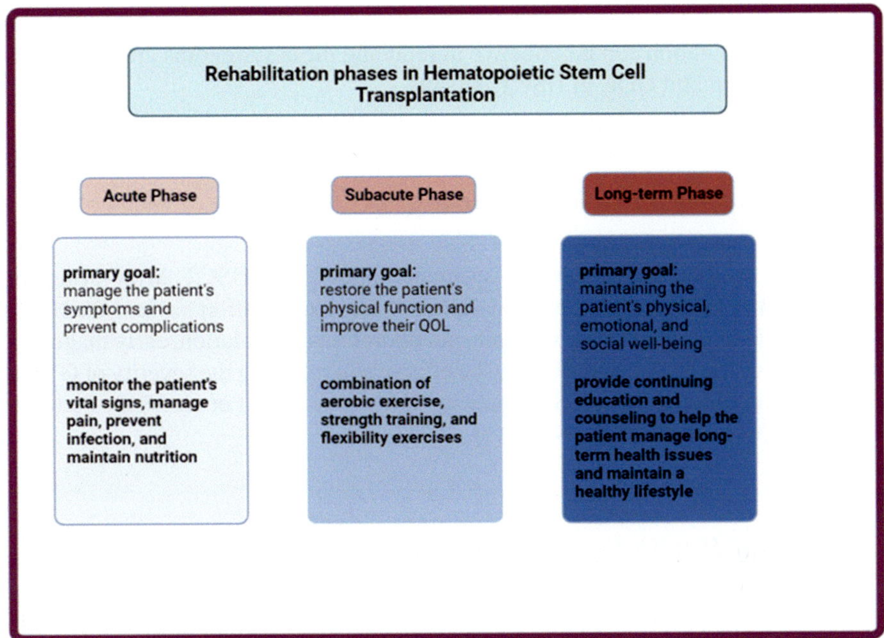

Fig. 1 Rehabilitation phases in HSCT

during this phase is maintaining the patient's physical, emotional, and social well-being. The rehabilitation team will work with the patient to develop a maintenance exercise program that addresses their ongoing needs and goals. The team will also provide continuing education and counseling to help the patient manage long-term health issues and maintain a healthy lifestyle.

In conclusion, the acute phase focuses on managing symptoms and preventing complications, the subacute phase focuses on restoring physical function and improving QOL, and the long-term phase focuses on maintaining well-being and preventing long-term complications. The rehabilitation team plays a critical role in helping patients achieve their rehabilitation goals and improve their overall health and well-being.

2.3 Management of Physical and Emotional Symptoms Such as Fatigue, Pain, and Weakness

Management of physical symptoms such as fatigue, pain, and weakness is essential to the several rehabilitation phases for HSCT patients (Stein et al. 2008; Esser et al. 2017b; Eriksson et al. 2023; Mosher et al. 2009).

Noteworthy interventions such as Exercise therapy, Cognitive Behavioral Therapy (CBT), and medication can be effective in reducing these symptoms and improving physical function and QOL in HSCT patients.

2.4 Monitoring and Management of GVHD

GVHD is a common and potentially severe complication of HSCT. Monitoring and management of GVHD is an essential aspect of the rehabilitation process for HSCT patients. Treatments such as Ruxolitinib, gut microbiota modulation, early diagnosis, and aggressive treatment, and ECP can be effective in reducing the severity of GVHD and improving outcomes in HSCT patients (Novitzky-Basso et al. 2023; Perreault et al. 2020; Mozo et al. 2021; Nair et al. 2021).

2.5 Mental Health Support for Patients and Caregivers

Mental health support is an essential component of rehabilitation for HSCT patients. These Patients and their caregivers often experience a significant psychological burden due to the treatment process and potential complications. Although depression and anxiety are common among HSCT patients and can persist even after treatment, one study found that up to 45% of HSCT patients experienced depression or anxiety symptoms during the first year post-transplant, and up to 20% experienced symptoms beyond the first year (Applebaum et al. 2016; Lee et al. 2005; El-Jawahri et al. 2016; Liang et al. 2019).

Comprehensive supportive care programs such as Cognitive-behavioral therapy, Family-based interventions, and peer support groups can play a beneficial role in helping patients and caregivers cope with the emotional stress of HSCT, However, they improve their psychological well-being and improves overall QOL.

2.6 Nutritional Support and Management of Gastrointestinal Symptoms

Nutritional support and management of gastrointestinal symptoms are essential components of several rehabilitation phases for HSCT patients.

Due to the conditioning regimen and GVHD, HSCT patients may experience gastrointestinal symptoms such as nausea, vomiting, diarrhea, and mucositis, affecting their ability to eat and absorb nutrients properly. In addition, HSCT patients may have increased nutritional needs due to the demands of the transplant process and the need for tissue repair. Various approaches may be used to manage gastrointestinal symptoms in HSCT patients, commonly the results of Studies have shown that nutritional intervention in the process of treatment, can improve outcomes for HSCT patients, including reducing the risk of infections and improving overall survival rates (Williams-Hooker et al. 2015; Andersen et al. 2020; Beckerson et al. 2019).

2.7 Pulmonary Rehabilitation for Patients with Lung Complications

Pulmonary complications are common in patients undergoing HSCT and can significantly impact morbidity and mortality.

Pulmonary rehabilitation programs may include Exercise Training (ET) and interventions such as breathing exercises, airway clearance techniques, and education on self-management of symptoms. Studies have shown that, the pulmonary rehabilitation to be an effective intervention for patients with lung complications in HSCT, helping to improve lung function, exercise capacity, and overall QOL (Kelsey et al. 2014; Smith et al. 2015; Choi et al. 2020; Bacigalupo et al. 2012).

Overall, the studies highlight the Positive effects of pulmonary rehabilitation in managing lung complications in HSCT patients. They suggest, it can significantly improve pulmonary function, exercise capacity, and QOL.

2.8 Incorporation of Complementary and Alternative Medical Therapies

Complementary and alternative medical therapies (CAM), such as acupuncture, massage, homeopathy, herbal supplements are becoming increasingly common in managing symptoms and side effects in hematopoietic stem cell transplantation patients (Lindsay et al. 2016; Ratcliff et al. 2014; Tsitsi et al. 2014; Dóro et al. 2017; Mohty et al. 2021; Beri et al. 2022). However, Due to the limited number of studies, the prevalence and extent of the usage of CAMs in HSCT patients is uncertain.

Nonetheless, more research is needed to determine the effectiveness of CAM therapies in HSCT and to guide their incorporation into rehabilitation plans.

2.9 The Multidisciplinary Approach to Rehabilitation

A multidisciplinary rehabilitation team approach with a comprehensive rehabilitation program is essential to restoring the physical functioning and QOL in hematopoietic

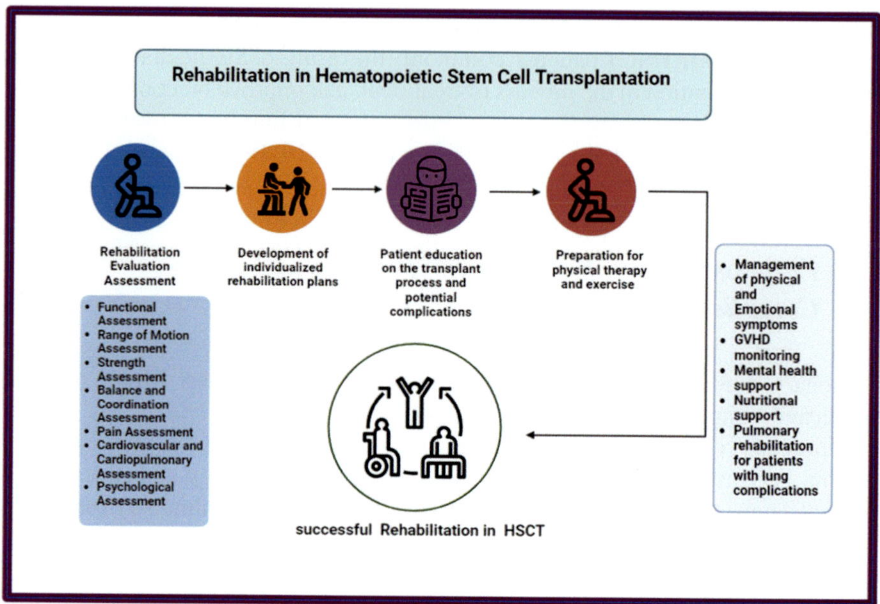

Fig. 2 Multidisciplinary approach to rehabilitation

stem cell transplantation patients. The rehabilitation program consisted of PT, OT, speech therapy, psychological counseling, psycho-oncological support, nutritional counseling, and social work support. However, Effective communication between medical professionals, physical and occupational therapists, psychologists, and dietitians is crucial to improve the outcomes of HSCT patients (Mohammed et al. 2019b; Roberts et al. 2020; Ngo-Huang et al. 2023; Obaisi et al. 2022; Molés-Poveda et al. 2021) (Fig. 2).

3 Economic Considerations and Insurance Coverage

Communication between patients, providers, and insurers is much in demand for the treatments process of HSCT patients. Studies have reported that, Insurance coverage varied widely, with some patients having full coverage and others having limited coverage or being denied coverage for certain services and emphasized the need for improved communication between patients, providers, and insurers to address financial concerns and ensure patients have access to necessary treatments and support (Aljurf et al. 2020; Gratwohl et al. 2015; Chamoun et al. 2021; Snowden et al. 2017).

4 Conclusion

HSCT is a challenging and intensive treatment. Rehabilitation plays a crucial role in the treatment process of HSCT patients by improving physical function, QOL, and overall well-being.

Several studies have shown the effectiveness of rehabilitation interventions, including individualized rehabilitation plans, patient education, preparation for PT and exercise, management of physical symptoms, monitoring and control of GVHD, mental health support, nutritional support, and management of gastrointestinal symptoms, pulmonary rehabilitation, multidisciplinary approach, a collaboration between medical professionals, the timing of rehabilitation interventions, patient compliance and motivation, economic considerations, and incorporation of complementary therapies such as acupuncture and massage.

Therefore, it is essential to incorporate a multidisciplinary approach to rehabilitation early in the treatment process of HSCT patients and maintain ongoing communication between team members. By doing so, patients can achieve better outcomes, reduce hospitalization time, and improve their overall QOL.

Overall, ongoing research is needed to investigate the improving rehabilitation outcomes for HSCT patients and emphasis the most effective strategies for optimizing physical, psychological, and social recovery after transplant.

References

Aljurf M, Weisdorf D, Hashmi S, Nassar A, Gluckman E, Mohty M, et al. Worldwide network for blood and marrow transplantation (WBMT) recommendations for establishing a hematopoietic stem cell transplantation program in countries with limited resources (Part II): clinical, technical and socio-economic considerations. Hematol Oncol Stem Cell Therapy. 2020;13(1):7–16.

Amonoo HL, Massey CN, Freedman ME, El-Jawahri A, Vitagliano HL, Pirl WF, et al. Psychological considerations in hematopoietic stem cell transplantation. Psychosomatics. 2019;60(4):331–42.

Andersen S, Banks M, Brown T, Weber N, Kennedy G, Bauer J. Nutrition support during allogeneic stem cell transplantation: evidence versus practice. Support Care Cancer. 2020;28:5441–7.

Applebaum AJ, Bevans M, Son T, Evans K, Hernandez M, Giralt S, et al. A scoping review of caregiver burden during allogeneic HSCT: lessons learned and future directions. Bone Marrow Transpl. 2016;51(11):1416–22.

Bacigalupo A, Chien J, Barisione G, Pavletic S. Late pulmonary complications after allogeneic hematopoietic stem cell transplantation: diagnosis, monitoring, prevention, and treatment. Semin Hematol. 2012;49:15–24.

Beckerson J, Szydlo RM, Hickson M, Mactier CE, Innes AJ, Gabriel IH, et al. Impact of route and adequacy of nutritional intake on outcomes of allogeneic haematopoietic cell transplantation for haematologic malignancies. Clin Nutr. 2019;38(2):738–44.

Beri A, Pisulkar SG, Bansod AV, Dahihandekar C, Beri A, Bansod A. Alternative prosthodontic therapies: a multifaceted approach. Cureus. 2022;14(9):20.

Bertz H. Rehabilitation after allogeneic haematopoietic stem cell transplantation: a special challenge. Cancers. 2021;13(24):6187.

Bjelland I, Dahl AA, Haug TT, Neckelmann D. The validity of the hospital anxiety and depression scale: an updated literature review. J Psych Res. 2002;52(2):69–77.

Brotelle T, Lemal R, Cabrespine A, Combal C, Hermet E, Ravinet A, et al. Prevalence of malnutrition in adult patients previously treated with allogeneic hematopoietic stem-cell transplantation. Clin Nutr. 2018;37(2):739–45.

Chamoun K, Firoozmand A, Caimi P, Fu P, Cao S, Otegbeye F, et al. Socioeconomic factors and survival of multiple myeloma patients. Cancers. 2021;13(4):590.

Cheville A, Smith S, Barksdale T, Asher A. Cancer rehabilitation. Braddom's Phys Med Rehabil. 2021;2021:568-93.e7.

Choi HE, Lim S-N, Lee JH, Park S-H. Comprehensive pulmonary rehabilitation in patients with bronchiolitis obliterans syndrome: a case series. Resp Med Case Rep. 2020;31:101161.

Diep PP, Rueegg CS, Burman MM, Brinch L, Bø K, Fosså K, et al. Graft-versus-host-disease and health-related quality of life in young long-term survivors of cancer and allogeneic hematopoietic stem cell transplantation. J Adolesc Young Adult Oncol. 2023;12(1):66–75.

Dóro CA, Neto JZ, Cunha R, Dóro MP. Music therapy improves the mood of patients undergoing hematopoietic stem cells transplantation (controlled randomized study). Support Care Cancer. 2017;25:1013–8.

El-Jawahri AR, Vandusen HB, Traeger LN, Fishbein JN, Keenan T, Gallagher ER, et al. Quality of life and mood predict posttraumatic stress disorder after hematopoietic stem cell transplantation. Cancer. 2016;122(5):806–12.

Eriksson LV, Holmberg K, Hagelin CL, Wengström Y, Bergkvist K, Winterling J. Symptom burden and recovery in the first year after allogeneic hematopoietic stem cell transplantation. Cancer Nurs. 2023;46(1):77–85.

Esser P, Kuba K, Scherwath A, Schirmer L, Schulz-Kindermann F, Dinkel A, et al. Posttraumatic stress disorder symptomatology in the course of allogeneic HSCT: a prospective study. J Cancer Surv. 2017a;11:203–10.

Esser P, Kuba K, Scherwath A, Johansen C, Schwinn A, Schirmer L, et al. Stability and priority of symptoms and symptom clusters among allogeneic HSCT patients within a 5-year longitudinal study. J Pain Sympt Manag. 2017b;54(4):493–500.

Fiuza-Luces C, Simpson RJ, Ramírez M, Lucia A, Berger NA. Physical function and quality of life in patients with chronic GvHD: a summary of preclinical and clinical studies and a call for exercise intervention trials in patients. Bone Marrow Transpl. 2016;51(1):13–26.

Gratwohl A, Sureda A, Baldomero H, Gratwohl M, Dreger P, Kröger N, et al. Economics and outcome after hematopoietic stem cell transplantation: a retrospective cohort study. EBioMedicine. 2015;2(12):2101–9.

Hacker ED, Mjukian M. Review of attrition and adherence in exercise studies following hematopoietic stem cell transplantation. Eur J Oncol Nurs. 2014;18(2):175–82.

Hamada R, Arai Y, Kondo T, Harada K, Murao M, Miyasaka J, et al. Higher exercise tolerance early after allogeneic hematopoietic stem cell transplantation is the predictive marker for higher probability of later social reintegration. Sci Rep. 2021;11(1):1–10.

Ishikawa A, Otaka Y, Kamisako M, Suzuki T, Miyata C, Tsuji T, et al. Factors affecting lower limb muscle strength and cardiopulmonary fitness after allogeneic hematopoietic stem cell transplantation. Support Care Cancer. 2019;27:1793–800.

Jessop H, Farge D, Saccardi R, Alexander T, Rovira M, Sharrack B, et al. General information for patients and carers considering haematopoietic stem cell transplantation (HSCT) for severe autoimmune diseases (ADs): a position statement from the EBMT autoimmune diseases working party (ADWP), the EBMT Nurses Group, the EBMT patient, family and donor committee and the joint accreditation committee of ISCT and EBMT (JACIE). Bone Marrow Transpl. 2019;54(7):933–42.

John MJ, Chandy M. Setting up a hematopoietic stem cell transplantation unit. Contemp Bone Marrow Transpl. 2020;14:1–19.

Kanter J, Liem RI, Bernaudin F, Bolaños-Meade J, Fitzhugh CD, Hankins JS, et al. American society of hematology 2021 guidelines for sickle cell disease: stem cell transplantation. Blood Adv. 2021;5(18):3668–89.

Kelsey C, Scott J, Lane A, Schwitzer E, West M, Thomas S, et al. Cardiopulmonary exercise testing prior to myeloablative allo-SCT: a feasibility study. Bone Marrow Transpl. 2014;49(10):1330–6.

Kuehl R, Schmidt ME, Dreger P, Steindorf K, Bohus M, Wiskemann J. Determinants of exercise adherence and contamination in a randomized controlled trial in cancer patients during and after allogeneic HCT. Support Care Cancer. 2016;24:4327–37.

Kvinge AD, Kvammen T, Miletic H, Bindoff LA, Reikvam H. Musculoskeletal chronic graft versus host disease—a rare complication to allogeneic hematopoietic stem cell transplant: a case-based report and review of the literature. Curr Oncol. 2022;29(11):8415–30.

Lee S, Loberiza F, Antin J, Kirkpatrick T, Prokop L, Alyea E, et al. Routine screening for psychosocial distress following hematopoietic stem cell transplantation. Bone Marrow Transpl. 2005;35(1):77–83.

Liang J, Lee SJ, Storer BE, Shaw BE, Chow EJ, Flowers ME, et al. Rates and risk factors for post-traumatic stress disorder symptomatology among adult hematopoietic cell transplant recipients and their informal caregivers. Biol Blood Marrow Transpl. 2019;25(1):145–50.

Lindsay J, Kabir M, Gilroy N, Dyer G, Brice L, Moore J, et al. Epidemiology of complementary and alternative medicine therapy use in allogeneic hematopoietic stem cell transplant survivorship patients in Australia. Cancer Med. 2016;5(12):3606–14.

Manettas AI, Tsaklis P, Kohlbrenner D, Mokkink LB. A scoping review on outcomes and outcome measurement instruments in rehabilitative interventions for patients with haematological malignancies treated with allogeneic stem cell transplantation. Curr Oncol. 2022;29(7):4998–5025.

Mohammed J, Smith SR, Burns L, Basak G, Aljurf M, Savani BN, et al. Role of physical therapy before and after hematopoietic stem cell transplantation: white paper report. Biol Blood Marrow Transpl. 2019a;25(6):e191–8.

Mohammed J, Aljurf M, Althumayri A, Almansour M, Alghamdi A, Hamidieh AA, et al. Physical therapy pathway and protocol for patients undergoing hematopoietic stem cell transplantation: recommendations from the eastern Mediterranean blood and marrow transplantation (EMBMT) group. Hematol Oncol Stem Cell Therapy. 2019b;12(3):127–32.

Mohananey D, Sarau A, Kumar R, Lewandowski D, Abreu-Sosa SM, Nathan S, et al. Role of physical activity and cardiac rehabilitation in patients undergoing hematopoietic stem cell transplantation. Cardio Oncol. 2021;3(1):17–34.

Mohty R, Savani M, Brissot E, Mohty M. Nutritional supplements and complementary/alternative medications in patients with hematologic diseases and hematopoietic stem cell transplantation. Transpl Cell Therapy. 2021;27(6):467–73.

Molés-Poveda P, Comis LE, Joe GO, Mitchell SA, Pichard DC, Rosenstein RK, et al. Rehabilitation interventions in the multidisciplinary management of patients with sclerotic graft-versus-host disease of the skin and fascia. Arch Phys Med Rehabil. 2021;102(4):776–88.

Mosher CE, Redd WH, Rini CM, Burkhalter JE, DuHamel KN. Physical, psychological, and social sequelae following hematopoietic stem cell transplantation: a review of the literature. Psycho Oncol J Psychol Soc Behav Dimens Cancer. 2009;18(2):113–27.

Mozo Y, Bueno D, Sisinni L, Fernández-Arroyo A, Rosich B, Martínez AP, et al. Ruxolitinib for steroid-refractory graft versus host disease in pediatric HSCT: high response rate and manageable toxicity. Pediat Hematol Oncol. 2021;38(4):331–45.

Nair S, Vanathi M, Mukhija R, Tandon R, Jain S, Ogawa Y. Update on ocular graft-versus-host disease. Indian J Ophthalmol. 2021;69(5):1038.

Ngo-Huang A, Ombres R, Saliba RM, Szewczyk N, Adekoya L, Soones TN, et al. Enhanced recovery stem-cell transplantation: multidisciplinary efforts to improve outcomes in older adults undergoing hematologic stem-cell transplant. JCO Oncol Pract. 2023;22:520.

Novitzky-Basso I, Schain F, Batyrbekova N, Webb T, Remberger M, Keating A, et al. Population-based real-world registry study to evaluate clinical outcomes of chronic graft-versus-host disease. PLoS ONE. 2023;18(3): e0282753.

Obaisi O, Fontillas RC, Patel K, Ngo-Huang A. Rehabilitation needs for patients undergoing CAR T-cell therapy. Curr Oncol Rep. 2022;24(6):741–9.

Osborn RL, Demoncada AC, Feuerstein M. Psychosocial interventions for depression, anxiety, and quality of life in cancer survivors: meta-analyses. Int J Psych Med. 2006;36(1):13–34.

Perreault S, Schiffer M, Zhao J, McManus D, Foss F, Gowda L, et al., editors. 577: incidence and outcomes of positive outpatient surveillance blood cultures in hematopoietic stem cell transplant (HSCT) patients with graft versus host disease (GvHD) on high dose≥ 0.5 mg/kg/day (HD) and low dose< 0.5 mg/kg/day (LD) steroid therapy. Open Forum infectious diseases. Oxford: Oxford University Press; 2020

Ratcliff CG, Prinsloo S, Richardson M, Baynham-Fletcher L, Lee R, Chaoul A, et al. Music therapy for patients who have undergone hematopoietic stem cell transplant. Evid Based Compl Altern Med. 2014;2014:3605.

Roberts F, Hobbs H, Jessop H, Bozzolini C, Burman J, Greco R, et al. Rehabilitation before and after autologous haematopoietic stem cell transplantation (AHSCT) for patients with multiple sclerosis (MS): consensus guidelines and recommendations for best clinical practice on behalf of the autoimmune diseases working party, nurses group, and patient advocacy committee of the European society for blood and marrow transplantation (EBMT). Front Neurol. 2020;11:1602.

Rupnik E, Skerget M, Sever M, Zupan IP, Ogrinec M, Ursic B, et al. Feasibility and safety of exercise training and nutritional support prior to haematopoietic stem cell transplantation in patients with haematologic malignancies. BMC Cancer. 2020;20(1):1–9.

Santa Mina D, Dolan LB, Lipton JH, Au D, Camacho Pérez E, Franzese A, et al. Exercise before, during, and after hospitalization for allogeneic hematological stem cell transplant: a feasibility randomized controlled trial. J Clin Med. 2020;9(6):1854.

Schmidt M, Breyer S, Löbel U, Yarar S, Stücker R, Ullrich K, et al. Musculoskeletal manifestations in mucopolysaccharidosis type I (Hurler syndrome) following hematopoietic stem cell transplantation. Orphanet J Rare Dis. 2016;11(1):1–13.

Silva TC, Silva P, Morais DS, Oppermann CZ, Penna GB, Paz A, et al. Functional capacity, lung function, and muscle strength in patients undergoing hematopoietic stem cell transplantation: a prospective cohort study. Hematol Oncol Stem Cell Therapy. 2021;14(2):126–33.

Smith SR, Haig AJ, Couriel DR. Musculoskeletal, neurologic, and cardiopulmonary aspects of physical rehabilitation in patients with chronic graft-versus-host disease. Biol Blood Marrow Transpl. 2015;21(5):799–808.

Snowden JA, Badoglio M, Labopin M, Giebel S, McGrath E, Marjanovic Z, et al. Evolution, trends, outcomes, and economics of hematopoietic stem cell transplantation in severe autoimmune diseases. Blood Adv. 2017;1(27):2742–55.

Stein KD, Syrjala KL, Andrykowski MA. Physical and psychological long-term and late effects of cancer. Cancer. 2008;112(S11):2577–92.

Steinberg A, Asher A, Bailey C, Fu JB. The role of physical rehabilitation in stem cell transplantation patients. Support Care Cancer. 2015;23:2447–60.

Stiff P, Erder H, Bensinger W, Emmanouilides C, Gentile T, Isitt J, et al. Reliability and validity of a patient self-administered daily questionnaire to assess impact of oral mucositis (OM) on pain and daily functioning in patients undergoing autologous hematopoietic stem cell transplantation (HSCT). Bone Marrow Transpl. 2006;37(4):393–401.

Syrjala KL, Martin PJ, Lee SJ. Delivering care to long-term adult survivors of hematopoietic cell transplantation. J Clin Oncol. 2012;30(30):3746–51.

Takekiyo T, Dozono K, Mitsuishi T, Murayama Y, Maeda A, Nakano N, et al. Effect of exercise therapy on muscle mass and physical functioning in patients undergoing allogeneic hematopoietic stem cell transplantation. Support Care Cancer. 2015;23:985–92.

Thygesen KH, Schjødt I, Jarden M. The impact of hematopoietic stem cell transplantation on sexuality: a systematic review of the literature. Bone Marrow Transpl. 2012;47(5):716–24.

Tsitsi T, Raftopoulos V, Papastavrou E, Charalambous A. Complementary and alternative medical interventions for the management of anxiety in parents of children who are hospitalized and suffer from a malignancy: a systematic review of RCTs. Eur J Integr Med. 2014;6(1):112–24.

van der Lans MC, Witkamp FE, Oldenmenger WH, Broers AE. Five phases of recovery and rehabilitation after allogeneic stem cell transplantation: a qualitative study. Cancer Nurs. 2019;42(1):50–7.

Williams-Hooker R, Adams M, Havrilla DA, Leung W, Roach RR, Mosby TT. Caregiver and health care provider preferences of nutritional support in a hematopoietic stem cell transplant unit. Pediat Blood Cancer. 2015;62(8):1473–6.

Wood WA, Weaver M, Smith-Ryan A, Hanson E, Shea T, Battaglini C. Lessons learned from a pilot randomized clinical trial of home-based exercise prescription before allogeneic hematopoietic cell transplantation. Support Care Cancer. 2020;28:5291–8.

Wu Y-H, Hsu Y-J, Tzeng W-C. Correlation between physical activity and psychological distress in patients receiving hemodialysis with comorbidities: a cross-sectional study. Int J Environ Res Public Health. 2022;19(7):3972.

Index

MIX
Papier aus verantwortungsvollen Quellen
Paper from responsible sources
FSC® C105338

If you have any concerns about our products,
you can contact us on
ProductSafety@springernature.com

In case Publisher is established outside the EU,
the EU authorized representative is:
**Springer Nature Customer Service Center GmbH
Europaplatz 3, 69115 Heidelberg, Germany**

Printed by Libri Plureos GmbH
in Hamburg, Germany